ECOLOGY AND ENVIRONMENT IN THE HIMALAYAS

ECOLOGY AND ENVIRONMENT IN THE HIMALAYAS

K.S. Gulia

ISHA BOOKS
Delhi - 110009

ECOLOGY AN DENVIRONMENT IN THE HIMLAYAS

Rs. 900

ISBN: 81-8205-423-0

Published in 2007 in India by
ISHA Books - B-63, 1st Floor, New Gupta Colony,
Near DDA Market,
Delhi-110009
Phone: 9212142040
E-mail: ishabooks@hotmail.com

Laser Typesetting at: Best Computers, Delhi
Printed at: Salasar Imaging System, Delhi

Contents

Introduction

Beauty is everywhere. But nothing to beat the natural beauty of the hills. It lifts us to a height from where we can see a little farther over the dry, dusty plain. Perhaps that is the reason we need to go to hills to seek beauty.

The Himalaya is the world's loftiest and most majestic mountain chain. It is a global heritage of all mankind. Divine nature of its snow-covered peaks and its icy glaciers have attracted the sages and saints since time immemorial. The places of pilgrimage have attracted the devotees from every corner of the country. Its life-sustaining rivers, luxuriant valleys, climate, terrain enhance the Himalayan beauty.

At the sametime Himalaya is a powerful magnet for those who are interested in adventure and mountaineering. It has attracted a large number of them from all over the world. But this, along with modernization and development have adversely affected the natural balance of the Himalaya. The super intelligence of mankind, their rapid advances in the field of science and technology and the exploitation of natural resources to meet the ever increasing demands of the rising population have caused ecological degradation in the Himalayan region. Not only that, they have seriously threatened the wildlife of the region too.

Each chapter of this book deals with the existing ecological and other problems in the different parts of the Himalayan region and also how they forced other forms of life on earth at bay. In a couple of chapters, the protective measures of conservation initiated by the Government and some other organizations have also been discussed.

The first chapter, "Eco-Stress Around Himalaya" deals with ecological degradation around Himalaya. Once beautiful forest hills are gradually becoming ugly. It is loosing its environmental richness. Himalaya could be the source of wealth in the shape of timbers, forest produces, minerals etc. But man from plains is fast invading the calm atmosphere of the highlands, thus disturbing its natural balance. These problems are discussed in this chapter.

The second chapter is "Himalaya's Strain Environment Calls for Quick Action". This range of mountains is covered with green dense forests and surrounded by snow-covered mountains. This fairyland attracts tourists from the every corner of the world. How Himalaya, the abode of celestial snow, is losing its environmental beauty and what are the steps to be taken to check it is discussed in this chapter.

The third chapter—"Himalayan Expedition By A British Climberc—A Review"—is the extraction from the diary of Sir Chris A. Bonington, a renowned British climber. It was his first Himalayan expedition about thirty years ago, at the age of twenty-five. How the trekkers and climbers make the base camps and camp spots impure by leaving their mark in the shape of rubbish is explicitly expressed by him in this chapter.

"Himalaya—Its Geology, Meteriology and Hydrology" is the fourth chapter. This chapter tells us about the formation of the Himalaya. It also deals with the changing climate of Himalayan region and how it determines the climate of the Indian subcontinent. The impact of human intervention, rising population, expansion of geological, meteriological and hydrological condition of the region is highlighted in this chapter.

The next chapter, "Constitutional Provision to Save Himalaya," deals with the natural calamities and loss of natural resources around the Himalaya. How this is adversely affecting the rich wildlife resources of the region, to save the alarming situation what are the major actions the Government is taking and establishment of tiger reserves are discussed in this chapter.

Modernization has adversely affected the human life. Being tired of hectic daily routine man wants a periodical interval in some isolated place, in the midst of nature, away from the busy city life. This was given birth to tourism. How far tourism and pilgrimage have devastated the scenario of the Gangotri-Gaumukh has been revealed in the sixth Chapter—"Effect of Pilgrimage in Gangotri—Gaumukh Region."

The pilgrimage movement to the Himalayan shrines has increased tremendously. Environmentalists around the world are worried about the deteriorating ecological system of Gangotri. The seventh chapter, "Gangotri Conservation Project," discusses the steps the Government is taking to check the alarming situation.

The eighth chapter—"North Eastern Frontier Region in Jeopardy"—attributes the deteriorating ecosystem of the North Eastern

Frontier region to the rising population deforestation, expansion of agriculture, construction of dam etc. How the local people, including women folk have come forward with their own land laws to check the situation is discussed in this chapter.

The ninth chapter is "Human Interference into Nature Accelerates Natural Hazards". In order to meet the personal needs man has caused havoc in the Himalayan region. Some construction works as well as deforestation have affected the environment adversely. But are they the sole factors that are responsible for the deteriorating environmental condition? The readers will come to know about it in this chapter.

Biotic pressure on forests is the major problem in the Himalayan region. Its reasons and effects are highlighted in the tenth chapter titled—"Effect of Environmental Problem in Himalaya". Besides that, it is also reviewed in this chapter how open cast mining is not only affecting the environment of the region but it is also a threat for the human, animal and aquatic life.

The eleventh chapter, "Threat to Wildlife in Himalaya—A Regionwise Review", tells us how some of the rich species of wildlife in the vast Himalayan chain are on the verge of extinction. It is a major matter of concern for the world. Zonewise review of the problem and implementation of Wildlife Protection Act as well as creation of national parks and sanctuaries are explained in details in this chapter.

"Flora and Fauna of Kumaon Region" is the next chapter. In this chapter readers will come to know about the famous Corbett National Park, situated in the foothills of Kumaon region. Despite being effected by the poachers, it has a very good cross-section of animals. Jim Corbett, a famous naturalist and a shikari, stayed in Kumaon region for sometime. During that period he intensively travelled through the forests of the region studying the character and behaviour of the wildlife there. This chapter is mainly based on his assessment about the region and its wildlife.

The next chapter—"Nepal's Protected Areas and Their Management"—deals with the history and current practice of park and wildlife conservation within the Himalayan Kindgom of Nepal. In this thirteenth chapter attention has been focused on the illegal fur trade in Kathmandu. Since the execution of the 1973 National Parks and Wildlife

Conservation Act, Nepal has implemented a number of conservation programmes successfully. All these have been discussed in this chapter.

Chapter fourteenth is "Rich Flora and Fauna of Arunachal Pradesh." It given an idea about the wildlife resources of Arunachal Pradesh. One can learn from this chapter that the vegetation cover of Arunachal Pradesh stretches from foothills to snow covered peaks. Due to differences in geological, topographical and climatic conditions it has a great variety of plants and animals. These aspects are highlighted in details in this chapter.

The next chapter of this book deals with Red Panda. Rightly titled as "Beautiful But Little Known Species: Red Panda on the Brink of Extinction," this chapter tells us about the animal which was once commonly found all over the Eastern Himalaya. Generally found at an altitude of 2200-4800 in the temperate forests of the Himalaya, Red Panda has presently reached the brink of extinction. Their features, food habits and various other such factors have been explained in this chapter.

Chapter sixteenth, "Eastern Himalaya—Habitation of Salamanders" deals with the highly endangered species, Himalaya salamanders. It is very sensitive to human disturbances. This chapter tells us about its structural features, habitat as well as conservation measures.

Trade in rare insect species has grown worldwide. How thousands of butterflies are captured for commercial benefit is discussed in the next chapter, i.e., chapter seventeenth—"Illegal Trade Threat the Butterflies in Darjeeling."

The Lloyd Botanic Garden—Its Aesthetic and Recreational Value," the next chapter focuses on the heritage status and conservation value of the Lloyd Botanic Garden which underwent a severe setback in 1987. It also explains how a visit to this garden is educative and provides a lesson on the importance of conserving nature.

The nineteenth chapter—"Changing Process of Convervation Around the Great Himalayan National Park"—draws our attention towards changing perceptions of conservation and degradation around the Great Himalayan National Park. In making of a conservation project how state agencies and various groups disagree as well as collaborate is highlighted in this chapter.

Chapter twentieth is titled as "Valley of the Lost Horizon." It tells us how human beings, by a false sense of self glorification, disfigured rocks and trees of Nanda Devi, which was once considered as a unique masterpiece of nature's wilderness areas. It also reveals about the endangered species of the region.

The twenty-first chapter is about Garhwal Himalaya titled as "Ecological Disturbances and Conservation of Wildlife in Garhwal Himalaya" this chapter deals with geography of Garhwal Himalaya and its wildlife with special reference to endangered and threatened species and its conservation.

The last but one chapter, "Jigme Dorji National Park"—"Largest Protected Area of Bhutan," reveals about the wildlife, floral aspects and vegetation of Jigme Dorji Natural Park, Bhutan.

Transboundary wildlife trade has been a major matters of concern. The last chapter of this book "Global Trade of Wildlife" draws readers' attention towards wildlife trade realities from both the global and Indian scenario. It has become a matter of concern for conservationists all over the world.

1

Eco-Stress Around Himalaya

Introduction

Ecological degradation has become a major problem the world wide. Deforestation has spoilt the environment around us, thus making it unhealthy. Once beautiful landscape of the Himalaya with its green forests is losing its environmental richness. Scientists and environment experts are very much concerned about this problem as it is spreading fast and affecting the world like cancer.

"There is a saying, the earth has cancer, and the cancer is man:" The cancer of pollution and poverty is extending itself all over the face of the earth. Gradually it is spreading over the once beautiful, forested hills including Himalaya. They are becoming bare, ugly landscapes. Ecological degradation is setting in across the whole range.

'The Himalayan arc', says Eric Eckholm 'forms an ecological Gibraltar whose fate looms over the well-being of hundreds of millions'—about 50 million in the mountains, and about another 300 million in the Indus, Ganges, and Brahmaputra valley systems. The deserts of Afghanistan are in the late evening; Bhutan and Arunachal Pradesh may still be in the early morning; and the Central Himalaya is in the late afternoon of the Himalayan day. Wildlife is being fast exterminated, and both man and mountain are becoming poorer.

Impact of Man on Mountains

Since Stockholm, 1972, and the spread of environmental consciousness, an awareness has been dawning that the mountain ranges of the world, the Himalaya and Andes in particular, are in a state of danger and that their biospheric problems begin with the impact of man on the mountains. UNESCO-sponsored conferences have been held in the European Alps, then in the Andes, and latterly for the Himalaya-Hindu Kush area at Kathmandu in September 1975. Over forty scientists and mountain experts from many countries assembled in

Munich in December 1974 to discuss this problem. At the end of the conference, they issued the following Statement of Concern:

STATEMENT OF CONCERN

Mountains and the Human Future

1. Man has been the maker of deserts in past ages, even with his bare hands, and often in the name of progress. Now that he is multiplying fast, and has more powerful technology, his capacity to endanger the balance between Man and Nature ('Only One Earth') has been enhanced many times.
2. About a third of the earth's land surface, the mountain regions, seemingly permanent, are among the most fragile once deforestation and soil erosion begin. Yet their potential for human well-being, for agriculture, forestry, water and power systems, and recreation, can affect half the human race, for better or for worse. In many places the misuse of mountain land causes economic losses far in excess of any gains from development endeavours.
3. Eco-stress in the mountains has widening destructive impact in the plains with floods; the siltation of dams, reservoirs and ports; the loss of agricultural produce and of homesteads leading to irreplaceable human and economic losses.
4. The combined eco-stress on mountains and plains affects the world's two major problems, food and energy, especially in the developing world. Eco-stress undermines national and international efforts in economic development.
5. Yet, there is hope, and there is vast potential for human benefit, given awareness of the problem; and if conservation-oriented development techniques are explored. Many are already known. They need implementation. In the developed mountain countries there are many useful examples of timely economic and ecological development to serve as a guide to prevent an avalanche of disastrous consequences.
6. Eco-development also needs new levers for resource and employment generation, especially in poor developing

mountain lands, which are in both an ecological and a resource trap. In many areas important possibilities exist especially where recreational potential and hydrological resources exist but are unavailed of.

7. New sources of energy for mountain peoples are the key to the problem of saving and spreading forest cover, the shield of agriculture in both mountains and plains. There is promising potential for such new sources of energy from hydro-power, fuel forestry, and even solar energy.
8. Urbanization and population increase are driving millions of plainsmen to seek recreation and peace in the hills. This is a mounting environmental threat, but one which can be turned into potential benefit if ecologically planned tourism, which can respect landscapes and cultures, could provide alternative employment for populations whose subsistence technology is rapidly depleting their resources of land and forest.
9. We who are concerned with the implications of ecological degradation of mountain environments, believe that the dimensions of this problem call for an immediate trans-national, multidisciplinary response including the joint use of all available scientific, technical, economic and political possibilities.

Source of Wealth

In ages the Himalaya seemed everlasting but man's ferious attack has rendered them among the most fragile eco-systems of the earth. Most people still treat the Himalaya as some kind of Shangrila, or as the Abode of Gods; of little economic use to man and with limited geophysical implications for the lands and peoples around it. True, the Himalaya, the highest and most majestic mountain range of the world, has a beautiful and spiritual quality which makes it almost a symbol of the Super Conscious; but it is more than an altar to worship at, an arena for human sublimation by yogi or mountaineer. It is the weather-maker of Asia.

The interaction of the annual monsoon and the eternal Himalaya makes the weather for the whole of South Asia, and it also influences the continental climate of the whole of Central Asia. Without the Himalaya there would be major climatic changes in most Asian countries. They are also the source of rivers and of hydroelectric power—of the Indian subcontinent. These rivers are watering Pakistan, India, Nepal, Bangladesh and Bhutan. The whole range is like Bhagiratha's bent bow which strikes arrows at glacier and snow slope, and releases hundreds of streams of water. The hydroelectric potential is incalculable and only partially tapped; the potential of a single valley—the Karnali valley in Nepal—is expected to be the same as the entire installed capacity of India or about 18 million kW. There is tremendous hydroelectric potential in the 70 km Brahmaputra bend, in which its waters come crashing down nearly 3000 m in so short a distance. Its estimated hydroelectric potential is about 30 million kW. In *India's Water Wealth* K.L. Rao has estimated the hydel power potential of the Indus basin to be 4630 mW at 60 per cent load factor; that of the Ganges at 2395 mw, and that of the Brahmaputra at 1315 mw. The Himalaya could also be the source of much wealth in the shape of timber, forest produce, minerals, and a major source of fruits and wool. But for that there is a need for an environment-oriented planning, not blind exploitation. But unfortunately the people of the subcontinent have turned their backs on the Himalaya for centuries, except for spiritual purposes. They have turned to them as places of pilgrimage. People in other parts of the world have responded to them as challenges to adventure upon the world's highest mountains; and names like Everest, Annapurna, Kanchenjunga, Nanda Devi, and Nanga Parbat have beckoned heroic hearts from every continent.

Alarming Consequences

Much of the western Himalayan region from Afghanistan through Ladakh to Lahul and Spiti has been mountain desert for centuries; but the cancer is spreading fast eastward, again, the carrier is man—both Himalayan and non-Himalayan man. Mounting populations in the inhabited hill regions in many cases multiplied as fast or even faster than the population of the Gangetic plain or the coastal plain of India. Statistic shows that in the 50 years since 1921, the population of 18 sample districts below 3000 m elevation in the Himalaya have risen 170 per cent, whilst those of selected districts in the Gangetic plain have risen 127 per cent, and those in the coastal plain only 99 per cent.

These mounting numbers of poor hill peasants are pushing up the valleys to the last rocky slopes which can be tilled for the most meagre subsistence farming. The wearing away of vital forest cover, which should be at least 60 per cent of the land area in hill regions for ecological stability, has unfortunately led to alarming consequences in the shape of soil erosion, the siltation of dams, and the scourge of annual floods.

The hillman has very limited cultavable land, on which he can barely produce 50 per cent of his food. For example, there are 17.6 people per ha of cultivated land in the UP hills; the figure ranges from 2.3 to 5.8 for the rest of the state. The apparent population densities of 75 per sq km in the UP hills, and 370 per sq km the UP plains is deceptive and misleading. With erosion fast removing the hills' thin irreplaceable top soil and its nutrients, the percapita food production with further diminish.

Unless immediate and effective remedial steps are taken. There will be widespread consequences of this dangerous impact of multiplying man and his multiplying livestock on the Central Himalaya, in Garhwal, Kumaun and Nepal, in the next stage of the extension of the western mountain desert. Nepal provides the most dramatic example of the speed of desertification of the most beautiful part of the range, the Central Himalaya, which Frank Smythe described only thirty years ago as its 'cream'. Two-thirds of its people live in the hills, and they cannot feed themselves. Its people have, therefore, had to flow down and spread themselves over the entire *tarai* belt, extending into the hills of Garhwal, Kumaun, Darjeeling, Sikkim and Bhutan; and into the plains of India. This disastrous phenomenon is not limited to the Central Himalaya alone. Everywhere the treeline is receding and the hills are browning. The brown disease is spreading fast like cancer and raising an alarming situation.

Ecological Devastation

Man from the plains equipped with higher technology and larger resources is fast invading these remote and calm highlands. In their pursuit of the glittering summits, mountaineers need to pause, and ask themselves how well-founded their love of mountains is; how long the sanctity of environments below the snowline will will last or it is lost for ever. When expeditions destroy vegetation at these altitudes, it may take generations for the regeneration of plants. Nature conservancy experiments in such high-altitude regions as Snowdonia in North Wales have shown that when mountain lands are preserved from human beings, it takes over 20 years for grass to grow about 30 cm. Nature must be given time, if man is to replenish intelligently what nature so sedulously creates in such harsh conditions.

Huge dams have been built in India, Nepal and Pakistan at a cost—averaging between Rs. 1000 to Rs. 2000 million each with life expectancies of between a hundred and two hundred years. Within ten years of the construction, their life expectancies of these dams have shrunk to a half or a quarter; and enormous siltation in the treeless catchment areas behind the dams. This has led to a huge waste of resources. The mere erection of concrete walls in narrow defiles cannot produce irrigation water and hydroelectric power for long, if the cancer behind the dam is not taken care of.

Sedimentation rates of the reservoirs of such Himalayan dams as Bhakra and Ramganga are given below, in years:

TABLE 1.1

	Year of first impounding of water	*Originally estimated life to fill dead storage*	*Revised estimated life to fill dead storage*	*Number of years needed to fill 50% of live storage*	*Balance life to fill live storage*
Bhakra	1959	88	47	147	105
Ramganga	1973	185	45	223	177

In spite of huge expenditure on major dams, the benefits of irrigation have gone entirely to the downstream command areas. There has been an almost total neglect of the hydrology of springs, which sustains most of the vegetation and human life of the hills, and the survival of which is seriously threatened by the increasing denudation of forests. The preservation of hill springs is a matter which should be given the first priority in any ecological planning of the hill areas. Because if they dry up, the organic life around them will perish; and it is beyond the combined effort and resources of governments and international agencies to bring water back to the plant and human communities in the mountains.

Forestry practices have been a major ecological threat to Himalayan environments. The afforestation plans of the governments of the Himalayan states have not been able to cope with the rising environmental problem.

Actual afforestation has not exceeded 2 per cent of the area deforested. It takes minutes to fell a tree, but 50 to 100 years for it to mature at these altitudes. So, it is clear that devastation is upon us already, between the infinitely slow vegetative process of afforestation, and the rapid mechanical sweep of deforestation. On the one hand, this rapid deforestation has created a serious firewood crisis for Himalayan populations, and, on the other, it has led to massive depletion of the scarce top soils upon which human populations and organic life depend.

R. Hoegger has drawn our attention to the alarming figures of soil depletion in the Himalaya as compared with the Alps. It was calculated that the Karnali river is moving some 75 million cu m of silt and debris every year, an amount that corresponds to a 1.7 mm soil cover of the whole Karnali Watershed. As a comparison: these 75 million cu m of solid material are two and half times more than the Kander, a river in the Swiss Alps, has deposited in the lake of Thun during the last 250 odd years, since 1713 (Kander Delta).

If this trend continues, it will be impossible for mounting human populations to survive in such regions; not to speak of the radical transformations of entire eco-systems for the rest of organic life. This depletion of soil and energy potentials is absolutely basic to all developmental planning in such regions. But because of our mistakes, neglect, and shortsightedness, today we stand at a critical point in Himalayan forestry policy; many past assumptions both explicit and implicit, no longer hold good. When forestry policies were framed more than 50 years ago in India and Pakistan, the assumptions then were of low populations, low demand, and unlimited forests. We now have high population densities per unit of arable land, and limited forest resources.

Earlier policy assumption was that it was enough for governments to give priority to commercial forestry in heavily-forested areas yielding the maximum revenue for the department. There was hardly any regard for the ecology of the region, or for soil conservation. The policy in earlier years also led to a relatively low plough-back of revenues into afforestation measures to match deforestation.

Another policy assumption was that, if the forest department give up proper forestry management responsibility for the class II forests (not the best commercial forests) these, and the civil forests, could be suitably managed by local *panchayats*. In UP, there was a hope that forestry education might be given to the local people, though the forestry department did not take responsibility for it. As a result, 8000 sq km of civil forest lands in the UP hill areas are now virtually treeless.

But now there is need for a total revision of forestry policy on the basis of changed factors. As there are very few forests left, the forest priority should be given to the forest-produce needs of increasing local populations for their firewood, fodder, agricultural implements, and house construction. In Nepal it has been estimated that the wood needs of the average hill family amounts to between 1 and 1.5 tons a year. The second priority is the soil conservation, widespread afforestation, and ecology-based land-use planning. The revenue needs of the forestry department must now come third, after having enjoyed first priority for over fifty years. Only such drastic change in the forestry-policy objectives can help the hills and the people of the hills.

In 1947, Karponang, 15 km up the track from Gangtok to the Nathu la on the Tibet border was dense with one of the world's finest forests, dripping with 500 cm of rainfall a year. It was shocking to discover ten years later, that not a single tree stood there. Inadequate technology in road building has led to this result. Authority and its contractors were the carriers of this cancer. Even in economic terms, the recurring maintenance cost of the new road in an eroded landscape would amount to more than the capital cost of the construction of the road. Such road-building technology has led to serious damage to the hills in the Alakananda Valley in Garhwal, and in the upper Satluj Valley in Kinnaur. Besides that the traditional nomad communities, *gujjars, gaddis* and so on have overgrazed vast tracts of the. Himalaya with mounting livestock, to keep pace with human population. There are other 'nomads' too—the tourists, trekkers and mountaineers. This new nomads in the last fifteen years has shown its destructive potential for both landscapes and cultures. New tourist resorts have been built without any serious thought on alternative sources of energy. Trees, the only source of local energy, have been cut down. This is the pattern from Gulmarg to Kulu to Sola Khombu, with tourism expanding its own nest. The classic example is that of Sola Khombu, the land of the Sherpas. Ironically, the attraction of the Everest region is proving to be Sola Khombu's ecological and perhaps cultural undoing.

The destructive potential of unplanned tourism is apparent. The Singalila ridge dividing Darjeeling district and Sikkim from Nepal was denuded of its forest cover long ago. In Kulu, the tree-line has been visibly receding in the last thirty years, and it is now near the ridges. The Alakananda valley, dividing Garhwal and Kumaun, is in an advanced state of ecological degradation. These are only a few illustrative examples of the blight creeping over the Himalaya.

The number of trekkers and climbers rose from only 20 in 1964 to 3400 in 1974. They need to acquire an ethic of respect for their natural environments; the planning of tourism, mountaineering and trekking needs appropriate ecological guidelines. Especially at altitudes above 3000 m numbers need to be kept low so as not to overburden the limited carrying capacities of the fragile eco-system; and parties of over six persons need to carry their own fuel, until alternative sources of energy are made available.

Useful International Techniques

There are two interesting models of conservation and development of mountain regions present themselves in the European Alps and the Japan mountains. In central Europe in places like Switzerland, Bavaria and Austria, similar conditions prevailed a hundred and fifty years ago as in the Himalaya today. In the early phase of the industrial revolution, central Europe almost lost its forests; fortunately coal, and later hydel energy came to its rescue. The first was land-use planning, with the help of land-tenure systems which, in some cases, originated in the fourteenth and fifteenth centuries. These were not always enforced, but at least there was a structure of law and public opinion, around which land-use systems could evolve. There was none such in the Himalaya. The second factor in Europe was the development of hydroelectric power as a major alternative source of local energy for development, preventing the destruction of forest cover. The development of planned tourism was the other factor. These three made possible a happy blend of conservation and development.

In Japan there is the remarkable phenomenon of only 2 per cent of the large population of over 100 millions living in the mountains, which cover 75 per cent of the land area. This has been achieved, partly, by a policy of conserving the mountain regions as reservoirs of resources of forestry, hydroelectric power, and tourism, and partly, by the attraction of economic development in the plains, especially in the last thirty years. A price has admittedly been paid for this in the acute congestion and pollution of the remaining 25 per cent of the land area in the plains where 98 per cent of the population lives. The same basic principle of treating mountains as reservoirs of natural resources, with the same combination of conservation-oriented development factors has been applied here, as in central Europe. Both these are useful models for the Himalaya too, despite the -difference in climate, terrain and people.

To assess the mnagnitude of the problem and the adequacy of our response to it, a macro survey of the Himalayan environment is needed. Presently there is not enough hard data upon which to base a sound planning response. One of the first requirements is data from aerial photography, or from remote-sensing techniques, to present a clear picture of both land-use and land-misuse patterns, the actual forest cover, and the extent of denudation and degradation. This may call for an international effort, as well as national and bilateral efforts.

A brief reference has been made in above paras to the success in the central European and the Japanese highlands, which point to solutions from three major directions: (1) land-use planning; (2) alternative energy sources; and (3) environment-oriented planning for the development potential of hill regions. But, to be able to do this in practice, it would be necessary to design and implement sub-catchment areas, in the three major agro-climatic regions which have most felt the impact of mounting human and livestock populations. These three regions may be described as the tropical and subtropical regions up to about 1000 m elevation; the lower temperate and higher temperate between 1000 and 3000 m, and the tundra or subarctic region between 3000 m and 4500 m; above which would be the permanent snow line with arctic conditions.

Multidisciplinary study and planning is easy to suggest, but hard to plan and impliment especially in these complex areas with variables of climate terrain, and man, apart from the social and political problems which often obstruct development.

One thing is certain—we can look at developmental problems surrounded with the problems of agriculture, animal husbandry, forestry, roads, power, etc. The fundamental problem of planned land-use demands an understanding of land capabilities and an analysis of the trade-offs between its major uses, e.g., cultivation, livestock rearing, forestry—and now tourism and industry. The interdependencies of land, energy, food, water, man and livestock are a basic factor.

The concept of plains development cannot be implimented to the hill regions. The physical features of the hill areas and their problems are to be considered. Be it in soil conservation, or in road construction, or in tourism, and all these have to take into account. Mostly, the problems of man and his livestock are also different in the hills, and therefore require different responses too.

Hans Rieger's remark, 'There is only one Himalaya to lose' is a major reality, not only for the 50 million people living in the Hindu Kush-Himalayan system, but also for the hundreds of millions living around that system in Asia, in the hills and in the plains. The lands in the plains of three major river systems south of the Himalaya, those of the Indus, the Ganga and the Brahmaputra, are the most directly affected.

There are certainly similar problems in other parts of the world, especially in the Andean region in South America and in the Central African highlands. The Himalaya are, therefore, the most spectacular manifestation of what is a global problem demanding global attention. It is not just a matter of science or organization or money. We need to care.

DEGRADATION OF MOUNTAIN ENVIRONMENT

Ecological Disturbances in the Himalayan Region

Hans Christoph Rieger

For quite some time Environmental problem in the Himalaya is a matter of concern for scientists of varying fields—botanists, zoologists, geographers, geologists, anthropologists—and the literature on mountain ecosystems is now quite lavish. However, most of it is highly specialized and thus narrow in scope. Whatever data is available does not fit together very well. So, attempts to link the different aspects of mountain ecosystems together into a general picture of the Himalayan environment suffer from the fact that empirical research up to now has been sporadic. The result is that the available attempts at a more general view of the Himalayan environment are rather journalistic.

In our attempt to describe the process of environmental degradation in this chapter, we will employ a rather more systematic approach. He will discuss the various macro-ecological ecosystems involved and to determine the relationships between them and the processes involved in the present rapid degradation of the environment. Our approach in this chapter is likely to apply on a wider scale, but we will be restricting ourselves to the central Himalaya,— part of the mountain chain belonging to the Ganga catchment (essentially Kumaun to east Nepal). At the same time our analysis will take into account not only the processes in the mountains themselves but also of their impact on the Gangetic plain. Starling with the natural factors, we will discuss man and study his influence on the ecological system in the hills. Besides that, we will describe the effects of ecological disturbances in the hills on the plains, and finally look at some other factors that make the situation worse.

Climatic Erosion

Natural erosion occurs in the Himalaya, due to massive downpours during the monsoon season (July to September). The south-west monsoon winds cross the Arabian Sea and the Bay of Bengal, picking up large quantities of moisture on the way, turn north-westwards at the head of the Bay of Bengal and then advance along the Himalaya from east to west. Here the moisture-bearing winds strike the longest, highest and one of the youngest mountain ranges of the world, causing precipitation levels unequalled elsewhere. The steepness of the terrain, the structural instability of the area, as well as the relatively young age of the mountain, all contribute to the erodability of the slopes, and a natural weathering process results.

In the high mountains of perpetual snow and ice, glaciers strike the mountain slopes and transport rocks and boulders to the lower valleys. The huge amount of water run-off caused by precipitation and snow melt, and the considerable speed of run-off result in a natural erosion process. It has been estimated that if heavy rains double the water flow, 'scouring capacity is increased four times, carrying capacity thirty-two times and the size of particle carried sixty-four times.' (Tempany and Grist, 1958, p. 88).

The climatic distress, topographic and geological conditions of the Himalayan mountain range lead to natural erosion. Without such a natural weathering process there would be no fertile river valleys or plains, or, for that matter, a fertile delta area in Bangladesh.

Depletion of Forest

When man interferes in natural processes, systems become more complex, and erosion in the Himalaya is no exception. The mountains serve as a source of resources for the population of the plains. To fetch timber for construction of railway sleepers, of resin for turpentine production, or of other forest products, excessive commercial exploitation of the mountain environment is carried on—as in some areas of Kumaun—thus forests are destroyed.

Although this form of environmental degradation is probably quite wide spread, it is, paradoxically, a less serious problem with regard to finding solutions. With growing awareness of the forest's importance for the mountain environment, deforestation should be controlled in the future. Hopefully the authorities in UP are realizing their responsibilities in this area of the Himalaya.

Another factor, which happens to be widespread in the Nepal Himalaya, concerns the destruction of mountain ecosystems by the resident population.

As the human population expands in the hills, forests are depleted or degraded. The water-retaining capacity of the natural vegetation is reduced and run-off is increased, both in quantity and speed. This, deforestation and the use of inappropriate agricultural techniques on unsuitable land leads to different forms of soil erosion and, ultimately, to the loss of cultivable land. Man's presence in the Himalaya makes itself felt in two ways: first, by increasing numbers, and second, by his behaviour.

To survive, man needs food, clothing and shelter as well as fuel for cooking and, according to location and climate, heating. The way in which he fulfils these needs depends on the traditional and culture-bound techniques as well as the norms and mores of the society in which he lives.

Productivity of Land

Cultivation requires land. Population growth results in pressure to expand the cultivable area by clearing forest lands. The land requirement depends, among other things, on the productivity of the soil, and this can be considerably increased—or reduced—by the agricultural techniques applied. Increasing population pressure also results in marginal land, i.e., less suitable and mostly steeper slopes, being brought under cultivation, thus leading to a reduction of overall productivity of the land. Finally, mention must be made of shifting cultivation which is still being practised in some parts of the Himalaya.

Terracing

The quality of terracing is subject to local variation. The classic rice terrace has an outer *bund* of about 30 cm height to retain the water, and thus provides good resistance to erosion. Maize terraces, on the other hand, are more susceptible to erosion, having no *bund* and being generally inclined outwards to prevent water retention. Irrigation of terraces can increase the productivity of the available land. But causes certain dangers. As the main rivers usually flow far below the terraces, the steeply flowing side-streams are tapped for irrigation at a level high enough to permit a flow along the slope of the side valley to the fields in the main valley. There are usually no facilities for limiting flow, so that floods caused by excessive rainfall may enter the canal and overflow at the point where the steepest slopes are traversed and where the canal is generally smallest, often causing big gully erosion (Tautscher, 1974, p. 8).

Extinction of Trees

In the Himalaya, livestock is usually held in excessive numbers partly for religious reasons, partly on account of the low yields of cattle, and partly to fulfil the need for animal manure. The forest is used for grazing all the year round, and the cattle feed on young plants and the leaves and twigs of small trees. Moreover, the leaves and twigs are cut off for cattle feed by the population. Gradually, whole branches are amputated and the trees are consequently considerably weakened. The thinning of the leaf cover reduces the protection of the forest soil, and induces the growth of grasses and thus poses a fire hazard for the forest. Chopping also leads to stunting and finally the extinction of trees. Only the trees which can survive such brutal treatment remain.

Existence of livestock affect the destruction of forests by fire. Many peasants set fire to the dry cover of grass and pine needles in order to improve the growth of fresh grass and to destroy the slippery pine needles which pose a danger to livestock grazing on the slopes. In areas where the burning of forest grass is practised regularly, a characteristic open forest of old pines without any young trees has resulted. Of course, self-incendiary fires also occur, but according to Donner (1972, p. 355), purposely-laid or carelessly-caused forest fires are the rule. Since yields on the primitively-cleared and cultivated land are low, the area burnt is correspondingly large.

Wood for Fire

Due to lack of alternatives the fuel needs of the population in the Himalaya are met by firewood. Branches of trees are used as firewood. According to Robbe, 1954, annual firewood consumption is of the order of 0.6 cu m per family or well over one million cu m for the whole of Nepal. However, this figure seems to be a gross underestimate, for FAO calculates the production of wood for fuel purposes in Nepal to be no less than 6.6 million cu m in 1967 (Donner, 1972, p. 354). In a study by Mauch, 1974, it is estimated that over 90 per cent of wood extraction from the forests in the Kalinchowk area is for fuel purposes.

Wasteful Use of Timber

The forests also have to meet the population's need for construction timber. This varies considerably according to climate and local building styles. As far back as 1931, Heske (1931, p. 571), criticised the wasteful use of timber in constructing houses in Tehri-Garhwal and concluded that an average family of five persons requires one tree trunk annually for construction purposes. Mauch (1974, p. 9), confirms that in the central Himalaya (eastern Nepal) about 70 cu m of valuable wood is logged per house, although less than 20 cu m would suffice if properly and efficiently utilized.

Timber and firewood extraction, forest clearance for cultivation, grazing, lopping for fodder and burning of the undergrowth, inefficient timber utilization these all together are causing a general degradation of the forests.

It is clear that the destruction of the forests is rapidly every year and that Nepal, for instance, is likely to be totally denuded by the end of the century. Without alternative sources of energy this process cannot be stopped (Mauch, 1974, p. 5). On the other hand, Donner (1972, p. 354), considers the provision of alternative fuel resources in the near future an unrealistic proposition.

Effects on Soil and Water

Present Condition of the forest and vegetation cover has definite effects on the water household of a given area. A good forest will provide relatively high infiltration-rates and a correspondingly low run-off of water, because of the relatively stable and porous condition of the soil and the protective layer of leaves and other organic substances.

Numerous experiments in Europe, America and Japan have confirmed the absorptive capacity of the forest.

However, the application of these results to Himalayan conditions is not without problems. Where rainfall is of long duration and intensive, as is frequently the case during the monsoon period, saturation may occur and the run-off-retarding effect of the forest may decrease rapidly.

Alternatively it has been proved that denuded soil soon becomes compacted to such an extent that run-off is accelerated and soil erosion results. The high kinetic energy of raindrops falling on the naked ground breaks up the soil and carries it away down the slope. The progressive drying out of the soil through overheating leads to a destruction of organic matter, reduced water-holding capacity, and increased evaporation of moisture.

To complete the picture of man-made erosion in the Himalaya, a number of additional factors may be mentioned. Thus we have seen that the erosion in the Himalaya is man-made. The cultivation of wheat in winter, propagated by several development assistance programmes, reduces the grazing area available to cattle. Recently, several authors have noted that the receding forests have made firewood procurement difficult so alternatively cattle dung is being dried for fuel. This again deprives the soil of valuable manure, so the productivity drops (Eckholm, 1975; Hogger, 1975). Finally, road construction also contributes to erosion. While some roads are aligned along the river valleys and are relatively secure, others are aligned higher up, thus changing the run-off of the slopes. The roads frequently have to cross geologically unstable slopes and thus cause numerous landslides. Where hill-side cutting is practised, the soil is simply dumped into the rivers, thus enhancing the sedimentation problems of the rivers for several years (Tautscher, 1974, p. 9).

Effects of Natural and Man-Made Erosion

The main interrelationship between natural and man-made erosion on the one hand and their effects downstream and in the plains on the other are explained in this section of the Chapter. Increased run-off makes itself felt in two ways: in terms of run-off quantity and in terms of run-off speed. To begin with, increased run-off quantities may temporarily exceed the capacity of downstream river beds and thus directly result in floods. In addition, reduced infiltration of water into the soil may reduce the groundwater level and thus lead to the drying-up of springs.

Mountain springs are important suppliers of drinking water and irrigation. According to Jahn (1975, p. 5), there seems little room for doubt that many springs have reduced their flow or dried up completely in the last decades because of the destruction of forests. In many villages visited by Donner the villagers confirmed that water now has to be fetched from farther off than a generation ago.

Heske (1931, p. 586), found a direct connection between the regularity of water flow of rivers originating in forested areas and those from areas in which deforestation was widespread. He predicted that detailed analyses would prove the relationship to be more pronounced in India with her long dry periods. However, such detailed analyses do not seem to have been made up to now. In any case, they would require long-time series of measurements, which are not available at present.

The continuous erosion of the bedrock due to increased speed of water flow reduces the support of the slopes, thus causing landslides. Slope portions right above the gorge, which are kept in place only by friction but have no support at the foot, slide into the gorge, removing the support for higher areas and increasing infiltration into the formations behind the steep rock face (Tautscher, 1974, p. 3). This type of erosion not only extends uphill but can also lead to the collapse of the steep face, causing the very dangerous damming up of big rivers. Sheet erosion, gully erosion and landslide erosion, together with the scouring effect of torrents and streams, produce sand and silt in huge quantities which is transported to the Gangetic plain.

Thus the Karnali shifts 75 million cu m of solid material annually, equivalent to a 1.7 mm layer for the whole catchment area (IBRD, *Nepal Agricultural Sector Survey,* Vol. III, Annex. 6, p. 2). The sand and silt are deposited when the rivers reduce their speed, i.e., when they break into the plains, thus constantly raising the river bed.

According to Nepali observers the beds of the *tarai* rivers are rising by 15 to 30 cm annually (HMG, *Draft Proposals of Task Force on Land Use and Erosion Control,* 1974). This leads to flooding as well as to quite considerable shifts in the courses of rivers. The Kosi in Bihar, for instance, has shifted its course 115 km westwards within two centuries, leaving 15,000 sq km of once fertile land buried under a mass of sand and rubble, and displacing 6.5 million persons.

In addition to these effects, excessive sedimentation silts up dams and irrigation canals, thus rendering such construction works useless after a short period of operation. It is interesting to note that this problem is in no way new, as quotations from engineering reports from the turn of the century prove (Heske, 1931, p. 581).

That flooding in the Gangetic plain and in the delta also serves to spread fertile soil in these areas. According to IBRD, Nepal Agricultural Survey, Vol. III, Annex 6, p. 2,240 million cu m of soil are 'exported' from Nepal every year. Newby *(Ganga,* 1974, p. 44) locates the end of the Ganga under the sea a hundred kilometres to the south, where the long tails of sand run down towards the deeps of the Bay of Bengal or even further 'Four hundred miles out in the Indian Ocean the sea is discoloured by the silt brought down by her.'

Pressure of Growing Population

Migration is creating new problems. We have already read in same previous paras that population pressure in the hills has led to deforestation and erosion, and the situation is becoming more critical every year. Many have no other choice than to look for alternative means of livelihood. The *tarai* at the foot of the Himalaya now presents such an opportunity since the main reason for the low density of population there has been removed by the eradication of malaria. In fact, the *tarai* belt is gradually filling up.

There are two types of settlement in the *tarai.* The Government of Nepal, together with international organizations like FAO and IBRD, is implementing organized resettlement schemes in various places.

By far the largest group of settlers is constituted by illegal and uncontrolled settling or 'squatting'. The World Bank estimates that 400,000 settlers migrated to the *tarai,* accounting for 180,000 ha of which 130,000 were illegally settled on. FAO quotes an even higher figure and estimates that about 36,000 ha of good forest land are lost annually through squatting (UNDP/FAO, Forest Development, Nepal, Land Use, 1974, p. 7). On the basis of ERTS-satellite imagery, FAO estimates that no more than 818,600 ha of forest area remain in the *tarai* (*Nepal, Land Use,* 1974, p. 7), and Eckholm (1975, p. 765), points out that less than half of this will be suitable for cultivation. If migration into the *tarai* continues at the pace of the last ten years, all the good farmland will be occupied in little more than a decade.

By now it is clear that the deforestation in the hills, is indirectly resulting in the destruction of the forests in the *tarai.* Both squatting and legal settlement are dependent on forest clearance, while a growing population is placing increasing demands on the dwindling forests for firewood and construction timber. This is not easy to stop, since Nepal depends on the export earnings for the import of essential items to meet the needs of a growing population.

But destruction of forests in the *tarai* does not have the same destructive effects as in the hills. However, another form of erosion, i.e., wind erosion, is induced where adequate precautions are not taken. The sandstorms in UP during June, for instance, bear adequate witness to this fact.

The Four Subsystems of Ecological System

In the previous sections we have depicted four subsystems of the ecological system of the Himalaya and the Gangetic plain. In the first, a natural weathering process in the hills and the transportation of fertile sediments to the plains constitute the natural erosion which cannot be curbed even by the complete restoration of the Himalayan forests.

In the second subsystem we focussed on the effects of population pressure on the existing forests, the soil and the water household. This was followed by an analysis of the downstream effects in the river valleys and the plains, which are likely to cause increases of floods and droughts both in terms of intensity and frequency.

In the fourth subsystem we have depicted the effects on migration of a destruction of the mountain environment and the way in which migration, a stopgap solution in any case, is likely to pose new ecological problems in the *tarai.*

Two points need special mention. First, several of the processes described are practically irreversible. Afforestation is a lengthy process with difficulties and setbacks. Where the soil has been eroded down to the naked rock, afforestation is not possible. Once the soil is washed away it is no longer usable except at lower levels. Secondly, in case of man-made erosion, there are a number of positive feedback loops which have the effect of accelerating the shift towards imbalance. Loss of soil through erosion vigorously attack the remaining forests, which only serves to further erosion and loss of soil.

Nearly all studies of Himalayan ecosystems have contented themselves with examining causal relationships of the type we have tried to integrate in the foregoing analysis. In general, these relationships take the form of the statement 'if event A occurs, event B will follow'. But there is no indication of the time periods involved, which alone would give us an idea of the magnitude of the problems. For that a detailed research is to be done.

Population growth is estimated to be of the order of 2 per cent per annum, taking into account a slightly higher natural growth-rate and the migration to the *tarai.* If the average size of households is assumed to be five persons, and the average size of agricultural holding is 0.5 ha, the amount of purely agricultural land required per capita is 0.1 ha. However, additional areas are required for housing, common grazing land, tracks, etc. so that a land requirement of about 0.2 ha per capita would seem fairly realistic. The extraction of firewood, construction timber, and leaves and twigs for cattle fodder, etc. is about 1400 kg per capita per year according to estimates made by Mauch (1976). The natural forest density is about 360 tons per ha of timber, twigs and leaves, while natural forest growth, given sufficient space, is about 5 per cent per decade. On the basis of these assumptions, we can conclude that Because of the exponential growth of population, extraction from the forest increases so rapidly that it exceeds the natural regeneration of the forests. Within a few decades, the remaining forests are depleted to the point of complete destruction.

It is indicated that in some areas of Nepal, particularly in the eastern middle-hills, the point has already been reached where the extraction from the forest is greater than natural regeneration. Mauch has calculated for an area of some 840 sq km with some 160,000 inhabitants between the Sun Khosi and the Khimti Khola that, from the stock of some 9 million tons of timber, about 170,000 tons are withdrawn per year, but only about 60,000 tons are reproduced (Mauch, 1976, p. 127). How this wide spread disaster can be prevented is a serious problem for the statement in the above portion of the chapter should not be taken as a statement or prediction of what will happen in the future. This techniques of agriculture and forestry are looked upon as constant although they are bound to change with the deterioration of the environment. The purpose of the above discussion, however, is to indicate the dangers inherent in a static view of the ecological system and to show the way in which future research could be geared could provide specific guidelines for policy-makers by indicating the ecological carrying capacity of a given area, e.g., a river catchment. It is in this direction that efforts to analyse the Himalayan ecosystems can eventually ensure that man and mountain come to terms of their natural future benefit.

Find a Solution

Possibly the degradation of the environment in the central Himalaya cannot be stopped and that the area will certainly deteriorate into a mountain desert. Before we submit ourselves to the fact that the Himalayan hills are going to be lost for human habitation once and for all, we must make all possible efforts to see whether and under what circumstances a solution might be possible.

Examine Individual Factors

First we should examine the individual factors of the ecosystem with a view to pinpointing those points that are capable of being influenced by policy measures of the government. At this stage it is neither necessary nor profitable to enquire into the practical feasibility of individual policy measures. The purpose is rather to identify those points of the system that can be influenced by man in some way; for instance, while it is impossible to influence the monsoon, it is not impossible to influence population pressure in a given region, even though the prospect of achieving this may look slim when examined in the light of present facts. It is clear that within the subsystem of natural erosion there is very little scope for influencing the course of events, since the forces of nature at work in the Himalaya are literally gigantic.

However, there are a number of marginal possibilities for reducing the run-off speed of water in some torrents and streams by man-made structures, such as check dams, river-bank consolidation etc., and these should not be overlooked in the integrated approach. On their own, of course, they pale into insignificance.

In man-made erosion, perhaps the most important in the long run is man's number, i.e. the pressure of population. Any measures to decrease the population or even to merely reduce its rate of growth must obviously have a beneficial effect on the ecological balance. Besides efforts to effectively reduce the overall birth rate of the country's population, this can also include measures to draw a proportion of the existing population into alternative employment and habitat outside the Himalayan ecosystem. Secondly it is necessary to influence man's behaviour as expressed in the techniques of forestry, construction, irrigation, terracing, etc. he employs, in order to mitigate its negative effects on the environment.

Further, livestock is obviously a central variable of the system. The number, type and productivity of the cattle and other livestock population can be influenced by policy measures in order to bring it more into line with ecological requirements. The provision of alternative sources of fuel for cooking could reduce the demand for firewood, while a reduction of demand for agricultural land could result from the provision of food from outside the system, or from an intensification of agriculture on the more suitable cultivable land.

To reduce the effects of excessive run-off speeds, some technical measures such as river training works, check dams, etc. are useful. An increase of controlled settlement would reduce the negative effects of illegal squatting. Finally, the preservation of border lines of forest trees in areas cleared for settlement or the plantation of hedges could reduce the dangers of wind erosion to a great extent.

Multi-Pronged Approach

Rather than listing a number of individual, isolated problems connected with the mountain environment, we described the system as a whole along with its various subsystems. All the problems are interlinked, and this has obvious implications for their solution: it is unfeasible to tackle analytically isolated variables of the system which, in reality, are not isolated but closely linked with other elements of the whole system.

Most attempts to tackle the problems of mountain environments have suffered from a too narrow approach in that they have focused on one or two variables only, e.g., afforestation programmes; erosion control works; river training works, etc. If our view of the Himalayan ecosystem is correct, it would follow that the approach to restore its balance must be multi-pronged and on a broad front, and must co-ordinate, integrate and properly phase a number of quite different but interrelated measures. This requires not only an intricate plan of action for a given area and a particular time, but also a sensitive and flexible controlling agency capable of adjusting quickly and smoothly to changes of the system with which it interacts.

Closely linked with the systems aspect is the multi-disciplinary character of the problem of the Himalayan environment. Reality seldom does us the favour of falling into the neat compartments we make for it at our universities, and the Himalayan ecosystem is no exception. An examination of the available literature shows that there is no dearth of specialized works on or about the ecology of the Himalaya by botanists, hydrologists, foresters, economists and anthropologists. The Himalaya do not, however, consist merely of a set of plant species arranged horizontally or vertically, or of ethnic groups scattered more or less at random among the hills. At the risk of leaving much to be desired for each individual discipline, we must try to cover all aspects sufficiently to do justice to reality. This means that we must combine the expertise of many different scholars from many different fields in order to tackle the problem on a sufficiently broad front.

Mechanisms of Social Behaviour

From our analysis of the Himalayan ecosystem one thing is clear that the problem we are dealing with is a social one. It is the man's influence that has upset the natural balance, and it is in the adaptation of his interaction with the environment that we must seek for ways to restore it. The system within which man operates has biological, hydrological, technical and climatological facets also. But unless it becomes possible to influence the behaviour of the population in the hills, all hopes of saving the fast-deteriorating ecosystem of the Himalayan region will be disappointed.

To avoid the unrealistic solutions, it is important to have a clear understanding of some of the mechanisms of social behaviour operative in the hills. It is probably true that nowhere else is the knowledge of the ecosystem and its rapid deterioration more widespread than among the hill population itself. It is delusive to suppose that the forests are being degraded and decimated merely because of the ignorance of the population in the hills. On the other hand, destruction of the ecosystem is taking place despite they are quiet aware of the situation.

Priority of Present Over Future

Hill people often act without considering the future effects of their actions. Steep slopes are cultivated, although he knows that he may thereby endanger the future of the hillside. Or slopes are denuded, although the villagers know that the water household on which they depend for their very life may be disturbed.

There are two distinct causes for their such ignoring behaviour, about the future, one involving long-run considerations, the other of a short-range nature. The first refers to actions whose future reaction will be felt only by the next generation. Donner (p. 351) reports that when he asked Nepalese hill dwellers why they did not plant trees for firewood, the answer was: 'Trees grow slowly, and we will not live to see the day on which they can supply us with firewood.' This attitude seems to be closely bound to a cultural and religious philosophy that accepts the environment as given and looks upon developments within it as an expression of an external being's will. In this view of the world the next generation will be supplied with its own resources and cannot depend on those available to this one.

But the second, i.e., short-range only has a completely different background. Whenever an individual is in acute distress with regard to the fulfilment of his primary needs, his actions are directed towards sheer survival. He does not bother even if this entails new and perhaps greater dangers for himself. This priority of the present over the future expresses itself, for instance, in the cultivation of marginal land although it endangers the surrounding area, for the simple reason that no feasible alternative to starvation exists.

Ignoring the Social Losses

Costs which do not affect the cost calculations of those individuals who incur them but which have to be borne by the society are called social costs. Production or technology decisions which ignore the costs to society are individually profitable although they may be a loss for the society.

In some areas of eastern Nepal tree trunks are burnt in order to speed the drying process in the production of paper. The fact that the producers look upon and utilize trees as a free good with no market value, leads them to continue this practice. But it is great loss for the society. Trees at one time were so plentiful that they could be looked on as free; now the practice has reached a point where each additional tree felled represents a high price to society, and society must pay in the absence of methods to recover it from the feller.

Needs a Social Organisation

But heve we ever thought why this practice is followed? It is because one dose something what no body wants him to do for the fear that some body else will. It is evident in the lopping to death of nearby trees rather than walking a greater distance to an area where the trees are in a better state and able to be lopped without danger to their regenerative powers. If the tree nearby is spared by him another may come and cut its leaves. This very expectation directs the first man to take the first step for personal profit.

Same is the reason behind indiseriminate grazing. Nobody makes the first move by restricting his own cattle from grazing in the area, because by doing so he would lose a grazing opportunity while his neighbours profit from the reduced grazing density.

In both cases it requires to form a collective social organization to bridge the gap from a stable equilibrium at a sub-optimal level to another equilibrium at a higher level of satisfaction. This collective organization can take the form of mutual agreement or compulsion or a mixture of both.

Project to Influence Human Behaviour Needed

As this form of social organization has been created at the national level, there is danger of a lack of identification by the rural population with the public purpose. This is caused by the fact that at the village level the national state and its central government appear as anonymous agents rather than as an expression of the individual's own efforts to further the welfare of society. This danger is by no means limited to developing societies. It is common in all societies in which the central government assumes powers to further the public good and exercises these powers through legislation, administration of justice, bureaucratic procedures, and the institution of public property.

In Nepal, where the villagers have always made use of the forests around them, the institution of reserved forests, created in the national interest, is frequently viewed as an unwarranted restriction on their traditional rights brought about by an allen power.

Effort to preserve and protect these national forests are often hindered by the actions of the local population, precisely because their individual rationality diverges from collective rationality or, in this case, national rationality.

To conclude that where socially irrational or seemingly irresponsible behaviour is observed, the individuals concerned are irrational is incorrect. Precisely because they act in an individually rational manner, socially irrational behaviour results. It is important to bear these principles in mind when designing a project strategy with the intention of influencing human behaviour.

Appropriate Policy Measurers and Awaikening of World-Wide Interest

The problems outlined are obviously tremendous in several ways: throughout the length and breadth of the Himalaya—in terms of the many aspects of the environmental system of which the problem is a part, and in terms of the many aspects of the problem which must be taken care of in tackling them. Obviously it is not feasible to propose and implement a programme to tackle the problem in all regions of the Himalaya at the same time. On the one hand, the finance and expertise that can be arranged are not enough with regard to the geographical magnitude of the problem; and on the other, presently there is no fool-proof methodology available to tackle the problem at present. So any project proposed is to some extent only experimental in nature.

Initially the project to tackle the problems should be small enough to permit an intensive input of all the required for controlling the ecosystem. This will permit a certain variation of input intensities so that conclusions on the best mix of policy measures can eventually be drawn. This can have important cost-saving implications for future projects in other areas.

To control such large and complex ecosystem efficiently, international cooperation plays a major role. This has two advantages. First and foremost, such co-operation would ensure the availability of the experience gained in other countries, and could bring to bear on the problem of the Himalayan ecosystem the best expertise available in the co-operating countries. Secondly, the knowledge gained in the Himalaya could be propagagated on a wider international scale, thus raising worldwide interest in the Himalayan problems. This may awake the world-wide readiness to fund the large-scale programmes that may emanate from the initial pilot projects could be greatly enhanced.

Bibliography

Donner, Wolf, 1972, *Nepal, Raum, Mensch und Wirtschaft.* Band 32 der Schriften des Instituts fur Asienkunde in Hamburg. Wiesbaden.

Eckholm, Erik P., 1975, 'The Deterioration of Mountain Environments; Ecological Stress in the Highlands of Asia, Latin America, and Africa Takes a Mounting Social Toll. *Science* Vol. 189.

———, 1976, *Losing Ground, Environmental Stress and World Food Prospects,* New York: W.W. Norton.

Heske, F., 1931, *Probleme der Walderhaltung im Himalaya.* Berlin.

HMG Nepal, 1974, National Planning Commission Secretariat. 'Draft Proposals of Task Force on Land Use and Erosion Control.' Kathmandu (mimeo).

Hoeoger, Rudolf, 1975, *Die Schweiz in Nepal—Erfahrungen und Fragen aus der Schweizerischen Entwicklungszusammenarbeit mit Nepal.* Bern und Stuttgart: Verlag Paul Haupt.

IBRD/IDA, 1974, 'Nepal Agricultural Sector Survey, Vol. III. Washington DC.

Jahn, Hans C., 1975, 'Forstwirtschaft im Distrikt Almora, Provinz Uttar Pradesh, Indien'. Munich (mimeo).

Mauch, S.P., 1974, 'The Long Term Perspective of the Region's Forest Resources—and the Associated Availability of Firewood, Cattle Fodder and Construction Material for the Local Population.' Integrated Hill Development Project. Zurich (mimeo).

2

Himalaya's Strain Environment Calls for Quick Action

Introduction

Himalaya—a fairyland surrounded by snow covered mountains with green, dense forests around it—is gradually losing its beauty. Only debate across the table cannot help to find the solution. Is tourism allow us responsible for this? But we cannot keep men away the beauty of Himalaya. So tourism has to be there. Then? What is the solution? This aspect has been discussed in this chapter.

Man-mountain interaction-Venture Tourism is spoiling the Himalayan environment. It is already under stress. Now the experts establish it as a alarming fact. What makes us worry is the fact that once most attractive place to draw the attention of hordes of visitors—Himalaya—the abode of celestial snows, is losing its environmental beauty.

Man is of course known for doing nothing beneficial for the environment wherever he goes. Why, the whole world is ringing with the alarm bell! But imagination goes haywire with the thought that man could destroy a thing of beauty that has been a joy for ever for him. That's what shatters something inside one's soul. The situation seems to be going out of hand and calls for quick action.

Effect of Tourism

Think of the Himalaya, and we visualize the entire spectrum of a fairytale world—embracing high mountains, mighty glaciers and ferocious rivers. Blue hills, green valleys sporting forests, meadows and pastures throbbing with floral wealth, a veritable cascade of colours. It's indeed here that our study begins. In these highlands is spread a constellation of cultures spanning like a rainbow the entire 2500 km stretch, East to West, of the Himalaya. The people inhabiting the hills are indeed the heartbeats of the Himalaya. A study of environment of the Himalaya has to be related to the whole—the land and the people. Over the ages, the people there have developed a relationship of considerable intimacy with the nature around them. So much so that disturb one and you disturb the other. Tourism had done exactly that—disturbed that balance. How? That will be discussed in the rest of chapter.

To debate an issue across a table is different from listening to the eye-witness accounts. Definitely the wordy images weave and carry an instant impact on one's mind. We have already heard Sir Edmund Hillary sketching scenarios of the Nepal Himalaya, as he saw, then (in 1940s) and now (in the 1990s).

To achieve objectivity, we would listen to an eye-witness, a perceptive lover of the wilderness. Jeoffrey Campbell dreamed of Ladakh as a kind of Shangri la James Hilton had envisioned in his celebrated work *The Lost Horizon.* It says: "The incredible beauty of the vast, dry, trans-Himalayan landscape, thinly populated with small villages perched on cliffsides or nestled in narrow gorge, combine the wonder of a lonely 'untouched' wilderness, with the novelty of an 'unchanged' civilisation". He went there in search of his Shangri la after Ladakh had been opened to Tourism in 1974, by which time of course many visitors had travelled through the length and breadth of the land. His dream turned into a nightmare. The idyllic image on the mirror of his mind was shattered; he had the frustration of a person who had idealised the real but failed to realise the ideal. As he proceeded along the points on his itinerary, he discovered something vital had been lost.

Here is the image he transmits about the landscape: "The most immediate and glaring examples of the debasement are found along the major trails and campsites in the profusion of undisposed litter and garbage. It's sad reality that tourists representing highly advanced material cultures can leave behind them, with apparently little guilt, trails of their industrial detritus. Candy wrappers blow along the dusty trails, and imitations of dried rhuberb leaves and rose petals. Smashed bottles lie amidst the rocks. Tins, soup packets, cardboard boxes and chicken feathers dominate the surroundings at camp sites. And of course the supreme insult to the environment is the ubiquitous crinkled toilet paper. An eye-sore everywhere, toilet paper in Ladakh is the more nauseating as the extreme dryness of the atmosphere makes disintegration and decomposition a process of years ... Even rubbish like tins of half-finished tuna fish and boxes of spoiled noodles which are given a cursory burial a few inches below the ground, are soon dug up by prowling village dogs ... Water channels and irrigation canals, often several kilometers in length, may serve as the only source of water for villages high up on a mountainside. Imagine the unhealthy effects upon a villager used to drinking this perfectly clean water, when a large group chooses to bathe, defecate, wash dishes or clean their chickens in it just upstream...".

And now some flashes to touch the cultural aspect of the landscape: "The invariable clashes of differing cultural concepts which occur between trekkers, operators and locals complete the picture. Most saddening amongst the cultural faux pas is the baseless degradation of religious monuments, in particular *mane* (prayer stone) walls and *chortens.* It is shocking to see, side by side with European signatures, the names of many Kashmiris and Ladakhis scrawled on rocks and even over prayer stones, and in the mortar walls of *chortens.* Surely a more respectful attitude towards these monuments should be encouraged, not only among foreign tourists but among the local operators as well. In a land where both men and women cover themselves with adequate clothing, the tendency towards almost total nudity on hot windless days, on the part of many trekkers, is also in direct contrast to social mores and certainly changes the landscape. Further problems arise from the indiscriminate distribution of 'bon bons' to local children which is fast turning the younger villagers into persistent beggars, urged on often enough by their parents."

Jeoffrey did not mention, perhaps he was not aware of the disappearance of some religious relics of immense cultural value from the treasures preserved for ages in the monasteries where prayer-wheels whir, day and night, and whisper of the Jewel in the Lotus.

What venture tourism has done to Himalayan environment is expressed in the above statements. The environment there had enabled the highlanders to overcome the poverty with patience, poise and piety in that Godly society. But after the contact with lowlanders from the materialistic society the environment these has totally changed.

Now the writ see "What we've done to Venture Tourism in the Himalaya." Sir Ed said in the conference: "One of the great joys of venture tourism is getting away from the mob, enjoying the peace and contentment of areas far from the public pressure—bubbling streams, empty forest paths, and lonely mountain passes. We all need to escape from the concrete jungle and experience Nature at its best. We can then return refreshed to our city environment and dream of sparkling waters and icy mountain tops." Viewed through the prism of imagination, this escape theory appears so exciting in the Himalayan context that we feel tempted to wedge in our own.

Surely, it was no tectonic freak, as the geologists would have us believe, that two continents clashed causing the earth to shoot up huge rocks out of its belly, hanging them so high in the sky that clouds could not cross over to the other side stealing water they carried and, at the same time, the cold Siberian winds could not blow in over to this side. The fantasy spun on for eternity, to imbue the abode of snows with magic milieu. This abode of snow was made for man with no boundaries, no barriers laid. So, keeping man away from the hills made no moral solution. Tourism had to come.

Over the years venturism has become a significant segment of tourism, ever burgeoning as man is informed by a new spirit in search for leisure, finding more pleasure in not killing time but making time come alive. Why, it has grown into an industry, a new developmental activity. Remember the old adage, coined by some sage, that Time and Tide wait for no one. You have to be in tune with the times and swing with the tide. This brings us to the crucial stage of our study. You would no doubt argue that the value of Venture Tourism is alright but what about its adverse impact on environment. To find an answer to this we will have to do a bit of self-analysis and dive into depth to discover some harsh home truths unperceived by us.

Extreme Insensitiveness to Environment

Mass tourism is a world phenomenon. No doubt about that. The crash is not only here in the Himalaya, it's elsewhere too, everywhere. Take for example the European Alps. These mountains are shared by several nations as we do the Himalaya. They are smaller in size and stretch, comparatively speaking, but are thrashed by throngs in season, out of season, all the year round. There's a virtual explosion of numbers—people come from all parts of the world to climb, picnic, ski, skate or simply suck the scenery. There are cafes, chatelets, hotels, inns and huts dotting the entire landscape. What's more, as Dr. Tacke, Director

ISIMOD told us in the conference, the Alps are not only overused, they are also over-mechanised.- cabins, ski-chairs, ski-lifts, telepheriques and what not. And yet there's no aesthetic degradation of the kind we find in the Himalaya, Why? Have we ever asked ourselves this question? A forthright Englishman rightly asked: if we can keep our mountains clean, why can't they? Firstly, it is because insensitiveness to environment is extreme. We call Himalaya as the abode of gods and gabble about godliness but that's about all. Nothing is farther from our mind, that the fundamental precept of living that cleanliness is next to godliness. Dirt does not disturb us, trash does not trouble us. They have become a part of our environment. Look at the holy spot like Badrinath, the pilgrim's paradise, surrounded by the most soul-sustaining scenery, you would find ugliness rearing its head around in the shape of slums gaining ground. And the hill stations—Srinagar, Shimla, Mussoorie, Nainital, Darjeeling—once the health resorts are turning into health hazards. Got the point?

No Management Plans

Secondly absence of ethics all along the track, from top to bottom. Take the men at the helm in the entire Himalayan regions. Tourists were sought for the money they brought without any thought or impulse of a host. No adequate infrastructure was laid, no management plans worked out. The visitors came, they spread and their number spiralled. From the moment they arrived, they were left to fend for themselves. Nothing was more obvious than the fact that the far and remote trails the tourists will tramp would not enable them to live off the land, so they brought their own things to eat and drink. They could not find fuel for cooking, heating and warming purposes, so they had to find fuel-wood from the forests as the only alternative. There were no huts or holes to worm in for shelter and toilet. On top of that they were enjoined not to litter their left-overs and try to take away the trash.

Atmosphere of Avarice and Indifference

The visitors alone are to be blamed. Behind them trailed many others. Men from the mushrooming travel agencies, contractors and go-betweens, porters, guides, animals with men who tended them, errand boys to boot. They too eat and drink and defecate. Being hazy about hygiene, they were the least troubled by trash. In such a motely crowd one finds that not only the sights are ugly but the behaviour is uglier.

The pollutants are not debris and deforestation but the human behaviour. All through there is an atmosphere of avarice and indifference.

Proper Demarcation and Basic Facilities

A proper masterplan for management is the need of the moment appears to be the only solution. Demarcation of areas are needed.

Some to be demarcated as sacred place like the Nanda Devi Basin, others can be turned over to tourism with arrangement for strict management like the Kanchanjunga National Park, the Valley of Flowers National Park and the proposed Gangotri National Park. This would not only help control and channel traffic but also conserve forests and wildlife. New areas are to be opened, hitherto unused, for tourism so as to relieve pressure on areas over-used. Besides demarcation definite infrastructures will have to be built, layer upon layer, offering basic facilities to the visitors they need. Stores, to provide articles of food, drink and fuel, have to be situated at convenient points on the track. Some shelter has to be there all along the trails with minimum amenities to lie down, do toileting, collect trash, and keep them clean. Then, strict guidelines have to be worked out delineating the environmental responsibilities for all groups who are also to be educated on the code of conduct. Above everything else, the authorities have to get activised to play their part more effectively.

Duty to Preserve Beauty

In the context, the Himalayan Adventure Trust has come not a day too soon. Looking at the aims and objectives and the plans of action of the Trust, one hopes that the problem lies in reliable hands.

We want that adventure tourism must stay and flourish in the future. At the same time we also want that the beauty and purity of the abode of snow must be protected and preserved for all time to come. In fact the two are complementary to each other. It's the magic of environment and the natural attractions that lure tourists to the land and heighten their enjoyment. If we understand this, then we must also understand that if we do not manage our mountains, two awful possibilities stare at us. One: increasingly controlled wilderness. Two: no wilderness at all.

The flecks of philosophy to follow were visible in the message Sir David Wilson, Governor of Hong Kong. In his message to the Himalayan Adventure Trust on the occasion of its inauguration Sir David Wilson

Governor of Hong Kong, said: "Time has not changed the dramatic beauty of the mountains but it has made them more accessible. With increased opportunity to see the beauty of the Himalaya goes a duty to preserve that beauty." "Duty to preserve beauty" is the new environmental ethic Himalaya call for. That should be our motto and motive whatever our dream and destination. Let's shout this slogan and carry on the silent revolution with resolution.

3

Himalayan Expedition By A British Climber—A Review

Introduction

A renowned British climber, Sir Chris A. Bonington had his first Himalayan expedition at the age of twenty five. It was the first time that he had been out of Europe. His first sight of Himalaya, His first encounter with its people. Here is an extraction from his diary written at the start of his first Himalayan expedition:

"Here's a smell of wood smoke in the air and the rounded form of hills blur in the evening haze. My body tingles with the feeling of love for the sounds and shapes around me—a feeling of voluptuous excitement that just wants to absorb everything that is so new and strange."

In his diary Sir Chris A. Boinington has given a detailed pen picture about his expedition. He states: Before the Second World War, the country had been closed to foreigners and although there had been a less number of expeditions, there had been few tourists and no trekkers. The pressures that have since made themselves felt, partly through the growth of the tourist trade, but mainly through the increase in population had barely made themselves felt in 1960. In Kathmandu, there was only one western style hotel and it certainly wasn't either modern or five star. The Royal Hotel was situated in a moldering Rana palace, managed by a Russian called Bons. There were hardly any tourists and comparatively few cars or lorries in the streets. The road from Raxaul on the frontier to Kathmandu, had only recently been completed and this was the only road in Nepal.

He was a member of an international expedition, comprising members of the armed forces of India, Nepal and Britain, attempting the first ascent of Annapurna II, a mountain just below the magic height of 8000 metres. He was not the first to attempt it. Two small expeditions, one of them led by Bill Tillman had already been there in the early fifties, but their impact was very light. They started walking on the outskirts of Kathmandu and did not see a single trekker or climber throughout the approach. The route they followed has become one of the most popular treks and today many hundreds follow the same trail. Their expedition comprised ten climbers and nine Sherpa high altitude porters. It was traditional in every respect, for Jimmy Roberts the man who gave birth to trekking industry in Nepal with the founding of Mountain Travel, had climbed in the Himalaya before the Second World War. He was Gurkha officer and serving as military attache in Kathmandu. He knew and understood Sherpas well, looked after their interests and received their complete loyalty. This expedition was of special value for him. If gave him a learning experience of how to go about leading an expedition and in establishing a good relationship with our Sherpa employees.

Next year, he joined another expedition to Nepal but with a smaller team. This time to Nuptse, the third peak of Everest. This time he saw three other Europeans on the entire approach march to Solu Khumbu. He met Autschnarter Heinnch Harrer's companion in Tibet in their famous escapade and sojourn just outside Khathmandu and then two members of Ed Hillary's Scientific Expedition. Because it was a major trade route, there were some litt͜le tea shops by the wayside. Namche Bazar was about half the size it is today with none of the hotels that have sprung in the last fifteen years.

Effect of Tourism and Mountaineering

It is easy to regret the changes that have taken place over the last thirty years. The travellers love to find places that are untouched and unspoiled by the impact of the industrial world. Tourism and mountaineering have undoubtedly brought many benefits to Nepal and Solu Khumbu in particular—the schools started by Ed Hillary, the clinic in Khumbu and the overall prosperity that can be seen in changes in the architecture with large glass paned windows, cowls over the cooking fires to reduce the smoke that filled the traditional Sherpa living room, a wider variety of crops and a better balanced diet.

But the growing deforestation throughout Nepal is the most serious problem. The prime cause of this is the overall increase in population, the result of greater prosperity and better medical care. This has led to the clearing of more forest for cultivation and increase in cutting of wood for fuel. The development of tourist accommodation has undoubtedly aggravated the problem with cooking going on throughout the day and in the evening, the promise of hot showers on all the main trekking routes.

The environment problem has become so accute the solutions are not very easy. Effective birth control can start limiting population growth. More efficient farming methods can increase yields without putting more land under cultivation and perhaps most important of all, fuel can be conserved by using more efficient cooking methods—simple enclosed wood stoves rather than open fires favoured throughout most of the Himalayas. The greatest saving of all though would be by making greater use of the vast power that is available. Small hydro-electric schemes are already being developed and there is need for more. This of course requires a heavy investment, followed by effective support for the essential follow-up maintenance of the schemes.

How to Find A Solution?

Can Tourists Help in Solving the Problem?

And what of the tourists? Certainly. All trekking groups and climbing expeditions could help the fuel problem by being self contained, bringing in their own food and cooking stoves and fuel. Not just for themselves but for their porters as well. This might not be popular with the local people, who are developing their business by catering for tourists. The ideal long term solution would seem to be one where alternative fuel resources can be found but also where there is some kind of control by giving key areas national parks status to preserve the very beauty and integrity of the society, which of course is also its greatest long term assets. This is already being done in Sagarmatha National Park and also in Annapurna Sanctuary, a project that is run and controlled by the local people themselves. The problems of tourism are not as acute as the problem of population growth and deforestation. In 1987 and 1988 when Chris visited the Rongshar Gorge and Menlung Chu on way to Menlungtse, he was shocked to observe an intensive logging operation of timber for building. The process was bringing a local boom to the economy of the villages of Rongshur George. This could destroy some of the finest unspoilt forest in the Himalayas, if they carry on at their present rate. In doing so, they will loose a valuable source of income unless it can be carefully planned with restrained logging.

The threat of extensive deforestation across the entire range must be the most critical problem in the Himalayas. It is true that trekkers and climbers can only influence in a small way but this must not deter us. Each small effort contributes both in its own effect and also as an example.

In Nepal the situation is more complex since the local economy is undoubtedly benefited by tourists and their provision for food and accommodation. If they were unable to offer this service they will suffer a decline in revenue. Another solution is the alternative source of power which has already been introduced in Solu Khumbu with a number of small hydro-electric schemes. Definitely there is a problem in financing and then effectively maintaining them. The projects need to be heavily subsidized, not only in their installation time but also in their running costs since it is unlikely that the local people would be prepared to pay for electricity when they have been accustomed to getting their firewood free of charge.

Another impact that the trekker and climber have on the environment is the mark he or she leaves in the shape of rubbish that makes the base camps and the camp spots pmpure. This is something we can and must try to stop about. Most of us have been guilty in the past, however careful we have tried to be. If the rubbish has been dropped into the crevasse, the crevasse will open at some later stage, leaving the rubbish exposed. The only solution is to burn as much possible and to carry the rest out. There still needs to be some kind of effective rubbish disposal system, ideally at the last village on any major trail.

An environmental pressure group in Europe—'Mountain wilderness'—has come up with is an expedition to clean up the waste combined with the introduction of waste recycling plant. It will be set up in the Skardu and Gilgit. An American group is planning an expedition to the North side of Everest with the same aim, but is placing their recycling plant at the first village so that the villagers can profit from the enterprise.

To improve the code of practice of mountaineering expeditions and particularly the trekking agencies the Himalayan Environment Trust with its representation both from the host Himalayan countries and among the visiting climbers has set the realistic aims concentrating

Ideally the expedition should clear its fixed ropes away, but in practice this is rarely feasible because of the risks involved. The same is true of the Everest Ice fall, where removal of ladders from the dangerous positions is often risky. It is important to get the problem in perspective. A few ropes or perhaps ladder sticking out of serac tower are not visible from distance and are not threatening the environment. The impression is cosmetic. If there is an excess from an expedition, it becomes aesthetically unaccordable and perhaps even danger to healthy environment but in small quantities it can be lived with. I believe that we should do away as much as we possibly can but at the same time be realistic about what is possible.

The biggest eyesore of all is on the South Col of Everest, the rubbish and the discarded oxygen bottles of every expedition since the Swiss first reached it in 1952. It represents the syndrome created by custom. The sight of rubbish already there, the fact that the climber or Sherpa is exhausted makes it all too easy to leave one's own rubbish as well. We must start acting in a more responsible way repairing the damage of the past and caring for the mountains in the future. We are not only polllutingh the environment of the Himalaya, but we are polluting the world as a whole.

What is happening to the mountains of Himalaya is a microcosm of the way person is polluting the world as a whole. We need to start treating the mountains with greater care, try to solve the problems of deforestation, pollution and rubbish before it is too late. It's only if we act right now that we can avoid irreparable damage to the place we love so much.

4

Himalaya—Its Geology, Metereology and Hydrology

Introduction

In this chapter we will know how two drafting earth plates caused the earth's crust to buckle and rise, thus forming the Himalaya. It is said that this process started less than 40 million years ago and is still the process is going on. It is the determinant of the Indian climate. Its glaciers support some of our mighty rivers.

The effect of rising population, deforestation and agricultural expansion on the geological, meteriological and hydrological condition of the region is also discussed in this chapters.

Human intervention, resulting in deforestation is aggravating floods in the plains, as greater quantities of soil and water now rush down the Himalayan valleys to raise the river beds at an even faster rate. Himalayan rivers, like Ganga and Brahmaputra, are also carriers of enormous quantities of water, causing havoc in the Indo-Gangetic and Assam plains.

To find out the mechanism at work an overview of the region's physiography, tectonics, geology, meteorology, hydrology and the process of erosion is necessary.

Geography of Himalaya

The Himalaya Comprises of three parallel mountain belts spreading along a length of about 2,400 km with a width ranging from 200 km to 400 km. The northernmost belt is known as the greater, Himalaya with a number summits exceeding 7,500 m to 8,500 m. With the average height of about 6,000 m they are among some of the highest mountains in the world.

The greater Himalaya's southern flanks merge with the middle mountains, whose average height is about 2,000 m to 3,300 m. The famous ranges here are Pir Panjal in Jammu and Kashmir and Dhauladhar in Himachal Pradesh. The width of the middle mountains which is also called lesser Himalaya varies from 60 km to 80 km. These mountains have traditionally a high population density. The northern portion of the lesser Himalaya which lies between 2,900 m and 4,000 m still remains under upper mountain forests. Below is a belt of intensive agriculture.

Adjacent to Indo-Gangetic plain are the Siwaliks which constitute the outer Himalaya. Rising abruptly from 200 m to 500 m they attain an height of 900 m to 1,200 m. Here occur the famous tectonic depressions known as the dun valleys supporting the richest agricultural land in the region. The duns range from Dehradun in India to the Rapti valley in Nepal. The width of the Siwaliks varies from 10 km to 50 km and they extend from the Brahmaputra valley in the east to the Potwar plateau in Pakistan.

Birth of Himalaya

A massive collision of two drifting earth plates—the vast continental mass of Asia to the north and the smaller continental mass of the Indian Peninsula to the south. This collision caused the formation of Himalaya. The uplift has taken place in three phases.

The first phase took place about 38 million years ago. This gave rise to the greater Himalaya. The area in the north of it is known as Transhimalaya.

The second phase took place between seven and 26 million years ago and this formed the ranges of the lesser Himalaya.

The third which formed the Siwalik hills started about seven million years ago.

The Himalayan mountains still continue to rise as the mountain building activity continues even today as the Indian plate thrusts itself further under the Asian plate. During the last 1.5 million years, the Himalayan mountains have risen at least 1,000 m. Nanga Parbat, the tenth highest mountain in the world, is still rising.

The junction of the greater and lesser Himalaya regions is marked by a gently inclined thrust plane called Main Central Thrust. And the junction of the lesser Himalaya and the outer Himalaya is defined as the Main Boundary Thrust. Its geological condition is even worse.

Complex Landscape

The Transhimalaya and the greater Himalayan regions constitute an extremely complex landscape, heavily sculptured by glacial erosion. The rocks range from hard granites and metamorphic rocks to those formed out of the sediments of the ancient Tethys sea. It separated the Indian landmass from the Asian landmass which disappeared with their collision.

Lesser Himalayan rocks are extremely varied and include sedimentary, metamorphic, and granites. However, there are also extensive areas of phyllites and slate rocks which split very easily. Because of the steep slopes of the Himalayan mountains, there rocks are highly susceptible to erosion. The most easily eroded rocks of the entire Himalayan region is found in Siwalik hills. A drastic increase in soil erosion is found where there ranges have been extensively deforested.

Himalaya—The Determinant of Indian Climate

The Indian subcontinental climate is determined by Himalayan mountains. In the winters, they serve as an effective barrier to the intensely cold continental air blowing southward into India. This helps to keep India warm in the winter. In the monsoon months, the Himalaya force the rain bearing winds from the south to deposit most of their moisture on the Indian side. A monsoon simulation experiment conducted in the United States found that in the absence of the Himalayan ranges, South Asia would possess a desert like climate because the dry continental air flowing in from the northwest would push the rain belt far away to the south.

Within the Himalayan region the weather changes considerably over the year. October and November are typically dry, but beginning December winter rains start, reaching a maximum around February. Snowfall sometimes continues up to April. In winter, the Himalayan ranges receive hail from the mid-latitude westerlies. In the higher mountains, hail is typically in the form of snow, whereas in the lower valley it occurs as rain. Winter hail in the Himalaya exhibits a decreasing trend from west to east. The phenomenon is reversed in the summer. The eastern end of the Himalaya receives considerably more rain during the monsoon, where it also lasts longer.

The monsoon commences in the first week of June in the eastern half almost a month earlier, and ceases in September end, almost three weeks later, than it does in the western half. Rainfall at the microlevel varies even more. Local mountains strongly modify its distribution. With minor ranges, spurs and valleys running in different directions, rain bearing winds can often enter only through narrow gaps from particular directions. Northern slopes and east-west oriented valleys usually receive less precipitation because of the rain shadow effect than southern slopes or north-south oriented valleys.

In the hilly portion of the Yamuna catchment, there is not even one raingauge for every 300-400 sq km of the catchment area as recommended by the World Meteorological Organisation. In fact Indian meteorologists believe that given the extreme variability of rainfall in the Himalayan catchment, the density of raingauges should be one for every 150 sq km. Rainfall also varies with altitude. Analysing data obtained from the Kosi basin, meteorologists O.N. Dhar and B.K. Bhattacharya found that maximum rainfall occurs at two different elevations. The first occurs near the foothills and then starts decreasing, reaching a minimum at an elevation of 600-800 m. This region is located on the other side of the Siwalik hills, that is, in the valleys between the Siwalik and the lesser Himalayan ranges. Short duration, high intensity rains are common over the Himalaya.

The Mount Everest region in the greater Himalaya, for instance, is almost a semi-arid area. The 3,000-4,000 m high Pir Panjal range literally acts as a barrier to monsoon winds entering the Kashmir valley, which lies between the lesser Himalayan and the greater Himalayan mountains. The area to the south of the range receives four times more rainfall than the area to its north during the monsoon. The greater Himalayan mountains themselves stop almost all the monsoon winds from reaching the slopes which face Tibet. These slopes tend to be semi-arid to extremely arid because of the rain shadow effect. Ladakh, for instance, is a cold arid desert.

The Monsoon

During monsoon breaks, the axis of the monsoon trough tends to lie at the foot of the Himalaya instead of over the Indo-Gangetic plains. The trough axis is the line along which the two branches of the monsoon, namely, the Arabian Sea and the Bay of Bengal, meet. At such times, central India experiences near drought conditions. But in sharp contrast, the mountains of the eastern Himalaya tend to get heavy to very heavy rains. Because of this, a peculiar situation is created on a 'break day' in the catchment of Himalayan rivers, particularly those located in Sikkim and Nepal. These rivers often go in spate while downstream plains are experiencing almost drought conditions.

On a 'break day', the entire Himalayan region does not experience an increase in rainfall. The region to the west of longitude 79°E, that is, western Himalaya actually witnesses a general decrease. But rainfall increases to the east of longitude 79°E and maximum increases—200 per cent to 300 per cent over the average—are located in the region between longitudes 85°E and 87°E, where the catchments of Teesta, Kosi and Bagmati lie. Rain also increases in the catchments of the more western rivers like Gandak, Rapti and Ghaghara but not to the same extent.

Contribution of Glaciers

As compared to 2.2 per cent in the Swiss Alps Himalayan glaciers cover about 17 per cent of the mountain area. The 15,000 odd Himalayan glaciers form a unique reservoir which supports such mighty perpetual rivers as Indus, Ganga and Brahmaputra.

Unfortunately, data on the greater Himalayan environment is incomplete. Himalayan glaciers are spread over a chain of mountain ranges with different orientations and elevations, which makes the task of data collection extremely difficult. The glaciated area, including the area north of the Himalaya which drains into Himalayan rivers, is estimated to be about five mha. About 10 times the glaciated area gets snow covered during the winter months. The annual water yield from glacier-melt is about 15 mham per year. The snow melt contributes another 50 mham. Thus, on a very rough estimate, snow melt and glacier melt together contribute a water volume of 65 mham to the northern rivers. About 70 per cent to 80 per cent of all these contributions come during June to September. These figures are comparable to the total volume of water being used in India at the moment.

A study recently conducted on the Dhauliganga river, which originates in the lesser Himalaya in Pithoragarh district of Uttar Pradesh, shows that glaciers cannot be major causes of floods. Himalayan floods are mainly due to rainfall. The river's catchment is studded with 23 small and medium glaciers. About 45 per cent of the annual river flow comes from glacier and snow melt. The average daily glacier melt from April to September remains fairly constant, about 45-55 cumecs. The peak occurs after noon but it affects, the downstream reaches of the river only around midnight.

The Catastropic Events

When glacier dammed lakes suddenly burst—it is known as jokulhlaup—to causes severe floods. This type of major catastrophic are not uncommon in the Himalayan ranges though little has been known about them until recently. A moraine dammed lake drained catastrophically on August 4, 1985 in a tributary valley of the Bhote Kosi in Nepal. As scientific curiosity grew, it was found that at least three jokulhlaups, and possibly five, seem to have occurred in the region within living memory.

Dudh Kosi valley had at least two major floods caused by glacial dam bursts—one about 60 years ago and another in 1977 when an ice dammed lake at Mingbo burst. Both floods caused severe havoc and extensive erosion in the Dudh Kosi valley, depositing vast amounts of sediment.

In 1981, a glacial lake outburst occurred in the headwaters of the Sun Kosi river, which damaged the Sun Kosi barrage and a large number of old landslides were reactivated. A glacial dam burst on the Tamur river in Nepal resulting in a flood surge that for 30 km downstream stripped all trees 20 m above the river bank. Indian hydrologists have also recorded a jokulhlaup on the Dhauliganga, but fortunately its peak was not very high.

The outburst of a moraine dammed lake behind the Machapschare mountain between 600 and 800 years ago was a sensational event in the Nepal Himalaya. The resulting flood surge down the Seti Khola which deposited 5,500 m cum of debris in the Pokhara valley, damming large Phewa. This is the largest known glacier lake outburst. A recent study has concluded that these glaciers have been in a general state of retreat since 1850. The rate of retreat varies from 4.6 m to 23.5 m per year. Glaciers in the Kosi watershed have also been thinning and retreating. The progressive retreat and thinning of Himalayan glaciers during the present century is resulting in the formation of new moraine dammed lakes.

Disaster Caused in Nepal

On August 4, 1985 a moraine dammed glacial lake, Dig Tsho, burst in the Khumbu Himal area of eastern Nepal. Swiss scientists Daniel Vuichard and Marcus Zimmermann studied the catastrophe in detail. The Dig Tsho lake had been impounded by the moraines of the Langmoche glacier probably within the last 25 years. The lake surface was about 50 ha in area, one-third was covered by an ice mass about 300 m wide and 500 m long, the maximum depth of the lake was 18 m.

Before the burst, the lake was full to its brim. Just then an ice avalanche from Langmoche glacier fell into the lake sending a five metre impulse wave across it. The Dig Tsho burst and within four to six hours, the lake emptied into the Langmoche valley, one of the tributary valleys of the river Bothe Kosi, which flows along many Sherpa settlements.

Water plummeted along the Bothe Kosi and then into then Dudh Kosi in four hours and the initial peak discharge was at least 2,000 cumecs. The flood caused significant damage for more than 80 km of the river course till the confluence of the Dudh Kosi with the Sun Kosi. Eye witnesses report that the flood surge moved down the valley like a huge "black" mass of debris.

Trees and large boulders were dragged and bounced around, emitting a loud noise "like many helicopters" and a foul muddy smell. Where the river flowed over bedrock, the topsoil cover was washed off and the bedrock surface was polished at river bends. Slopes, which were made up of loose material, were subject to heavy undercutting, even if they had a good tree cover. This caused landslides several metres deep and several tens of metres high. Some houses were toppled by the vibrations caused by the flood surge. Lateral erosion destroyed several houses even if they were not situated immediately along the river bed. Nearly complete Namche hydroelectric power plant was destroyed.

Birth of Holy River

Indus, Ganga and Brahmaputra. All these rivers have their sources very close to each other in the Central Himalaya. Four of the world's great rivers — Karnali, Indus, Sutlej and Brahmaputra — rise in the watershed of one mountain, mount Kailash. The river Ganga, emerges at Gangotri, about 60 km away. Indus and Brahmaputra probably came into being 40 million years ago as the greater Himalaya was formed to drain the new highland near mount Kailash. The Sutlej and Indus flow westward from their sources to flow on into the Arabian sea through the dusty plains of Punjab.

Along the boundary between the Indian and Asian plate Brahmaputra flows eastward.

There it forms a loop around the Himalayan mountains to flow through the narrow Assam valley and finally to link up with Ganga before entering the Bay of Bengal. Together, Ganga and Brahmaputra form a garland around central and eastern Himalaya. These awesome mountains and the mighty, perennial rivers they give rise to, have been the subject of more myths, legends and prayers than any other mountains or rivers in the world.

Largely fed by the summer monsoon, these rivers remain perennial because of the huge Himalayan glaciers. Without them, the Indo-Gangetic plains, which support one of the highest densities of human population in the world, would be a desert.

But while the Himalayan rivers have been a source of life, they have also given birth to member of destructive floods too. River discharges increase considerably during the monsoon. Flows also vary sharply from one year to another. Even in a large river like the Brahmaputra, annual stream flow can be more than double in *one* year as compared to another. As one scientist puts it, even major Himalayan rivers often behave like mountain torrents.

As a result, averages of data collected over a short duration are often extremely misleading in the Himalayan context. Recent studies conducted on Beas, show that the flood data used to build the dam at Pong—one of the biggest in the country—in 1974 was an underestimate. When the dam was being built, little hydrometeorological data was available. The data after construction shows that the severest flood of 1947, which was used to determine the design inflow flood for the dam was exceeded in 1976, 1977 and 1978.

The Himalayan Hydrology

The hydrology within the Himalayan mountains is also extremely complex. Recent studies show that most of the rain that falls on Himalayan slopes gets turned into subsurface flows. Kumaon university scientists recently measured surface flow from eight forested and five non-forested sites. They found that only 0.3 per cent to 1.3 per cent of rainfall got turned into surface flows. The non-forested sites had greater surface flow but their magnitude was still very small.

The Scientists concluded that most water that flows into Himalayan streams comes from subsurface flows and the surface flows are too small to play a significant role in either causing floods or soil erosion.

But because soil cover in most areas is shallow and lies over soft, quick weathering rock subsurface flows make the mountains greatly susceptible landslides.

Also the Scientists in Nepal carried out a study. They concluded that even heavy storms should not produce much surface flow including areas that have been deforested and converted into heavily grazed and trampled grassland. The team measured the infiltration rates at five sites with different types of forest cover. The most degraded site revealed the lowest infiltration rate in the top layer of the soil — or 39 mm/hr — and a site with a well protected pine forest. It showed the maximum rate of 524 mm/hr. But available dam revealed that few rainstorms would exceed the infiltration rate even at the most degraded sites. Surface runoff from these slopes would, therefore, be rare and small, and the presence or absence of a forest would do little to increase or decrease it.

Some other scientists have, however, argued that higher surface runoff has been seen in higher rainfall areas than those chosen by the above team. The controversy notwithstanding, the available evidence does show that subsurface flows play an important role in Himalayan hydrology.

Denudation

The erosion in the Himalayan mountains can be gauged from the denudation rates prevalent in the region. All processes of soil erosion—surface erosion, erosion of rivers and mass movements—are at work in the Himalaya. One study even shows that surface erosion rates are usually very small, except in overgrazed pastures, very sandy soils. The low surface erosion can be explained by the rapid spread of vegetation with the onset of the monsoon. Mass movements include dramatic processes as such landslides, mudflows, rock falls, and rockslides.

When surface erosion is primarily a product of reduced vegetation cover, mass movements are the result of gravity. While the first can be influenced greatly by human action, the second is more the result of natural forces. Studies carried out by Indian researchers in the lesser Himalaya also show very low rate of surface erosion even including deforested sites. Studies conducted by Central Soil and Water Conservation and Research Training Institute (CSWCRTI) in the Siwalik hills do show high rates of erosion, especially in cases of extreme deforestation and heavy human disturbance. They also show that a grass cover or ever natural fallow with grasses and shrubs greatly reduce soil loss.

The high denudation rates in the Himalayan mountains indicate a very dynamic environment in geological terms. According to geographers Jack Ives and Bruno Messerli, in their recent book *Himalayan Dilemma* the denudation rates in the Himalayan mountains are estimated to be "up to five times or more than general estimates of erosion even in areas of steep relief", Several denudation rates are available for different parts of the Himalaya. Despite their wide range from 0.5 mm/yr to 20 mm/yr, these figures are very high when compared with data from other parts of the world. The Alps, for instance, have an overall rate of one mm/yr. Figures above six to seven mm/ yr are amongst the highest ever recorded or estimated.

Due to of this Himalayan rivers bring down enormous quantities of sediment. Ganga carries 340 mt of sediments per year, Brahmaputra 650 mt and Indus 320 mt. Ganga and Brahmaputra together have sediment yields higher than any other river in the world except for Hwang-Ho. Only one large river in Africa—the Orange—has a sediment yield exceeding one t/ha and most north and south American rivers have yields around one t/ha.

It is estimated that the world's oceans receive about 8,000 million tonnes of sediments per year, of which over a quarter comes from the Ganga and Brahmaputra alone. Though undoubtedly high, data on sediment yields of these rivers is extremely unreliable. Often, measurements are made during dry seasons, when loads are low, while rainy season floods carry maximum loads but are extremely dangerous to measure.

Moreover, the load measured is often only the suspended load. The bedload is now regarded to have been grossly underestimated throughout the Indian and Nepal Himalaya. Sand waves and dunes on the Brahmaputra bed in Assam reach heights of 15 m and lengths of 200 m to 900 m, and can migrate downstream by as much as 600 m per day. The bed load alone can amount to 10-20 million tonnes (mt) a day—almost equal to the suspended load during a flood. Not surprisingly, sedimentation rates of reservoirs built in the Himalaya have invariably been higher than those predicted prior to construction.

The Effects of Landslides

According to somescientists mass movements of landslides and catastrophic events like jokulhlaups make more important causes of soil loss in the Himalaya mountains than surface erosion. For that reason, annual averages practically of no use in the Himalayan context. The silt tends to move more in explosive waves whenever there is a catastrophic event than by averages. Landslides are the main source of soil loss in the central Himalaya, according to the study by Kumaon university.

Geologist K.S. Valdiya points out that massive landslide events like those of July 1970 in the Alaknanda valley and of August 1978 in the Bhagirathi valley can bring frighteningly high quantities of sediment in a very short period and upset all averages of data collected over previous years. Catastrophic events in the Teesta valley, increase average denudation rates several times claims another expert. Landslides are common all over the Himalayan ranges.

According to R.K. Bhandari, a landslide expert, Himalayan landslides continue to throw a challenge of such a magnitude that all tools and techniques of analysis and control in the armoury of a landslide specialist pale into insignificance. Vital roads, river valley projects and urban and rural settlements have been affected by landslides in the western Himalaya. Landslides are common also in the Northeast. The Geological Survey of India has prepared a map of the Northeast to depict the distribution and intensity of slide processes. The map shows that large parts of Arunachal Pradesh, Negaland, Manipur and Mizoram suffer from severe susceptibility.

Factors Inherent in Himalayan Ecology

Mainly three factors inherent in Himalayan ecology—weak geology, intense rainstorms and earthquakes. They make these mountains prone to landslides. It is intense rainfall that usually acts as the trigger. Heavy or prolonged rains often reactivate old landslides. A study in a watershed in Nepal found that the maximum number of landslides occur during the monsoon. As water generally flows down Himalayan slopes in the form of subsurface flows, and because the soil mantle is usually shallow and rests on soft, fast weathering rock, Himalayan slopes are inherently susceptible to landslides.

Experts of the Central Road Research Institute claim that experience in the Himalaya has taught them that a carefully designated drainage system has invariably proved to be the most effective measure in preventing or correcting a landslide. Besides bringing down large quantities of sediments, landslides become a major cause of devastating floods because they block the narrow gorges. Maps and aerial photos can be used to locate sites where such natural dams can form. They are normally marked by a sudden change in the gradient of a tributary stream, constriction at the point of confluence and weak geological conditions in the catchment of the tributary. The formation of landslide dams is common in the Himalaya.

In 1968 a landslide dam at Labubeusi in Nepal on the Bruni Gandaki river broke and thus caused devastating flooding downstream. Again in September 1981 river Tinneau in Palpa district, Nepal, was blocked by a landslide during continuous heavy rain. The breaking of the dam killed about 200 people in the Tarai region.

The high sediment flows resulting from landslides can be totally eclipsed jokulhlaups in the greater Himalaya which are now being recognised as of considerable importance for sediment transfer to the Ganga plains over longer geological periods.

The 1985 moraine dam outburst in Nepal removed about 0.9 mcum of material most of which was redeposited within about 25 km. Subsequent rainstorms will carry away the coarse material further downstream. According to geographers Jack Ives and Bruno Messerli, the enormous silt brought down by the Kosi could be the result of periodic jokulhalaups in its headstreams like Dudh Kosi, Sun Kosi and Arun.

Professor of geology at the university of Nebraska John F. Shroder Jr,, claims that all the 17 known major historical floods on the Indus were the result of breaking ice of landslide dams. In January 1841, an earthquake triggered massive landslide from the Hattu Pir spur of the Nanga Parbat mountain into the Indus. The landslide was 1.9 km high vertically and covered a horizontal distance of about four km. The resulting 150 m deep reservoir on the Indus backed up water 30 km upstream to the confluence of the Hunza and Gilgit rivers. As the impounded Indus rose, other rock slopes failed and made giant waves in the lake. When the landslide dam finally failed in June 1841, it sent a wave of water, mud and rocks roaring out of the Tarbela gorges.

"In the course of our recent search for the original 184 rockslide dam, we discovered many large dams," say Shroder Jr. Another major landslide blocked the Hunza river in 1858 at Pungurh. When the dam broke, it cause significant flooding downstream and large failures have continued on both sides of the valley ever since.

Man-made Desolation

Evidence clearly indicates that the Himalayas-Brahmaputra-Ganga-Indus system is one of the world's most dynamic mountain building and sediment transfer systems. The ongoing processes have combined to create a high mountain landscape of utmost complexity. Over this landscape, human activity has brought vast ecological changes over the last few centuries.

Large areas have been deforested, agriculture has expanded, vast networks of roads have been built and numerous dams have been constructed or are under consideration. According to many environmentalists this is increasing floods downstream. Let us see now.

The forests act like a sponge, absorbing large quantities of water during the wet season and gradually releasing it. Forests also reduce erosion and landslides. Therefore, if the Himalayan slopes can be afforested, the destructive flood-drought cycle that has been unleashed can be arrested.

Moreover forests influence the hydrological cycle by: (i) intercepting rainfall and allowing it to evaporate back to the air; (ii) transpiring large quantities of soil moisture and keeping the soil dry; (iii) creating soil conditions that allow greater infiltration of water into the ground; and, (iv) reducing peak water flows during a rainstorm.

On a forest canopy rainfall is caught by leaves and branches. Gradually it evaporates as it moves towards the forest floor. This is called interception loss—the amount of rain water intercepted by living and dead plant material and evaporated back to the atmosphere is influenced by wind, rainfall intensity and type and density of vegetation. Very light rainfall can be totally intercepted while heavy rainfall will rapidly saturate the forest canopy and most of it will get transmitted to the forest floor. Experts say that interception loss is governed more by rain intensity than tree species. In a forest of sal *(shorea robusta)* interception loss was 37.3 per cent of rainfall during a light shower (4.2

mm), and 4.1 per cent during a heavy shower (140.3 mm). As the quantity of intercepted water is small during a heavy rainstorm, interception by forests will have little effect upon major floods.

Evapotranspiration

The combined loss of water by evaporation from the soil and transpiration through plant leaves is termed evapotranspiration. While forests will normally reduce evaporation from the soil, they will increase transpiration. Evaporation from land covered by a forest usually ranges from 10 per cent to 80 per cent of that from bare soil. But in most cases transpiration loss from a good forest more than offsets the reduction in evaporation.

Various environmental factors and tree species determines the transpiration rates. Shallow rooting trees such as spruce and poplar have only 1.5-2 m deep roots, whereas deep rooting trees like oak and pine reach 4-6 m depths. Deep rooting trees transpire large quantities of water, sometimes as much as the entire amount of rain that falls over a year. Transpiration rates of well stocked forest lands can be over twice that of crop lands and five times that of meadows.

Studies on evapotranspiration in different types of forests have not been undertaken in India. But an evapotranspiration estimate for the Damodar catchment, which has 1.87 mha of dry deciduous forests, puts it at 560 mm/yr—about half the annual precipitation in the area. The high rate of evapotranspiration by forests increases storage space for moisture in the soil. Tree roots leave the soil in a more receptive condition to accept water during rainfall.

In other words, trees act almost like pumps and, hence, afforested catchments normally have reduced stream flows compared to deforested ones. Experiments on small watersheds across the world have shown that deforestation results in income annual water yield —usually between 20 per cent to 30 per cent.

Infiltration

Infiltration is the process by which rain water enters the leaf litter and greatly increase the organic content of the soil. The infiltration rates will be low during intense and protect rain, and afforestation will not make much different. Infiltration rates also depend on the nature of the soil instance. Infiltration can be total in a forest area when the soil is dry and deep and only a short, light rain occurs. But it is small, when there is heavy and prolonged rainfall, especially on wet, shallow soils.

CSWCRTI's Observation

How a forestimpacts on the hydrology of a particular site, depends on a number of factors like the type, number and density of trees, the state of leaf litter, soil characteristics, intensity and duration of rain, and various environment.

After deforestation dry season flows of Himalayan rivers have decreased. C.K. Sharma of the Water and Energy Commission of Nepal points out that "most Terai rivers, which previously contained water during the lean season, are now found to be dry during winter. Such effects are also seen in the rivers of the Middle Hills of Nepal. The low flow discharge of the Tamur river has decreased considerably in the last few years and this is mainly due to large scale deforestation in the upper valley". Indian observers claim that the chos of the Punjab Siwalik, once perpetnal streams, have become seasonal ones. Chos today seem to carry more mud and silt than water. They disappear after about 20-25 km from the foothills. Their beds silt up rapidly. The 55 km Swan river in Himachal Pradesh now has a width of two km and its muddy torrent cuts into hills, washing away fields. Cho beds which covered only 19,282 ha in 1852 in Hoshiarpur district of Punjab, now cover over 0.2 mha.

A study carried out by the CSWCRTI on small watersheds confirm that afforestation reduces total volume and peak rate of surface runoff; and streamflows last for a longer period after the cessation of the rains. It also found that in Dehradun the volume of runoff went down by 28 per cent after afforestation and the peak rate of runoff was reduced by 73 per cent.

Another study compared an agricultural watershed with a forest watershed subject to open grazing. Even this experiment showed that the volume of runoff was 15 per cent lower in the latter and the peak runoff rate 72 per cent lower. CSWCRTI's efforts to afforest and treat the eroded catchment of the Nalota Nala near Dehradun, which used to dry up soon after the monsoon, has resulted in streamflows for longer periods. In other words, a sponge effect does come into operation with afforestation which reduces peak runoff rates and prolongs dry season streamflows.

CSWCRTI studies have shown increased rates of surface erosion after deforestation. Surface erosion in non-forested sites in the central Himalaya a appears to be low.

Impact of Afforestation

The theory that afforestation of the Himalaya will prevent major floods in the Indo-Gangetic plains. They say floods occur during intense and prolonged rainstorms, when the watershed gets saturated and both interception and infiltration effects are minimised. Often major floods occur late in the monsoon season when watersheds are near saturation. No data exists to assess the impact of forests under such conditions compared to the impact of natural factors such as the intensity of the rain and size, shape and geology of the river basin, and human induced ecological changes in the flood plains themselves.

Moreover, most experiments have been conducted on small watersheds while several experts argue that hydrological responses of large watersheds can be different. Subsurface flows in large watersheds may be substantially different than in small watersheds. Data collected from experiments on the latter possibly explain local effects better than large scale regional effects.

A good forest cover in the Himalaya constitutes the safest land use to reduce floods but if deforestation is replaced by terraced agriculture and well managed grasslands, there should be little effect on downstream floods, said Lawrence Hamilton, watershed expert at the East West Centre in Hawaii. But, if deforestation results in an extremely abusive land use in the form of intensely overgrazed grasslands which lead to heavily compacted surface soils, eroded and gullied terraces, and roads and channels which speed water runoff on a sufficiently extensive scale across the river basin. This can aggravate downstream floods.

Lawrence Hamilton further argue that soil erosion rates found in small watersheds is not true for large watersheds. For instance, 90 per cent of the soil eroded in any year from a one ha watershed will be delivered outside the watershed by streams in that very yield itself. For a catchment of the size that of the Ganga, the sediment delivery ratio will probably be only 10 per cent. Most of the silt in the river comes from soil already deposited in the channels and from channel erosion. Even if all human activity was closed in the Himalaya and, as a result, all surface erosion hopefully came to a halt, the sediment load of the rivers in their lower reaches would show little change for a long time. The stand of eroded soil from the past would continue to feed rivers. In other words, if afforestation was undertaken to day it might improve local conditions within a few years, silt loads in the large streams will not decrease in decades.

There is no doubt that forests are being replaced by an abusive land use in many places in the Himalaya. Again there is no sufficient data available to show how extensive is this adverse land use and now far it affects the hydrological conditions. Studies carried out both in India and Nepal show that infiltration rates even on deforested sites, including those which have been converted into heavily grazed and trampled grasslands, are sufficiently high. Afforestation will, therefore, have a limited impact in terms of changing hydrological conditions. Existing studies do caution us towards beliefs that Himalayan deforestation has greatly exacerbated floods over and above natural causes.

An FAO publication summarises the impact of afforestation on floods in the following statement. It presents the probability of a flood of a certain size occurring under different levels of forest cover in the watershed. Major floods expected to occur once in 15-20 years will not be affected whether the watershed has a forest cover or not. But the peak discharge of floods that occur more frequently (say, once in two years) could more than double. In other words, forests can moderate minor and medium floods. But human society will have to learn to live with the major floods.

Can the soil erosion be reduced and the rapid silting up of river beds in the Indo-Gangetic plains be arrested is a big question. Forests have a significant influence on surface erosion and sedimentation of rivers. The existence of a forest canopy and, equally important, litter on the forest floor breaks the impact of rain drops on the soil. Moreover tree roots and increased organic matter in forest soils impart stability to them.

According to geographers Jack Ives and Bruno Messerli, all that silt loads in major rivers may not decrease substantially at all. The increased erosion rates due to deforestation do not match the levels of soil loss reached in the Himalayan maintains because of mass movements like landslides and catastrophic events. These processes are so predominant that human activity only has a limited impact in the dynamic environment of the Himalaya. They argue that forest will not be able to prevent major landslides, as they occur more because of steep slopes and other natural factors like weak geology, earthquake and intense rainstorms. They can postpone the occurrence of major landslides but not stop them. Forests, as compared to grazed and bare slopes, will get destabilised less frequently and, therefore small, shallow landslides will be prevented. But this will only facilitate the production of a thicker layer of top soil—and ensure a larger landslide in the future.

Thus, the overall soil loss over a longer time scale will be the same with or without forests. In other words, sooner or later major landslides will take place and Himalayan rivers will bring down enormous quantities of silt during intense rainstorms. Ganga and Brahmaputra will continue to carry a higher sediment load than most other major rivers worldwide and riverbeds will continue to silt up in the Indo-Gangetic plains.

According to a scientific data tree roots hold the tree and the soil itself, greatly increasing the shear strength of the soil and preventing it from slipping down. Loss of this shear strength by clearing forests can easily submerge shallow soils in areas prone to mass movements. Afforestation of landslide prone areas in the Himalaya will definitely increase slope stability.

Scientists also argue that forests can override the natural factors that cause landslides, especially in the Himalaya. A survey carried out in Kalim following the 1954 disastrous flood of north Bengal found that the primary cause of the landslides was deprestation. So human factors like deforestation, overgrazing and cultivation of slopes were substantial causes. Another study of landslips along the Sun Kosh (Tamur rivers) concludes that they have been mostly caused by unfavourable geology and heavy rainstorm though deforestation did play a role.

Generally landslides tend to come in cycles. Large scale landsliding is followed by stability presumably because time is required for the bedrock to weather and a layer of topsoil to accumulate before the next major slope failure can take place. Climax soils can sometimes never develop on steep mountain slopes because the slopes are too unstable to allow a mature soil cover to evolve.

Result of Steep Slopes

The current rate of erosion in Himalayan catchment areas is five times higher than the past says geologist K.S. Valdiya after a scientific investigation of the are. Geographer of University of Guwahati, D.C. Goswami, has also calculated that the present rates of erosion in the eastern Himalaya are far higher than the past rates of erosion. Goswami's studies show that about 20 per cent of the silt gets trapped in the Assam valley. The Brahmaputra valley in Assam is 5.62 mha in area and it has received a mass of sediments. 200-300 m thick in the last two million years which gives an average rate of erosion of 0.03 mm/yr over this geological period. But this is only about one-fiftieth of the rate of erosion observed in the last 30 years.

The present rates of erosion have increased because of deforestation and other human interventions. According Goswami the present high rate of denudation is mainly because of ongoing uplift of the mountains.

According to American geographer John F. Shroder Jr. modern rates of Himalayan uplift are, in general, about eight times greater than the average rate of erosion. The higher a mountain rises, the steeper will be its slopes. This will result in greater surface erosion, mass movement and erosion of stream valleys. When these steep slopes have to face heavy rainstorms and earthquakes, erosion will increase further. Therefore, scientists believe that there are major geological reasons to explain the increased erosion in the Himalayan mountains over geological history.

Result of Cultivation

The increased destabilisation of slopes, soil erosion and floods in the Indo-Gangetic plains is the result of expansion of agriculture in the hills. Geologist K.S. Valdiya, for instance, points out that the increase in population in the hills has driven people to greater heights and steeper slopes, even above 2,000 m. They cultivate over there which were once the forested tracts. Most of these villages and farms are located on the fan shaped debris of old landslides made of loose, fragmentary material and not very stable. Their cultivation has deprived these slopes of protective cover and thus increased erosion and waterlogging due to absorption of excess rainwater, especially in the absence of drains. So the poor farmers of the Himalaya are to be blamed for ecological destruction there.

Former Indian Prime Minister Sri V.P. Singh, in an interview to *The Illustrated Weekly of India* in 1989 said: "It is tragic to see the Himalaya now. I have flown over them many times in helicopters and seen the denudation of the forests. The mountains have been terraced by human hands, not bulldozers! You have to see it to believe it. The sheer human effort involved! The way they have gone about cutting down the trees and destroying the environment".

But many scientists do not believe that agricultural farms have become a major source of erosion. A team recently studied two mountain streams in Nepal to judge the impact of human action on them. They found that the only impact of human action on these streams was that the channel width of these in the uppermost eight km got effected because of the construction of agricultural terraces. The stream, therefore, adjusted itself to cutting a deeper channel. No aduerse effect could be noticed further down because of the activities of the local farmers.

Terraced Farming

A report by the Asian Development Bank (ADB) on Nepal, which states, "Terraces, especially on rainfed land, are often poorly constructed; they are outward rather than inward sloping and do not have a grassed bund on the edge". Jack Ives and Bruno Messerli are particularly critical of those who believe that terraced farming is leading to considerable erosion. They explain that bari or rainfed terraces in Nepal mostly support maize, millet, buckwheat and other crops. Situated usually on the upper, steeper slopes with no irrigation facilities, they slope outward from the hillside so that crops are not damaged by waterlogging. Local farmers are awared that accumulation of water on terraces—that would result from inward sloping terraces—would exacerbate landslides. The summer monsoon rain is intended to run down the outward sloping terraces. In contrast to the khet, or irrigated, terraces absence of a bund on the bari terraces is deliberate to rainwater runoff. Both bari and khet terraces are, for the most part, superbly engineered in Nepal, claim Ives and Messerli.

"Admittedly, during heavy monsoon downpours, available human energy is concentrated on repairing damage to the khet and irrigation systems, and the bari terraces may have to be left to collapse; this is because the much higher yielding khet terraces, usually under paddy rice, are more vital to the survival of the subsistence family. Any apparent neglect of the terraces may be due to shortage of available labour rather than a reflection of the ignorance of the farmer", claim Ives and Messerli.

Ives and Messerli made frequent visits to Kathmandu between 1987. Based on their study and experienced there they, in fact, argue that the subsistence farmers are highly knowledgeable and intelligent land managers with a wealth of accumulated, traditional wisdom. They recall the time when they were themselves not prepared to listen to them: "As somewhat representative of expatriate experts making short visits, we were able to observe the Middle Mountains of Nepal from brief road traverses out from Kathmandu in 1978 and 1979. Like many other visitors, we timed our presence to coincide with good weather—March, April and October—and were duly impressed with the large number of landslide scars and gullies that obviously had engulfed significant amounts of agricultural terrace land during the preceding summer monsoon periods. Since the project necessitated repeated visits of considerable duration to the field area throughout the agricultural cycle over a five year period, we were afforded the hitherto uncommon perspective of time. We were also able to develop communication with the subsistence farmers. This gave us a greater understanding of the landscape changes that have been occurring over several generations and an appreciation of local attitudes to dynamic slope processes and local responses to them."

H.W. Tilman, an English mountaineer, observed in 1952: "Whether it takes place little by little or in one swift calamity, soil erosion is generally attributed to man's careless greed, his idleness of neglect. It would not, I think, be fair to blame the people of these valleys on the Himalayan fringe for the frequent landslides which occur here. In turning the steep slopes into fruitful fields they have neither been lazy nor neglectful."

Human Contribution

Human beings contribute to the stabilising process of the slopes. In certain instances, local people perceived a landslide to be a beneficial occurrence because the more easily worked earth of the landslide scar actually facilitated terrace construction. Small landslides were sometimes deliberately triggered by water diversions in order to facilitate new terrace construction. The indigenous farmers have evolved an intricate set of coping strategies that include changes in land use to match changes in slope stability and subsequent reterracing of collapsed slopes. Agricultural deintensification often takes place as an adjustment to the threat of slope instability.

As human population numbers have increased, more forested land has been converted to agricultural terraces on steeper and more marginal slopes and more human energy is being spent to maintain a balance between stability and instability. The local definitely do not enjoy landslides which destroy houses, human lives, and livestock. During the days of severe rain farmers were spend fearful and sleepless nights in the small houses on the steep slopes. It is, no doubt, a serious problem indeed. But at the same time some of the most densely populated and extensively terraced land in the lesser Himalaya probably experience some of the lowest rates of soil erosion and land loss. A very real danger of soil erosion and slope collapse would arise if such areas were abandoned. It is, therefore, not correct to blame local farmers for all these ills.

So it is not always correct to say shat the conversion of a land from forest to agricultural land use always have a negative impact on either soil loss or water yield. This is also confirmed by CSWCRTI in one of their study. Soil loss in an agricultural watershed treated with soil conservation measures dropped from 2.4 t/ha/yr to 0.2 t/ha/yr, volume of runoff dropped from 44 per cent of rainfall to two per cent of rainfall, and peak rate of runoff also dropped dramatically. As compared to a neighbouring forest watershed, the treated agricultural watershed had only 28 per cent of volume of runoff and the peak rate of runoff was only four per cent. Before treatment, both volume and peak rate of runoff were higher in the agricultural watershed. Ecologically sound agriculture is, therefore, definitely possible in the Himalaya.

Intensification of Agriculture

It is believed that the agricultural practices the farmers in the Himalayan region have made Himalayan slopes more inclined to landslides. But the studies carried out by Sumitra Gurung of the International Centre for Integrated Mountain Development in Kathmandu and several other researchers show that local peasants are rich in environmental knowledge and contribute to the stability of the slopes.

Density of population per unit of cultivated land in the mountains is, in many cases, even higher than that of Bangladesh. In recent years the agriculture there has been expended and intensified.

Villagers in the Pokhara valley told US scientist Robert Schroeder, who studied the changing pattern of agriculture and pastoralism in

Nepal, that when they were children, upland fields were either single cropped or double cropped with low management crops. Currently, they are being triple cropped with high management crops and are also being intercropped to maximise yields. To minimise the risk of crop failure, the farmers in the Himalaya maintain a divarity in cropping.

Besides that the farmers often tested crop varieties brought from other areas and different crop management techniques. He found that one village in 1979 was growing over 50 distinct varieties of wet rice. If dry rice and other crops were included, they were growing over 75 varieties. In addition, they grew a wide range of vegetables, chili, oilseeds, fruit and lentil species.

Farmers utilise very developed land and plant classification schemes to match cropping patterns and crop varieties with each of the small fields that they farm. In the process, the peasants have transformed the middle mountains leaving little of the original vegetation.

Farmers in Nepal distinguish six different types of land use—*khet* (wet terrace), *bari* (dry terrace), *pakho* (unfilled and infertile land), *charan* (grazing land), *ban* (forest) and *gaun* (settlement). Khets are highly valued terraces. They yield two key staples, rice and wheat. They require heavy labour for upkeep and irrigation control. Baris are rainfed terraces on higher elevations. They yield maize, legumes and other staples. Its maintenance is easier. But due to weak soils these terraces are prone to erosion and sliding. Pakho is unfilled land, which is usually rocky or too infertile and steep to be cultivated. Charan is the commonly used grazing land, which often contributes landslides due to overuse and lack of upkeep.

Through the deintensification of land use farmers have developed ways of dealing with the problem. When khet terraces are threatened with erosion or sliding, they are left uncultivated till they can be reused. Deintensification often results is lower yields, but the farmers do it to save the land. When farmers are alerted to signs that their terraces may fail by development of cracks, slumping or undercut wall sections, they may shift to a lower intensity of land use. In the case of a khet, they may decide to discontinue irrigation in order to reduce soil saturation and plant rainfed crops. In the case of bari, the field is left uncultivated until the terrace either collapses or is judged to be out of danger.

Effect of Read Building

According to many experts the road building activities have been the worst form of human interventions in the Himalayan mountains in terms of promoting soil erosion and landslides. They believe that every human activity in the hills will directly or indirectly change existing slopes, slope forming materials and drainage system. As a result, all human activities can possibly increase soil erosion and landslides in fragile mountain areas.

The Indo-China war in 1962 prompted a massive road construction programme in the Himalayan areas. Prior to that, the Himalayan mountains were, for the most part, accessible only on foot. The few roads that existed led to hill stations like Mussoorie, Shimla, Nainital and Darjeeling, located not very deep within the mountains. Road construction programme and roads now go into some of the remotest Himalayan areas. More than 44,000 km of roads now exist in the Indian Himalaya.

Geologist K.S. Valdiya of Kumaon university claims that an average of 40,000-80,000 cum of debris is removed to build each km of roads and another 550 cum have to be removed per km every year to maintain them.

A six km stretch of the Nainital-Kilbury road built in the 1980s, which runs through a reserved forest, revealed that 38 per cent of the road had slumped, 39 per cent of it was affected by rockfalls and there were 24 landslides per km of the road.

Every year the landslide debris is pushed down the hill to keep the road clear, which damages the forest cover below and destabilises the hillside. The tree and vegetation cover below the road had been greatly reduced as compared to that above the road.

According to K.G. Tejwani, former director of CSWCRTI, 10 small to medium landslides occur on average over each km of Himalayan roads and the slope instability caused by road construction.

Road construction invariably involves deep cuts. This weakens the stability of the Himalayan slopes, say R.K. Bhandari and Chanchal Gupta of C.B.R. Road construction agencies are keen to complete their job rapidly. Few efforts are made to protect the cut slopes or provide effective drainage. As a result, during rainy season repeated landslides occur.

Changing Course of Rivers

Rivers have created new cllass of refugees constantly looking for a new home. In Bengal Changing courses by rivers is a constant problem especially the Padma and the Bhagirathi. The Padma, which flows along the Indo-Bangladesh border on the northern boundary of Murshidabad district, has washed away many villages. Some 30,000 ha of arable land have been destroyed. The rate of erosion is sometimes over two metres a week during the monsoon. Journalist Debjani Sinha, wrote on the condition of the people in the affected districts. For an example he narrated the case of Ismail Sheikh who once lived on a sand bank along the Bhairav river, a tributary of the Ganga, had to change houses six times due to the shifting course of the river during the last decade. Bindu Ghosh, a resident of Nayansukh, once owned six ha of orchards, but owns only a few cattle and some utensils now. The Bhagirathi has washed away the land and his house four times.

To protect the Padma's right bank, a vulnerable 94 km stretch downstream of the Farakka barrage. The state government has requested the Centre for money. According to some, the main cause of river erosion is the Rs. 160 crore Farakka barrage completed in 1975. Satyesh Chakravarti of the Indian Institute of Management is of the opinion that the barrage has increased erosion in the downstream reaches of the Padma. But it won't be correct to blame Farakka for all the disasters. As early as 1803, Major R.H. Colebrooke had found that the Padma was shifting westwards by 90 m a year.

The Geological Survey of India has found that the distance between the Padma and Bhagirathi in Murshidabad district, which was six km in 1944, had come down to 3.3 km by 1966 and 2.9 km by 1975. The Pritam Singh Committee, set up in 1978 to examine the post Farakka river ravages, also pointed out that the maximum distance between the Bhagirathi and the Padma is only a little over two kilometres. The committee warned that if the Padma breached its embankment, it could flow into the Bhagirathi, thereby, bypassing the Jangipur barrage and rendering it useless. At one time, the Teesta was not a tributary of the Brahmaputra but flowed into the Ganga, hugging the west side of the Madhupur jungle. Then the river migrated and joined the Atrai Gur to flow into the Meghna. But with the Brahmaputra moving to its present course, both the Teesta and the Atrai Gur have been captured by it and have become its tributaries.

Floods are common in the Bengal delta. Moderate floods occur at an interval of about four years severe, floods at seven years and catastrophic floods every 30 to 50 years.

This means that 1,760-3,520 mcum of debris had to be removed to make the Himalayan roads and another 24.2 mcum are removed every year.

Impact of Development Activity

Though no documented evidence is available several geologists believe that landslides are probably increasing because of increased human activity. A meeting of landslide experts held in Gangtok in 1975 had expressed fears that landslides will increase because of growing development activities. More roads will be required and, as these roads open up new areas, forests and other natural resources will be heavily exploited. Leading to an increase in landslides and erosion. Landslides and slope instability problems are now being noticed in numerous Himalayan towns.

Natural drainage often gets blocked or obstructed by human activity and surrounding hillsides get denuded because of increase human pressure. In addition, dynamite is often used to blast the hills for construction which further destabilises the slopes. There is no doubt that increasing construction activity in Himalayan areas is increasing landslides and soil erosion and leading to serious local problems. But it is not clear whether the increased erosion is large enough to increase sedimentation on the scale of the Indo-Gangetic plains.

Conclusion

Probably the impact of the increased erosion is small over natural factors. But there is no doubt that Himalayan environmental degradation is now widespread and the result is quite alarming. The forest cover has declined rapidly and its place has often been taken by overgrazed lands. Modern human activity like road building and commercial felling of trees have taken place in an ecologically destructive manner. It would, therefore, be wrong to say that Himalayan environmental degradation has no impact whatsoever on increasing soil erosion, surface runoff or landslides.

Considering the fact that a wholesale conversion of forests to overgrazed shrublands has taken place, it can be argued that the local effects of ecological degradation on flood flows are sufficiently widespread to produce a general and regional effect. But it would be equally right to say that these increases are probably far smaller than what environmentalists have implied—"perhaps in the tens rather than hundreds of percentages," as John Metz of the university of Wisconsin puts it.

For example,

Some archaeologists and scientists now believe that Himalayan rivers have constantly changed their courses long before deforestation began. At the time of the Indus civilisation, Yamuna probably did not flow into the Ganga nor the Satluj into the Indus. Instead they flowed into the Ghaggar—currently a small, seasonal river that emerges from the Siwalik foothills near Chandigarh. They believe that numerous floods and hydrologic changes have taken place in the Ghaggar plains over the last three to four millennia, possibly associated with tecton changes in the Himalayan foothills. Far more recently the Teesta used to flow into the Ganga until 1787.

In 1540, Portugue merchants described the very large, well fortified city Gaur on the banks of the Ganga. By 1575, the river has deserted the city and it was abandoned. In the lower course large islands appeared almost overnight only to be washed away with tremendous loss of life several years later when they had been densely settled. This process is still going on today in Bangladesh. Journalist Debjani Sinha recently reported that several river in West Bengal are changing their courses and causing immense riverside erosion leading to displacement people living in the Murshidabad and Malda districts.

5

Constitutional Provision to Save Himalaya

Introduction

Due to recurring natural calamities, loss of natural resources, the natural ecosystem around Himalayan region is at stake. Wildlife resources are highly sensitive to such environmental degradation. This has led to virtual extinction of some species in many cases.

Consequently, the conservation of environment and protection of wildlife has attracted the attention of people as well as the government. Considering the depth of the problem how far the government has come forward to save the situation is discussed in this chapter.

There is an old perception that man lives in close relationship with other living things like plants and animals. This concept of inter-relationship has been further strengthened over the last fifty years by the concept of ecosystem which provided scientific foundation to the interaction process operating between living and non-living components. If one attempts to alter any one of the components, all other elements are likely to be affected.

Reflects the quality and quantity of plants that can grow and survive on the surface of the earth, depend upon the diversity of environmental conditions. Every living being is limited by the non-living components which constitute the ecological system. Also every living being contributes to the formation of environmental system upon which all living beings are dependent for their existence.

The ecological system around us is balanced by wildlife in many ways. With the growing anthropogenic pressure on natural resources, various forms of deterioration in the environmental system have set. The most effected parts are the Himalayan and sub-Himalayan tracts which have resulted in severe erosion, serious floods and recurring landslides in the region. The process of desertification is more visible particularly in the arid and semi-arid regions. The natural vegetation cover is gradually disappearing.

Our natural ecosystems is at stake due to continuing natural calamities and gradual exhaustion of natural resources. Wildlife resources are highly sensitive to such environmental degradation since habitat destruction has in fact led to virtual extinction of species in many cases. The status of wildlife, therefore, acts as a good yardstick to measure environmental quality and health in a geographical region.

Protection to Wild Life

To safeguard the forests and wildlife of the nation fortunately provision has been made in our Constitution under the Directive Principles of State Policy. The role of common people in the protection and improvement of natural environment including forests, lakes, rivers and wildlife has also been defined in the Constitution was taken by.

The Union Government took the very first step in this, direction by constituting a Central Board of Wildlife in 1952. The Board was constituted under the chairmanship of the Inspector General of Forests of India, for suggesting necessary measures for the protection and conservation of declining wildlife of the country. The Board was later designated as the Indian Board of Wildlife (IBWL) which is the apex body to advise the Government of India in this matter.

Stringent Act Passed

In the field of wildlife conservation the Wildlife Protection Act played a vital role. It was passed in 1972. The Act has provisions for setting up wildlife sanctuaries and national parks for the complete protection, conservation and management of wildlife. Severe punishments have been provided for violation of the provisions of the Act. Prior to this Act, the protection of wildlife was governed by the rules of State Governments under the Indian Forest Act. Some special legislations were also enacted by different states. The prominent among these enactments are the Bombay Wild Animals and Wild Birds Protection Act, 1951, the Elephant Preservation Act, the Rhinoceros Preservation Act and the Game Act. Apart from a little control on shooting in the reserve forest areas, these legislations could not take any other action to protect the wild life.

The most important step taken under the wildlife Protection Act is to classify wild animals, birds, amphibians and reptiles into five different schedules. The Act has seven chapters dealing with extent, commencement and definitions of the Act; authorities to be appointed or constructed under the Act; hunting of wild animals; creation of Sanctuaries and National Parks etc.; regulation of trade on wild animals, animal articles and trophies; prevention and detection of offenses; and miscellaneous items.

For the first time the classification of wild animals into various schedules is brought under one legislative umbrella for complete protection of rare and endangered species. In the schedule first, rare and endangered species have beenb listed. Schedule second includes special game species which require more stringent protection and for which licences can be given only under special circumstances. Schedule third and fourth comprise big and small game, while schedule fifth lists vermin.

To build up a common organisation for the conservation of wildlife in all the states, a separate organisational system was also laid down in the Act for the first time. Trade in wild animals and animal products was also bought under legislative control in the Wildlife Protection Act, 1972. The Act lays down regulations governing the declaration of stocks, preparation of inventories, issuing of certificates of ownership, regulation of transfers of animals and their products from one state to another, and from one person to another, as well as registration of traders with provision for maintenance and inspection of records along with other necessary limitations and penal measures.

Amendment of the Act

To revise the provisions and schedules of the Act a sub-committee was appointed by the Indian Board of Wildlife after a couple of years. The most significant recommendations made by this expert committee were for bringing the hitherto neglected plant life under the provision of the Act, which in fact constitutes a vital element of the habitat of the wildlife. Thus the endangered species of plants were also considered for being included in the schedules of the Act. The committee also made recommendations for the control of trade in plants and plant products.

In addition to these recommendations, the committee also made suggestions for the creation of Biosphere Reserves. After realizing the gaps in the Wildlife Protection Act and considering the recommendations made by the sub-committee of Indian Board of Wildlife, the Government of India, brought out the Wildlife Protection Amendment Bill in the Parliament in 1991 and was enforced in the same year.

Protection to Wildlife and Forests

The term protected areas is referred to the places like National Parks, Wildlife Sanctuaries, Game Reserves, Closed Areas and Biosphere Reserves created for affording protection to wildlife and its habitat and genetic resources in different parts of the country (Wildlife Protection Act, 1972 and Wildlife Protection Amendment Act, 1991).

National park, is such an area considered to be of ecological faunal, floral, geomorphological or zoological significance and declared as protected through a specific notification issued by the state government after completing the due procedure such as enquiry and hearing for rights and concessions, under the provisions of the Wildlife Protection Act. There is a bar on all kinds of destruction, exploitation or removal of forest products, skins, trophies of wildlife or its habitat in a national park. Grazing by cattle, blasting and quarrying are strictly prohibited within the park. There is also provision for the shifting of villages existing in a national park to other areas. In case of a wildlife sanctuary, permanent residents of specified settlement around this area may interact with sanctuary area, but for the protection of the wildlife and forests of that area they are bound to perform certain protective duties. As a protective measure, regulation on grazing and fishing in the sanctuary area can be imposed. Commercial felling is not allowed in the sanctuary.

For special treatment, few of the national parks and wildlife sanctuaries are designated as Biosphere Reserves, Tiger Reserves, etc. Biosphere reserves are protected areas of respective terrestrial, coastal and marine environments that have been internationally recognized for this vital significance in the conservation of genetic resources. These areas are supposed to be completely free from all kinds of anthropogenic interference.

Rising Number of National Parks and Sanctuaries

The Wildlife Protection Act, facilitated the state governments to declare any such area of adequate ecological, faunal, floral, geomorphological, natural or zoological significance as national park and wildlife sanctuary, for the purpose of protecting, propagating or developing wildlife or its environment. Before the enactment, there were only eight national parks and 68 wildlife sanctuaries in the country. But in 1980 the number of national parks and wildlife sanctuaries was raised respectively to 19 and 204 (Saharia, 1982). The states of Andhra Pradesh, Himachal Pradesh, Karnataka, Madhya Pradesh, Orissa, Punjab and West Bengal took great interest in the protection and conservation of wildlife through the creation of national parks and sanctuaries under the provisions of the Act.

Since 1980, the numbers of national parks and sanctuaries and the geographical dimension of protected area have been increasing continuously. In 1987, there were 54 national parks and 372 wildlife sanctuaries. The number of national parks and wildlife sanctuaries was raised respectively to 147 and 633 in 1994 encompassing 12 per cent of the forest area and 4.3 per cent of the total geographical area of the country (Singh, 1994).

However, there exist great variations in the number of wildlife protected areas in the country as given by various sources. Some state forest departments and wildlife wings do not respond to queries pertaining to the details of wildlife protected areas in a proper way, and this causes great difficulties in assessing the status of wildlife conservation and management in different ecological regions of the country.

Some of the national parks are sanctuaries have been created specifically for the conservation and protection of the rare and endangered species. Notable among them are: the Gir National Park and Sanctuary for the Asiatic lion; Kaziranga National Park for the rhinoceros; Dachigam Sanctuary for the Kashmir stag; Keibul Lamjo for the brow-antlered deer; and Velvadar National Park for black-buck. Some other national parks are famous for the richness and variety of wildlife found in them. Amongst the most popular are the Corbett and Kanha national parks. The Bharatpur Sanctuary is world famous for its avifauna.

General Awareness and Important Projects

Today, the status of wildlife in India is alarming today. A large number of species are greatly endangered and are struggling for their very survival. It is good that a general awareness has developed among the intellectuals of the society and in some segments of the public about the necessity to conserve the degrading ecosystem and protect the wildlife of the country. For that specific conservation measures have been taken by both the central and state governments. The important conservation projects launched after the enforcement of Wildlife Protection Act, 1972 are: (i) Project Tiger, (ii) Gir Lion Project, (iii) Crocodile Project, (iv) Himalayan Musk Deer Project, and (v) Project Elephant.

Formation of Project Tiger

In view of India's fast vanishing wildlife, particularly the tiger which is now being considered as a symbol of country's wilderness, the Project Tiger was formed in early seventies by the special Task Force of Indian Board of Wildlife. The flow of nutrients through various trophic levels and complex food webs in the forest ecosystem culminates in the tiger. Plant and animal residues are decomposed by microorganism to form humus which is primarily responsible for the fertility of soils in wilderness areas. Besides providing food and necessary environment to various species of wildlife, it facilitate the conservation of soil and water resources. Among a host of animals that grow on this life support system, the tiger stands supreme in most of our forests. The well-being of tiger is thus synonymous with the health of the Indian wilderness.

Objective of Project Tiger

The special Task Force of Indian Board of Wildlife which fabricated the project, mentions in its report enunciating. The objectives of the Project Tiger were "To ensure maintenance of a viable population of tiger in India and to preserve, for all time, areas of biological importance as a national heritage for the benefit, education and enjoyment of the people". The Project Tiger is the most ambitious, prestigious and the biggest conservation project launched so far in the country. It was inaugurated as a central sector project by our late Prime Minister Mrs. Indira Gandhi in 1973. The objective of the project was strongly reflected in the message for the occasion by Mrs. Gandhi, who regarded it as a truly national endeavour and observed: "The tiger cannot be preserved in isolation. It is at the apex of large and complex biotope. Its habitat threatened by human intrusion, commercial forestry and cattle grazing, must first be made inviolate." (Report by the special Task Force of Indian Board of Wildlife).

The special Task Force which was then designing and formulating this ambitious national level project was fully convinced that Project Tiger must conform to an ecosystem approach.

Most of our potential tiger reserves were under great pressure due to increasing demographic pressure, exploitation of land resources for agricultural expansion, overgrazing, rampant fires, and soil erosion. Therefore, the areas to be declared as tiger reserves were to be free from al these hazards. The areas for such reserves were finally chosen so as to include as many biogeographic types of regions as possible in the distribution range of tigers in the country. The other important point of consideration for the selection of tiger reserves was the amenability of the specific area to concerted conservation action.

Based on following principles the management plans were prepared for each tiger reserve:

(i) Elimination of all forms of human exploitation and disturbance from the core area and rationalisation of such activities in the buffer zone.

(ii) Limitation of habitat management to repair damage done by man with the aim of restoring the ecosystem as close to its natural functioning as possible.

(iii) Researching facts about habitats and wild animals and carefully monitoring changes in flora and fauna

Initially nine tiger reserves were established in 1973-74, under the joint venture of central and state governments. Soon after villages situated in or near the reserves were relocated, cattle camps were shifted and halting of stock grazing and forestry operations in the core areas were effected in quick phases. Effective measures were taken for the conservation of critical environmental components, such as, land and water.

In the next phase a massive programme was started for the improvement of the original reserves by enlarging their core and buffer zones and upgrading their legal status. In order to enhance substantially the biogeographic coverage of the project, as many as 14 new tiger reserves have since been added between 1978 and 1995, bringing the total number of tiger reserves to 23. These 23 tiger reserves established under the Project Tiger now encompass a gross geographical land surface of 31,648 km^2 in 13 states of the country. Madhya Pradesh having the highest number of tiger reserves (5) has been designated as 'Tiger State'. The launch of Project Tiger was in itself a landmark in history of wildlife conservation in India. There was a steady rise in the animal population.

The population of tigers went up from merely 268 in 1972 in nine tiger reserves to 1227 in 1990 in eighteen reserves. By 1995 with the inclusion of five more reserves the population of tigers reached the figure as high as 1373. Besides, a number of other endangered species also benefited substantially. One significant achievement has been the considerable improvement in the preservation of overall floral and faunal diversity in such protected areas.

An urgent need to restore the productivity of the buffer zones of wildlife preserves and to increase the productivity of adjoining rural ecosystems through ecological development was felt. This ultimately resulted in the starting of eco-development programmes in the buffer zone of the Ranthambhor Tiger Reserve in Rajasthan in 1988-89.

A number of other tiger reserves including the Corbett Tiger Reserve have now started eco-development programmes in the reserve area in the areas around. The implementation of the Project Tiger is continuously monitored by the steering committee. This Committee, constituted under the chairmanship of the Prime Minister, regulation of tourism, management of buffer zones, interaction with local people through eco-development programmes, establishment of nature interpretation centres and veterinary care centres in the reserves. A high level meeting of Tiger Range Countries held in New Delhi in March 1991 to launch tiger and wildlife conservation programme through out the world.

Launching Global Forum

At present only five species of tiger are found in the 14 countries of Asia in the world. These countries are India, Bangladesh, Bhutan, Cambodia, China, Indonesia, Laos, Malaysia, Myanmar, Nepal, North Korea, Russia, Thailand and Vietnam. These together constitute the Tiger Range of the world. The membership of this forum is open to all other countries interested in the conservation of tiger. The major of this forum are to provide leadership to tiger conservation activities at global level, to develop uniform policy, to facilitate the exchange of information and experience among the Tiger Range countries and to maintain a constant vigil on the condition of tiger throughout the world.

Delhi Declaration

To facilitate the conservation of tiger within as well as outside the country, the Government of India has recently worked out a National

Tiger Action Plan. This strategy includes national as well as international policies for the effective conservation of tiger. The international policy was included in this plan at the 'International Symposium on Tiger' held in New Delhi in February 1993 and known as 'Delhi Declaration'. Establishment of the network of effective tiger reserves, appropriate management of tiger habitats outside the protected areas and tiger reserves, enhancement of the management efficiency of the tiger reserves, minimization of the biotic pressure of the people living in and around the tiger habitats on their natural resources and encouragement to eco-tourism are the major components of the strategy evolved for the conservation of tiger and its habitat for the coming ten years, under the national policy.

Other Important Projects

The Gir Lion Project was started in 1972 in the Gir National Park in the Saurashtra Peninsula of Gujarat, in order to conserve the remnant population of Asiatic lion. The Gir forest is unique and the last stronghold of the lion in Asia (Rashid, 1982).

The Crocodile Project which is specifically known as Project Crocodile Breeding and Management, was commenced initially in Tikerpada district of Orissa, in 1975, and later it was managed and monitored by creating a network of sanctuaries in different states (Bustard, 1982).

Habitat destruction, due to the removal of the forest by livestock-grazing and deforestation by villagers for timber and fuelwood are the couple of reasons for the disappearance of the once widespread throughout the Himalayan ranges, the Himalayan musk deer. (Green and Singh, 1982). The unique feature of this deer species is the presence of a musk gland in the made. This is chiefly responsible for its rapic decline. Once the threatened status of the Himalayan musk deer was recognized, the species was listed as vulnerable in the Red Data Book (Holloway, 1973) of the International Union for the Conservation of Nature and Natural Resources (IUCN), 1974. Then, in collaboration with the Government of India the world.

World Wildlife Fund Project on the Ecology and Conservation of the Himalayan musk dear was launched. The project is based in Kedarnath Sanctuary of Uttar Pradesh.

In 1993, on the recommendations of the Indian Board of Wildlife

(IBWL) the Project Elephant was launched by the Government of India. It was launched to protect the elephants in this country. The Project Elephant is being operated on a relatively smaller scale compared to the Project Tiger. In the initial stage two types of protected areas have been chosen for the implementation of this project. These areas are: (i) where the population of elephants is high, and (ii) where the population of elephants has reduced considerably during the recent years.

Programme for Biosphere Reserves

Of many environmental systems the biosphere system is the most fragile and highly sensitive from conservation point of view. the details of the existing and proposed Biosphere Reserves in India as released by the Wildlife Institute of India is as follows:

	Name of the state/ union territory		*Name of the Biosphere Reserve*	*Area (km^2)*
1.	Andaman and Nicobar	1.	Great Nicobar Biosphere Reserve	1045
		2.	North Andaman Biosphere Reserve	1376
2.	Arunachal Pradesh	1.	Namdhapa Biosphere Reserve	2500
3.	Assam	1.	Manas Biosphere Reserve	NA
		2.	Kaziranga Biosphere Reserve	NA
4.	Gujarat	1.	Rann of Kachchh Biosphere Reserve	NA
5.	Madhya Pradesh	1.	Kanha Biosphere Reserve	NA
6.	Meghalaya	1.	Nokrek Biosphere Reserve	NA
7.	Rajasthan	1.	Thar Desert Biosphere Reserve	NA
8.	Tamil Nadu	1.	Gulf of Mannar Biosphere Reserve	10500
		2.	Nilgiri Biosphere Reserve	NA
9.	Uttar Pradesh	1.	Nanda Devi Biosphere Reserve	630
		2.	Valley of Flowers Biosphere Reserve	3942
10.	West Bengal	1.	Sunderbans Biosphere Reserve	1700

NA = Not available.

The basic concept of Biosphere Reserves was evolved under UNESCO's Man and Biosphere Programme (MAB). The MAB is an interdisciplinary research programme to study exclusively the theme of management problems arising out of interactions and processes between natural environment and human activities. This programme emphasizes upon the ecological approach to evolve strategies for the most scientific and rational use and conservation of the earth's natural resources. It also emphasizes to improve the global relationship between man and natural environment, and to consider the future effect of present day human activities on the environment.

The Biosphere Reserve Programme was launched under the Project (8A),—'Coordinated Worldwide Network of Protected Areas,' and Project (8B)—'Conservation of Animals and Plants, including Microorganisms,' of the Man and Biosphere Programme. The main objectives of the Programmes are: (i) to conserve for present and future human use, the diversity and integrity of biotic communities of plants and animals within natural ecosystems and to safeguard the genetic diversity of species on which their continuing evolution depends; (ii) to provide areas for environmental and ecological research, including, particularly baseline studies, both within and adjacent to these reserves, such research to be consistent with the objective (i) above; and (iii) to provide facilities for education and training (Saharia, 1982).

Thus, conserve the genetic resources of flora and fauna in the country the concept of biosphere reserve is of immense value. This required a detailed survey and classification of natural ecosystems for proper identification of their biological diversity and pristine attributes. The areas identified as biosphere reserves should be large enough to be ecologically viable. The Biosphere Reserve Programme was launched at global level in 1971 and the first biosphere reserve of the world was created in 1976. Now there are as many as 300 biosphere reserves distributed over more than 75 countries of the world including India. In order to identify the potential areas to be constituted into biosphere reserves in the country, the Government of India set up the Indian National Man and Biosphere Committee in 1979. The committee created a core group of experts from various fields for submitting recommendations for the potential areas of critical importance and vital significance in genetic conservation. On the basis of the guidelines issued by the International Union for the Conservation of Nature and Natural Resources (IUCN) and UNESCO, 14 potential areas representing 10 biogeographic zones of the country were identified. But out of these 14 possible areas only a few have been assigned the legal status of biosphere reserves by the concerned state governments through issuing the notifications.

Protection and Conservation of Wildlife

For the effective protection and conservation of wildlife and its habitat a National Wildlife Action Plan was adopted in 1983. The main objectives of the Action Plan were: (i) establishment of a representative network of protected areas, (ii) management of protected areas and habitat restoration, (iii) wildlife protection in multiple use areas, (iv) rehabilitation of endangered and threatened species, (v) captive breeding of endangered species, (vi) wildlife education and interpretation, (vii) research and monitoring, (viii) domestic legislation and international conventions, (ix) national conservation strategy, and (x) collaboration with voluntary bodies and non-government organisations. In order to achieve these objectives specific programmes and projects have been launched in different biogeographic zones of the country from time to time and many of them have registered good progress such as the 'Project Tiger.'

To attain the objectives as laid down in the National Wildlife Action Plan, the Wildlife Institute of India carried out a survey of the protected areas of the country in 1984. It was observed that the protected areas are not evenly distributed over all the biogeographic zones of the country. Moreover, some rare, endangered and threatened mammal species had been protected in only a few specific protected areas. Thus the biogeographic diversity of the country was not being represented perfectly through the national parks and wildlife sanctuaries created on the basis of local considerations.

Thus considering the above view, the Wildlife Institute of India presented a workable biogeographic classification system. It also recommended for the representative network of protected areas based on this classification. The report of the Institute calls for creating a large number of new protected areas as well as enlargement, and up-gradation in the legal status of many of the existing ones.

The enlarged network of wildlife protected areas proposed to be constituted by 148 national parks and 503 wildlife sanctuaries encompassing a total geographical area of 1,51,342 km^2 which accounts for 4.6 per cent of the total geographical area of the country. Presengly, we have 12 per cent of the forest area and 4.3 per cent of the country's total geographical land surface under the protected area network.

Formation of Management Body

As forestry and wildlife are organisationally a unified subject under the control of states, the role of the Central Government is only advisory. It has no control on the management and administration of forest and wildlife. The setting up of organisational mechanism, formulation and implementation of management policies and making of legislations are done by state governments.

In India, most of the states are having department of forest under a separate ministry. The Principal Chief Conservator of Forests is the head of the forest administration in the state which constitutes both forests and wildlife. In most of the states separate wildlife wings are established under the provisions of Wildlife Protection Act. These wings are headed by the Chief Wildlife Warden. In a majority of states, Divisional Forest Officers have been notified as Wildlife Wardens for their respective forest divisions, whereas, important national parks are under the control of a Director assisted by Wildlife Warden.

At the central level the Department of Forests and Wildlife, is the apex body to look after the conservation and management of wildlife in the entire country. If functions under the direct control of the Ministry of Environment and Forests. The Inspector General of Forests is the chief of forestry and wildlife in India. He is assisted by a team of high level officers consisting of a Joint Secretary Forests, an Additional Inspector General of Forests, three Deputy Inspector Generals of Forests and two Assistant Inspector Generals of Forests. The Joint Secretary Forests and Wildlife is also designated as Director Wildlife Preservation.

Chief Advisory Body

The chief advisory body for advising the Government of India with regard to wildlife policy is the Indian Board of Wildlife (IBWL). Major achievements of the Indian Board of Wildlife are the constitution of specialized wings, such as the Zoo Wing, the Flora Wing and the Bird Wing for making appropriate recommendations. On its recommendation a Directorate of Wildlife Research and Education was established in Dehradun. In order to enhance the quantum and quality of scientific inputs into the field of conservation through better information and trained manpower, a National Institute called the Wildlife Institute of India (WII) was set up in Dehradun in 1983 by integrating two smaller training units initiated in the seventies. This Institute has been substantially strengthened and is now providing postgraduate education and training in wildlife biology and management. It has also undertaken field research on major wildlife problems in different parts of the country. To raise the general public opinion about the need of wildlife conservation the Board celebrates Wildlife Week every year.

The University Grants Commission has already initiated action to set up faculties on wildlife conservation in certain universities as per the directives and recommendations of IBWL. One of the major achievements of the Board is the evolvement of National Nature Conservation Policy in 1978. This policy statement was incorporated in the Sixth Five Year Plan.

To control the commercial exploitation which has been the prime cause of the rapid decline of wildlife in India, the Indian Board of Wildlife has been instrumental to enable the government in curtailing and restricting the export of wildlife and animal products to a considerable extent.

Moreover, to control the global trade in endangered wildlife species, India became a member of the International Convention on Trade in Endangered Species in 1976. This Convention has introduced an international system of licensing and legal procurement certificates to control the trade of all the species listed in the appendices to the Convention. On the recommendations of the IBWL, most of India's rare and endangered species have been included in appendices to the Convention to afford stricter international control on the illegal trade in wildlife and wildlife products. Now, State Wildlife Advisory Boards have also been set up in many states which have been playing a significant role in suggesting measures for wildlife protection, conservation and management, considering local conditions and management.

Voluntary Organisations Cropped-up

After 1955 some non-government, voluntary, national and international organisations have come up in India for the protection and conservation of wildlife. The Wildlife Preservation Society of India, the World Wildlife Fund of India and the Bombay Natural History Society are notable ones. The World Wildlife Fund (WWF) was started in India in 1969 at the time of the 10th General Assembly of the International Union for Conservation of Nature and Natural Resources

(IUCN) in New Delhi. It is an international organisation with its headquarters in Switzerland. WWF works in close collaboration with the IUCN. Financially as well as technically it is assisted by the United Nations Organisation and its other agencies, viz., UNESCO, UNEP, FAO, etc. The World Wildlife Fund has been supporting a number of wildlife conservation projects in India. The project, the Project Tiger, which is the single largest conservation project of its kind in the entire world, has been partially funded by the WWF.

The Wildlife Preservation Society of India was founded in Dehradun, in 1958 by some wildlife enthusiasts and naturalists. Since then the society has been attempting to promote interest in and impart knowledge regarding the preservation and conservation of all forms of wildlife, particularly among the youths of the country. It also assists the government, in the enforcement of the provision of Wildlife Protection Act, management of national parks, sanctuaries and other protected areas and conservation of wildlife.

Then in 1983, the Bombay Natural History Society was established by some residents of Bombay metropolitan. It has endeavoured in focussing public and government attention on the need for understanding, conserving and protecting the rich and diversified wildlife heritage of the country. It brought out some literatures, organised seminars, conferences and also carried out intensive research on wildlife problems. With the setting up of a National Institute of Ornithology under the Society, with support from the Ministry of Environment and Forests, Government of India, the Society became more active.

While conservation gains have been substantial, these are limited to the protected areas only. Even in these area a very small number of these animals are able to receive favourable protection. This is because of the low outlays for conservation and the lack of scientific inputs in the form of well investigated information and trained manpower. The ongoing depletion of forests outside the protected areas has led to degeneration and destruction of the habitats. This is another cause of concern.

6

Effect of Pilgrimage in Gangotri-Gaumukh Region

Introduction

Modernisation has adversely affected human life. Being tired of the hectic, monotonous daily routine, man wants relief. He wants a periodical interval to refresh his mind and soul in some isolated place in the midst of nature, away from the busy city life. He wants a physical rest.

This has raised the number of tourists to hill stations and in forests. Even Gangotri—Gaumukh, which lies at such an altitude, could not remain untouched. The pilgrimage to Gangotri is an old event. But the mixing of adventure with pilgrimage has changed the scenario of the place. It has created many economic, social, cultural and environmental problems there. How these are affecting the life of the people residing there and how these can be rectified is expressed in this chapter.

Beginning of Tourism Industry

Man is fond of wandering ever since the beginning of civilisation. Much of that was unconscious in the beginning. Later, in search of necessities of life. And today it is more in search of mental peace and physical rest. Human being are tired of Industrialisation, modernisation, spread of education and speedy communication systems. These have changed the human life and forced the people to go out of this environment for refreshment of mind, body and soul. Today, the city dwellers are looking for periodical intervals from their hectic schedule of monotonous work and are desirous for a retreat in isolated pockets of nature.

Gradually this has given birth to tourism industry. In this process out-door activities like climbing, mountaineering, trekking and kayaking etc., came into the front line of tourism. Man even leaped into Himalayan heights. Gangotri-Gaumukh area which lies in 78°55' to 79°10' E longitudes to 30°50' to 31°1' N lattitudes (with an altitude variation between 3000-4000 M) couldn't remain untouched with this new tourism. Of course, pilgrimage to Gangotri is an age old phenomenon, but this mixing of adventure activities with pilgrimage and sometimes simultaneous running of the two things in the area is changing the scenario. This religio-adventure tourism has initiated many economic, social, cultural and environmental changes in the region. Number of tourists and pilgrim to this area are increasing rapidly, every year.

In the year 1993 the number of tourists and pilgrims who visited Gangotri was 1,12,132 and 46,084 for Gaumukh (snout of 26 km long Gangotri glacier). This crowd of tourist/pilgrim to Gangotri starts with the door-opening ceremony of Gangotri temple in April/May, every year. May/June are the peak months for the tourist arrivals. Due to rains and fair weather roads, July-August remain slack months.

But the recent trend of *kawariyas* (pilgrims who walk all the way to collect Ganga Jal from Gaumukh and offer this to Shiv temples, in different parts of the country), is a new phenomenon. This has started giving crowd in July/Aug. (Shravan month of Hindu calendar), too. Later from the month of October to November, the crowd of trekkers come to the area. This ends on Dipawali day, when the ceremony of door closing for Gangotri temple takes place. Few mountaineering expeditions also come to this area in between April-September. After November, gradually the area gets covered with snow and the valley becomes isolated, and turns into a paradise for high wild animals living in that altitude.

Effect of Religio-Tourist Activity

No doubt this religio-tourist activity has boosted the economy of the region, but disturbed the ecology. It has changed the settlement and population pattern, sanitation system, general price structure and ecology. According to 1991 census, the total population of Gangotri was 128 and among them 7 were women. This temporary township which remains open for pilgrims/tourists, only for six months, had entertained 1,12,132 tourist/pilgrims. Earlier, for every tourist/pilgrim, who came upto Uttarkashi, Gangotri used to be the destination. But now-a-days, Gaumukh and Tapovan are the aim.

One has to pass through Chirbasa (9 km from Gangotri) and Bhojbasa (5 km from Chirbasa) to reach Gaumukh from Gangotri. The

18 km trek to Gaumukh, passes through the most beautiful natural rock garden. It lies in the reserve forest area of the forest department of U.P., where construction of any type is strictly prohibited. Tapovan is 4 km from Gaumukh at an altitude of 4,400 M is Tapavan. One has to cross Gangotri glacier to reach Tapovan. Beyond Gaumukh there are many snow covered peaks and glaciers which are great challenges to mountaineers and trekkers. Prominent peaks are Shivling, Meru, Bhagirathi group, Thelu, Bhrigupanth, Sudershan, Kirtistambha, Satopanth and towering above them all, the mighty Chaukhamba.

Due to pilgrimage/tourism activities local economy is developing, but few negatives to ecology are rapidly spreading like—deforestation and land slides, garbage menace, recession of Gangotri glacier and change in climate.

In between Gangotri-Gaumukh, the movement of thousands of tourist/pilgrims is going on. They are staying in this area but no proper system of collection and disposal is existing there. The throw away mentality of people is creating garbage heaps. Food is being cooked for them with forest fuel. Some of the pilgrims/tourists go to Gaumukh on horses and ponies. These horses and ponies graze freely in Alpine meadows. The biggest nuisance to the alpine flora are the horses and ponies of neighbouring villages, which are left there for two months (July-August) for grazing only. All these together is destroying the flora and initiating soil erosion deforestation, land slides, garbaga menace and vecession of Gangotri glaciers.

Research Project for Eco-conservation

To study the problems of in this area a research project 'Growth of Himalayan tourism and pilgrimage in Gangotri region: An impact on physical resources and economy", financed by Ministry of Environment and Forests, Government of India, was conducted in between 1989-93. Based on the research few recommendations and priority wise follow up actions were suggested.

Recommendations

- Supply and use of kerosene oil and LPG has to be made compulsory as fuel.
- Potential of non-conventional energy sources like—solar and micro-hydel power has to be explored.

- Proper garbage collection and disposal system has to be evolved. Installation of garbage bins in between Gangotri-Gaumukh area and icinerator at Gangotri is essential.
- Number of scavengers has to be increased at Gangotri and dry latrines should be banned.
- Sulabh complexes should be constructed at Gangotri, Chirbasa and Bhojbasa.
- Proper drainage system has to be evolved for Gangotri and strict ban should be imposed on putting drains into the Bhagirathi river.
- Nursery of indigenous flora has to be developed at Chirbasa and afforestation has to be done at a war footing.
- Grazing by horses and ponies should be immediately stopped.
- Awareness campaign for sanitation, ecology, hygiene, healthy pilgrimage and tourism among the hosts is required.
- Maintenance of tourist arrivals record, beyond Gangotri has to be done properly.
- Instead of free handling for the construction of hotels in Gaumukh area by the local people, Forest Department should erect the angle iron frame for the hotels which should be auctioned to the local people. This will stop the tree felling every year for the construction of hotels, give revenue to Forest Department, give the hotels an aesthetically sound look and will check the exploitation of the local people by the lower rungs of Forest Department.
- No mountaineeing or trekking expedition should be allowed to go straight away with any outside travel agency, except the travel agencies of Uttarkashi and from the travel agencies of Uttarkashi, a compulsory contribution (according to the business done by them), should be taken for the eco-conservation of the area. Even, Indian Mountaineering Foundation, New Delhi, should contribute (from the royalty it charges from the foreign expeditions to this area), in the eco-conservation fund. With the help of this fund, eco-conservation work should be undertaken.

Besides making paper recommendations for the eco-conservation of this area. But a new project has been started from the previous one. The task of eco-conservation, to be more specific—Birch afforestation in Bhojbasa, a place named after its birch forests, which is today devoid of birch trees, is in action. The target of the project was to prepare saplings of birch and other indigenous flora of the region, at Chirbasa. For the convenience of work and suitability of land, an area was selected for raising the plant nursery at Chirbasa, in the year 1993. Saplings of Birch and Blue pine are prepared with seeds and Salx and Popular saplings are prepared, with vegetative propogation method.

Approximately three thousand saplings of Birch, Blue pine, Juniper, Salex and Poplar are ready for planting and another crop of few seedlings of tens of thousands of birch are ready to be shifted in the polythene bags, very soon. The nursery work was financed by Uttarkashi seva Nidhi, Almora. Help was also extended up to certain extent by the Nehru Institute of Mantaineering.

It was decided that afforestation will be done on war footing. Apart from the hired labourers, neighbouring villagers (women, men, youth and school going children) will be motivated for afforestation and eco-conservation work. Later on pilgrims/tourists will be asked for more participatory type of eco-tourism and eco-conservation. This will be purely on voluntary basis.

Notes

Nursery work has been financed by "Uttarkashi Seva Nidhi", Almora and many a times help has been given by the Nehru Institute of Mountaineering, Uttarkashi.

Authors are thankful to Uttarakhand Seva Nidhi, Almora and the Nehru Institue of Mountaineering, Uttarkashi.

7

Gangotri Conservation Project

Introduction

Beside the increasing number of mountaineer and trekker, there is also a tremendous increase in pilgrimage movement to the shrines of Himalaya. Gangotri is the most populated area today. But due to lack of adequate civic amenities and sufficient facility to meet the daily needs of such a huge religio-adventure visitors there, the atmosphere there is getting polluted. What steps are taken by the state and the central governments to check the alarming situation is discussed in this chapter.

Gangotri—A Tiny Paradise

Environment is a global issue today. Leading organisations both government and voluntary, are concerned about the destruction of natural resources. This effect is more emphatic in the third world countries, where the intensity of human activity has accelerated more dramatically than the population growth.

As is clear from the previous chapter, in India, the pilgrimage movement to the Himalayan shrines has increased a hundred fold in the last decade. The Himalayas, which have a number of such shrines is getting drastically affected with the on rush of these pilgrims. Besides, more and more trekkers and mountaineers are venturing into this magnificent wilderness. Worried about the situation the Himalayan Environment Trust' (HET) previously known as the Himalayan Adventure Trust studied the effects of such activity in Gangotri, a very popular area for trekkers and mountaineers.

The study at the Indian Mountaineering Foundation came to the conclusion that 'Gangotri is the most polluted area in the Himalayas today' due to the influx of over a lakh and a half of the pilgrims, ten to fifteen thousand trekkers and three to four hundred mountaineers going there every year. The effect of this in a small place like Gangotri is mind boggling. This tiny paradise has no basic facilities existing and is accumulating all the garbage, which sadly is now being dumped into the very source of the Ganga.

Award for Rubbish Collection

More than 50 Indian expeditions and about 25 foreign expeditions are attempting various peaks in the Gangotri region. These are all approved by the IMF. Not only that, Indian expeditions are financially assisted by IMF. During 1991 IMF had asked the expeditions visiting Gangotri to include environment thrust as a special programme. They have been asked to collect as much rubbish as possible and bring it to Uttarkashi for depositing with the Nehru Institute of Mountaineering (NIM). Special awards will be given to the expedition which bring maximum rubbish from the region.

The effect of such a vast numbers of trekkers and mountaineers in the region included the destruction of flora and fauna of the area and the destruction of forests. The absence of basic amenities like latrines, baths and garbage collection facilities and the total lack of arrangements to meet the daily needs of human beings, for e.g. cooking media and rations resulting in the destruction of stately forests and polluting the Himalayan environment.

Himalayan Environment Trust's Recommendation

At the office we came to the conclusion that the following efforts could bring some balance of pollution control in this area:

- The government should declare Gangotri a—'National Park'.
- Private business houses should take part in this effort.
- Mass media could build a greater awareness in the people.

Local schools and colleges should come forward and help. This would have the dual advantage of young, eager and zealous help, besides teaching them when they are still young.

The Indian Mountaineering Foundation and the Nehru Institute of Mountaineering should monitor and control the movement of mountaineers and trekkers.

National Park

The concept of the Gangotri National Park was favoured by the

Government of India. The IMF is pursuing this case with the government persistently for early gazette notifications. This would restrict visitors and save the forests. In the National pick it should be forbidden to so much as even pick up a twig for use as firewood, thus saving the flora and fauna of the area.

"Keep Ganga Clean" Project

The Indian Government has special arrangements and funds for the project 'Keep Ganga Clean'. If the Gangotri Development Authority could only make use of these resources, then quite a few of the above-mentioned problems would automatically get solved.

The GDA should draw up plans for all future constructions in the restricted area of Gangotri. Haphazard constructions being done by different ashrams should stop. More importantly those constructions adversely effecting the ecological system should be demolished. Further the GDA should make immediate plans for construction of public conveniences well away from the river.

Role of Local Authorities

Depending on Gangotri's capacity for people at any one time, the local authorities should regulate the flow of pilgrim traffic there. Simultaneously in Uttarkashi a system to educate the pilgrims with the help of local video programmes and pamphlets should be started. They should be informed thoroughly about the damage being done to the shrines and how it could be averted. The facilities available in Gangotri and enroute to Gangotri should by then be well-known to the visitors. Student can be used during the season to educate these pilgrims.

Help Sought from Business Houses

Himalayan Environment Trust is in touch with some business houses who have agreed to help in the matter financially especially Shri Rattan Tata and Shri R.K. Birla. The proposed step to be taken are:

(A) Put up hoardings about the Do's and Don'ts for the public's information. These should start; from Uttarkashi right upto Gangotri and beyond to Gaumukh.

(B) Construct garbage bins throughout the route to Gangotri and beyond. This suggestion has been made by the Sir Edmund Hillary, on the lines of the practice being followed in the New Zealand National Park.

(C) Make arrangements to collect this garbage in bigger bins and then clear it down to Uttarkashi or any such place, which would have the facility to destroy the same.

(D) Construct incinerators close to the bigger garbage bins for burning part of the garbage.

(E) Have local people involved in the process. Post wardens chosen from them, to supervise this cleaning up process periodically. These wardens should be paid an incentive to do this job.

8

North Eastern Frontier Region in Jeopardy

Introduction

In this chapter we will discuss about the North Eastern frontier region of India-British administration reached there and imposed an alien market economy on localised autonomous system. It changed the simple, natural tribal life to an artificial one.

Before that the tribal people used to live in harmony with forests wound them. But with the rising population, deforestation, expansion of agriculture, construction of dams, the ecosystem of the region has been adversely affected. As a result the community forests are gradually vanishing Being alarmed by the situation the tribal people have worked out their own land laws. Even women have come forward to support the drug addiction compaign and anti-liquor movements. It is expected that involvement of the youth will help significantly for the success of the movement. Let us discuss in details.

Rapid Depletion of Tree Cover

Visit the North Eastern frontier region of India and you will observe the rapid depletion of tree cover. The hills have turned naked, trees have become fewer, scattered and younger. New human settlements have sprung up all around. During the dry days of summer one comes across even clouds of dust. A new problem of drinking water has been added to the list of the urgent issues to be attended in the abode of the clouds —the Indian state of 'Meghalaya'. Then one comes across 'the wet desert of Cherapunji.' The government forest officials claim not to be responsible for the alleged deforestation as they hardly control an appreciable acreage of forests. It is the community represented by the District Council, which controls and manages the extensive forests as per the constitutional provisions.

Elsewhere in India, the environmental activists are of the view that such a situation is the result of such a conflict between the state control and community interests. They demand that the forests should be handed over back to the community, as they were in the hands of the local communities in the pre-colonial days. Is it so that the community control has led to destruction of forests in this region? How come, in spite of the large scale deforestation, there is no environmentalists movement active in the region.

Effect of Population Pressure on Cherapunji

Cgerapunji certain part consists of fragile rocks which contain little top soil. This is one of the regions of the world which receives the maximum rainfall, causing extensive flood. Some other parts of the region has evergreen tropical rain forests full of rare varieties of flora and fauna. The populàtion pressure of the local communities coupled with demands of the expanding monetorized economy has emerged as a singular threat to the natural resources. Before the beginning of the colonialism most of the Third World societies consisted of a long settled and sophisticated agrarian culture. They had a finely tuned but delicately balanced relationship with their natural environment. Colonial and post-colonial capitalism had disrupted this relationship in many ways. They social consequences of this disruption are widely visible, what is less observed are the devastating ecological consequence. Due to lack of advanced knowledges was enjoyed by European colonists, the state-planned industrialization has to contend with a limited resource and rapid environmental degradation..."Urban and industrial development, though not even successful on its own terms, created tremendous havoc on the country-side, pauperizing millions of people in the agrarian sector and deminishing the forest, water and soil resources at a terrifying rate" (Guha, R: 1989: 195-196).

As in the case of this region, the absence of clear and simple rights of the small and localized communities over the natural resources in the pre-colonial setting was seen as ambiguous. This relates to the fact that the tribals knowledge of the forest and their ability to manage it for their needs is both primitive and not so effective. This very knowledge and capability, however, seems to have eluded the planners and developers. This situation has led to debate among intellectuals, policy-makers and grassroots organizations on two opposed notions of property and resources. The communal control over forests being paired with subsistence was on the one hand, and state control with commercial exploitation on the other (Guha, R:1988: 46).

The largest reserves of tropical forests in India are found in the North Eastern Himalayan region. One-seventh of the total Indian areas under forest. It plays host to a variety of birds and wild animals and rare species of plants. But the forest wealth of the region is disappearing at an alarming rate. Before the advent of roads, railway lines and so-called economic 'development', the region was covered with lush green sub-tropical forests. The hill people lived in harmony with the forests around them and used a form of agriculture that allowed them to cultivate the steep slopes.

Even today many of the tribes follow the same old method of cultivation—jhumming. An estimated 1.6 million people are engaged in it, but there is *a* disharmony between man and nature. The forest areas are decreasing, human habitations are increasing and the fertility of the soil is diminishing at such a rate that land is turning barren (Agarwal, et. al., 1982: 40). How was the paradise of the plentiful forests lost?

Effect of British Administration

The heartland of the hills was invaded by the British administration. They established a system of communication, imposed an alien market economy on localized autonomous systems All these disrupted the so far lively atmosphere of the region. The tribal people never accepted these impositions and rose in countless revolts throughout the British rule. But this was a loosing battle for the them. Even in the post-independence period, spirit of indigenous autonomy seems to have identified some steps of the Indian Union as continuation of the colonial trappings.

The original dense forests are located on the inter-national and inter-state borders, where relatively smaller ethnic groups practice their localized subsistence economy. These are the areas, where armed insurgency is on for decades and accordingly the 'state' regularly positions its counter-insurgency forces. For example, the Naga insurgents, the various force of Manipuri rebels, the Mizo National Front, United Liberation Front of Assam, the Boro Security Force, the insurgents in Tripura and Meghalaya station their hideouts in these 'impregnable' forests for the sake of logistic and as a strategy (*The Statesman*, October 1, 1996; *The Asian Age*, October 2, 1996).

Needless to say that beside defending the international borders, the defence forces as well deploy their units on these insurgency prone forested terrain for the same reasons. Besides locating their training hideouts, camps and shooting ranges for exercises, the movements of the armed forces necessitates construction of border roads undertaken by the B.R.O. for 'development' and 'communication' causing havoc to the fragile environment. Destruction of forest cover is very common in such a confronting situation where armed clashes occur quite frequently. These are also the areas, where various shades of ethnic movements against alleged outsiders are going on.

One may take two illustrations from the Assam-Nagaland borders on the southern Brahmaputra Valley and Assam-Bhutan border on the north, where Boroland movement remain unresolved. Both these areas have precious forest reserves. However, these forests had a tradition upon 1980 of 'forest villages' created by the state forest department. These villages invariably belong to a migrant ethnic group such as Nepali, Mymen singhia, tribals from Bihar or Orissa, or mixed migrant labour. While the counter-insurgency forces encourage their settlement, as a logistic and strategy in their favour, the insurgent local ethnic groups retaliate such moves. In the process the normal life and economic entangling of both the ethnic groups got disrupted. The only alternative for their survival left to them in such a chaotic situation has been to hack the forests and sell the fuel wood to the markets and urban centres for their upkeep. Unfortunately, the forest administration who is officially appointed for the preservation of these forest reserves are not taking much pain in their assignment. Travelling across the districts of Goalpara, Kamrup and Darrang one can witness the ugly half-burnt remains of the forests.

Man-made Destruction

Non-vegetarian food habit of the various ethnic groups of the region engaged them in hunting of the wild animals. In most of the hill districts such as Karbi Anglong of Assam, Nagaland, Manipur hills, northern Meghalaya and Mizoram etc., the hunting of the wild animals has almost been stopped because of the depleting forest and increased human population. As they have eaten up their all wildlife, the hill areas are supplied of a mutton, beef, pork, fish, eggs and chicken from distant Andhra, Bihar, Uttar Pradesh, Punjab and other states. The small wild animals have been totally eliminated; the ritual and ceremonial hunting of *mithan* is now restricted to isolated pockets of Arunachal Pradesh and even the big animals such as rhino, elephants, tigers, clouded leopards, golden languorous, etc., have turned out to be endangered species.

On the inter-national borders of the region there are two types of problems. Firstly, the unplanned and haphazard mining of coal, dolomite stones, limestones, etc. in the mineral rich hill states of Meghalaya and Bhutan have been polluting the adjoining agrarian plains in the northern Bangladesh and north Bengal (Buxa duar) respectively causing misery to the farmers and tea planters. On the other hand, the southern neighbours of these hill states have been living on the wildlife, timber, farm products and animals (*The Meghalaya Guardian,* September 28, 1996; *Kuensel,* Thimpu, 1993-1996) regularly.

Secondly, besides the smuggling of the factory produced consumer goods, states such as Mizoram are confronted with the large scale smuggling of their bamboo wealth from their irregular mountainous borders to Bangladesh (*The Asian Age,* May 28, 1996). It is alleged that these timber and bamboo cutters are procted and supported by the Chakma and Tripura insurgents across the borders.

Tipaimukh Hydel Project

Known for its rich hydel electrical potentiality, the region has attracted the proponents of big dams as well, the Tipaimukh hydel project on the river Barak costing as much as 1400 crore of rupees is proposed to be located on an ideal site among the Hmar tribes on the border of Manipur and Mizoram. This insurgency infected mountainous site is situated on an infrastructurally primitive locale. However, the river causes untold misery to the Barak valley and north-eastern Bangladesh because of the recurrent floods.

Further more the region is chronically starved of electricity. With a view to solving these two major problems this mega project was approved as a remedy. Instantly, the decision to construct the hydel project led to opposition from the immediately affected ethnic group, the Hmars. A team which was commissioned to study the impact of the project on the environmental and ethnic issues. It will be enlightening to quote from the report of the commission.

> "The Tipaimukh hydel project presents contradictions of various levels: between the needs of the plains and the hills, between the perceptions of the state administration and the micro-level tribesmen,

between the demands of the developed and the urbanized plains men and relatively isolated small communities and between the proponents of development and exploitation of resources and supporters of survival and preservation of the indigenous entities. The divided and fragmented frontier community of Hmar (on whose land the project is proposes to be located) is a poor match to the mighty state structure, capital investment, scientific skills and above all, the lack of national concern for a small vibrant community. Moreover, the Hmars have taken up arms against the state and spurned even award of amnesty. In the given situation, it appears that they have been taken for granted by the technocrats and policy makers" (Sinha, A.C. 1994: 207-208).

Forests Under Community Control

Hill states like Meghalaya, Manipur, Mizoram, Nagaland, Arunachal, Bhutan and Sikkim have worked out their own land laws. As per the with a few exceptions 'non-locals' cannot possess landed property there. Still there has been considerable transferring of land and other resources. All these states, with exception of Bhutan, do not have an appreciable percentage of their forests under the government forest reserves. Infact, the bulk of their forests are under communal control—either through the district council as in the case of Meghalaya, Mizoram, Tripura, Karbi-Anglang district of Assam, or within the control of the clan, village or tribes as in the case of Nagaland and Arunachal Pradesh.

Because of the guarantees provided to the customary laws in these states under the Indian Constitution, even their own state governments cannot enact forest laws affecting the basic customary practices. The State Forest Departments have a small patch of reserve forests and possess no land of their own to afforest and extend their activities. The forests are completely under the community control. Despite that forest resources are depleting, trees are cut on an unprecedented scale and wildlife is arbitrarily destroyed. Thus in case, the communal control over the forest resources against the state government control appears to have completely failed to safe-guard the forests. Here the question arises: Is the community control on the forests in itself a guarantee to the reasonable ecological preservation? Could the community control on the regional forests anticipate the commercial loot of the timber, killing of the wildlife and wilful destruction of other natural resources? Actually who controls the forest resources in the name of the communal control? Perhaps a mass movement for a balanced environmental preservation, a concern for the situation and an awareness is the need of the day.

Politically Influenced Tribal Elite

The conflict between the tribal forest users and the urbani industrial interest in not visible, because of the fact that the latter are not located within the region, so they remain unseen and hidden from the gaze of the public. Secondly, it is a group of politically-influenced, resourceful tribal people who ignore the local interest for supplying timber to the industries away from the states. The failure of the Meghalaya Government to enforce a law enacted by the state assembly in 1981 is a burning example of the fact.

Thirdly, the newly emerged middle class in these states is still embryonic in the sense that it has not been able to cause a split at societal level; there is little evidence of a conflict between the 'wretched of the earth' and the emerging middle class, and in any way, there is no upper class in the classical sense of the term. In fact, it is the middle class which articulates the various interests of these states on behalf of the masses against, the 'other'. Thus, they are not seen as exploiters and oppressors of their own community and communal resources.

Fourthly, this middle class-neo-elite combine has been instrumental in championing the cause of the local level 'under-dogs' vis-a-vis "non-locals exploiters" through various violent and non-violent movements for creation of the hills states. They are seen as the saviour of the local interests in the aftermath of the statehood. Many of them, really suffered politically and financially during the long drawn years of agitations. In fact, armed struggle was waged in the state of Nagaland, Mizoram, Tripura and Manipur and the insurgents were rehabilitated with grants in cash and kind and employment. When they use the community controlled forest resources to augment their private income in some way or other, it is just not taken note of.

Fifthly, in some way it is the common belief in the hill states that if the 'local' tribal elite do not exploit their resources and take advantage of it in their favour, it will be a matter of time in which the 'non-local' will fleece them away of the resources such as forest.

Sixthly, unlike Bastar (Madhya Pradesh), Singhbhum (Bihar) and elsewhere, where the local tribals and Dalits have suffered because of the excessive exploitation of the local resources by the non-tribals and upper castes, such a division between tribals and non-tribals as suffering masses and exploiters respectively, has not been established in terms of forest resources of the region.

And lastly, the middle-class-neo-elite combine has been able to focus on the various 'ethnic' demands on behalf of the community as a whole. So, they have been able to divert their attention from the internal contradiction within the states. Therefore, we do not find a noticeable awareness among them or a movement to safeguard their interests or the depleting forests.

Role of Local Inhabitant

We need to create an awareness among the people at the grassroots level for arresting the already threatened bio-mass of the region. Fortunately, community solidarity is still strong in the region. The various communities are still broadly attached to their land and traditions. This aspect may favourably be channelized through the following strategy of community involvement.

Reference to nature was considered undesirable religious act by the Church. So, after the introduction of Christianity the old practice of the sacred groves was surpassed. The Church leaders must be convinced that there is no conflict between the preservation of forests and practice of Christianity. They must be involved to view the forests and other national resources as divine gifts for posterity. Besides the Christianity, in order to have more community involvement, the positive approach of other religions like Buddhism, Seng, Khasi etc. should also be uncorporated into the new form of the movement against environmental degration.

Mostly women are the worst sufferers when the environment gets degraded. They have to strive hard for fuel, fodder and drinking water. Incidentally, the women of the region have shown exceptional energy in ritualistic social work. They also took part in drug deaddiction campaigns and anti-liquor movements. Efforts must be made to educate them about the endangered environment. Once they are convinced of its desirability, their energy may rightfully be utilized for the success of the movement.

Besides women the youth of the region can also play a vital role in second service. Their active participation in folk culture, games and sports and various hikes of political movements can help significantly.

Specially their dances, skits, songs and music, choir and orchestra performances are vigorous, energetic, and full of life. These talents need to be linked profitably with the environmental movements through audio-visual-adds using preferably local tradition. Preferably these should be in regional languages. This is of cardinal importance as the youth of the regional can feel the importance of the movement.

Courses on regional environmental resources must be taught along with extension work and some field exposure. The institutions must have a close liaison with the forest officials who are responsible for implementation of various government policies pertaining to environment. To boost their morals, the individuals who display exemplary expertise in terms of action, creative writings, research, motivation and audio-visual extension works must be recognized and their achievements should be highlighted.

References

1. Agrawal, Anil et al., 1982, *The First Citizen's Report: The Status of India's Environment,* Centre for Science and Environment, New Delhi.
2. Andersen, R.S. and W. Hueber, 1988, *The Hour of the Fox,* Vistar Publications, New Delhi.
3. *The Asian Age,* Daily Newspaper, Calcutta.
4. Guha, R., 1988, "State Forestry and Social Conflict in British India", Indian Institute of Science, Bangalore.
5. Guha, R., 1989, *The Ungueit Woods,* Oxford University Press, New Delhi.
6. *The Kuensel,* The Tri Weekly News Bulletin, Thimphu, Bhutan.
7. *This Meghalaya Guardian,* Daily Newspaper, Shillong.
8. Sinha, A.C., 1993, *Beyond the Trees, Tigers and Tribes Historical Sociology of the Eastern Himalayan Forest,* Har-Anand, New Delhi.
9. Sinha, A.C., 1994, *North Eastern Frontier of India Structural Imperative and Aspects of Change,* Indus Publishing Co., New Delhi.

9

Human Interference into Nature Accelerates Natural Hazards

Introduction

To meet the personal needs man has caused havoc in the Himalayan region. Deforestation, building of roads and dams and expansion of agriculture have created environmental problem.

Due to deforestation dry season flows of the Himalayan rivers have decreased. Even some of the streams which used to flow throughout the year have become only reasonal ones.

Increase in population has forced some of them to settle in greater heights which were once dense forest tracts. Their cultivation in the steeper slopes has destabilised the slopes thus resulting in erosion due to entry of excess rain water in absence of drains. This is common throughout the Himalayan region.

But a through survey reveals that human activities are not the sole factors responsible for the deteriorating environmental condition in the Himalayan region. In certain cases, human interference into the nature only accelerate the natural hazards.

This chapter deals with the environment related problems specially in the eastern Himalaya.

Widespread Degradation

Modern human activities have caused widespread environmental degradation in the Himalayan region. Road building and commercial felling of trees have resulted in ecological destruction. Forest cover had declined and its place has been taken by overgrazed lands. Runoff and silt tend to move out of the Himalaya in explosive waves. To construct roads and for its maintenance a large quantity of debris is removed from the Himalayan area. Besides that landslide debris is pushed down the hill to keep the road clear. As a result the forest cover below is damaged. It destabilises the hillside too.

Floods are common in the Indo-Gangetic plains. There have been many floods of gigantic proportions. Several geologists believe that landslides are probably increasing because of increased human activity, though no documented evidence is available. With the increase of population more forests and other natural resources will be heavily exploited resulting in an increase in landslides and erosion. It is now being noticed in numerous Himalayan towns. Natural drainage often gets blocked. Some are obstructed by human activity. Surrounding hillsides get denuded because of increased human pressure. Slopes are further destabilised by dynamite blasts, used for construction work.

Ecological Changes

The ongoing human activity has resulted enormous ecological changes. Large areas have been deforested, agriculture has expanded, vast networks of roads have been built up and numerous dams have been constructed or are under consideration. All these activities directly and indirectly are increasing floods downstream.

The forests act like a sponge, absorbing large quantities of water during the wet season and gradually releasing it throughout the year. Forest also reduce soil erosion and landslides. The destructive flood-drought cycle that has been unleashed can be arrested by a massive afforestation not only in the Himalayan area but also all over the world.

Other advantages of forest are: (a) intercepting rainfall and allowing it to evaporate back to the air; (b) transpiring large quantities of soil moisture and keeping the soil dry, (c) creating soil conditions that allow greater infiltration of water into the ground and (d) reducing peak water flows during a rainstorm. Rainfall on a forest canopy is caught by leaves and branches and it evaporates as it moves towards the forest floor.

According to a report of CSWCRTI, 1981, very light rainfall can be totally intercepted while heavy rainfall will rapidly saturate the forest canopy and most of it will get transmitted to the forest floor. The interception loss is governed more by the intensity of rainfall than wind velocity or the type of density and vegetation. On the contrary, Chandra S., 1986, stated that in a sal forest interception loss was 37.3 per cent of rainfall during a light shower (4.2 mm) and 4.1 per cent during a heavy shower (140.3 mm). Interception plays a vital role in controlling the minor floods.

Role of Forests in Reducing Evaporation

Forests will increase transpiration while reducing evaporation from the soil. Evaporation from land covered by a forest usually ranges from 10 per cent to 80 per cent of that from bare soil. Transpiration depends on various environmental factors and trees species. Deep rooting trees transpire large quantities of water sometimes as much as the entire amount of rain that falls over a year. Trees almost act like pumps and hence, afforested catchments normally have reduced stream flows compared to deforested ones. Leaf litter and vegetation greatly increase the organic content of the soil and improve its structure increasing its infiltration capacity.

Infiltration is large in a forest area when the soil is dry and deep and only a short, light rain occurs and it is small when there is heavy and prolonged rainfall, especially on shallow, wet soils. Most of the Tarai rivers, which previously contained water during the lean season, are now found to be dry during winter and disappear after about 20-25 km from the foothills. The river beds are silted up rapidly. On the contrary, afforestation reduces total volume and peak rate of surface runoff and streamflows last for a longer period after the rain ceases.

A study of CSWCRTI (1976) found that, the quantity of runoff went down by 28 per cent after afforestation and the peak rate of runoff was reduced by 73 per cent. In other words, aforestation acts as a sponage. It extends the duration of streamflows in dry season and reduces the peak run off rates.

Surface erosion and sedimentation of rivers significantly depend on forest cover. The existence of a forest cover and litter on the forest floor breaks the impact of rain drops on the soil. Tree roots and increased organic matter in forest soils bestow stability to them. Watershed expert Lawrence Hamilton further argues that soil erosion rates found in small watersheds will not hold true for large watersheds. For instance, 90 per cent of the soil eroded in any year from a one ha watershed will be delivered outside the watershed by streams in that year itself. But this sediment delivery ratio may be only 50 per cent for an 80 ha area and less than 30 per cent for a drainage area over 500 ha. For a catchment of the size of that of the Ganga, the sediment delivery ratio will probably be only 10 per cent. Even, if all human activity was closed in the Himalaya the sediment load of the rivers in their lower reaches would show little change for a long time. The stores of eroded soil from the past would continue to feed the rivers. On the other hand, tree roots anchor the tree and the soil itself. It increases sheer strength of the soil and prevent it from slipping down. Therefore, aforestation of landslide prone areas in the Himalaya will definitely increase slope stability. Because of their heavy evapotranspiration rates, forests will keep the soil relatively dry and lower the water table. Soon after the 1968 rainstorm in the Tista valley, a scientist found that landslide rates on non-forested tea slopes was 10-20 times higher than on forested slopes.

Effect of Human Activities

(A) Agriculture

All human activities can possibly increase soil erosion and landslides in fragile mountain areas by directly or indirectly changing the existing slopes, slope forming materials and drainage systems. Increased destabilisation of slopes, massive soil erosion and floods in the Gangetic plain is generally said to be the effect of agricultural expansion. Population pressure has driven people to greater heights and steeper slopes, even above 2000 m to cultivate which were once forested tracts. Most of these villages and farms are located on the fan shaped debris of old landslides made of loose, fragmentary material and not very stable. Their cultivation has deprived these slopes of protective cover and increased erosion.

Terraces, especially on rainfed land, are often poorly constructed; they are outward rather than inward sloping and do not have a grassed bund on the edge. Because the higher yielding khet terraces are usually under paddy rice, which is vital for the survival of their family, the labour, during the heavy monsoon downpours, mainly concentrate on repairing damage to the *khet* (irrigated terraces) and irrigation systems. The *bari* (rainfed terraces) terraces left to collapse.

(B) Road Construction

In the Himalayan mountains road building activities have been the worst form of human interventions in the Himalayan mountains in terms of promoting soil erosion and landslides. Before the Indo-Chinese war in 1962, the Himalayan mountains were mostly accessible only on foot. The few roads that existed led to hill stations like Darjeeling, Mussoorie, Nainital, Shimla, located not very deep within the mountains. But the Indo-Chinese war in 1962, prompted a massive road construction programme and roads now go into some of the remotest Himalayan areas.

Deserts Removal Due to Road Building

Length of roads (km)	44,000
Debris removed to make these roads (mcum)	1,760 to 3,520
Debris generated by each km of road/year (cum)	550
Debris generated by all Himalayan roads/year (mcum)	24.2

Source: Valdiya, 1985.

To build each km of roads an average of 40,000-80,000 cum of debris is removed. Another 550 cum have to be removed per km every year to maintain them which indicates that 1760-3520 mcum of debris had to be removed to make the Himalayan roads and another 24.2 mcum are removed every year, states, Valdiya (1985). Landslides are common in the Himalayan region. The landslide debris is pushed down the hill to keep the road clear, which damages the forest cover below and destabilises the hillside. The tree and vegetation cover below the road had been greatly reduced as compared to that above the road.

According to Tajwani (1987), 10 small to medium landslides occur on average over each km of Himalayan roads and most of them are the direct result of the slope instability caused by road construction. The stability of Himalayan slopes has been greatly jeopardised by roads. Road construction invariably involves deep cuts which set in a process of progressive failure. Because road construction agencies are keen to complete their roads rapidly, few efforts are made to protect the cut slopes or provide effective drainage. As a result, during rains, they are damaged by recurrent landslides and rockfalls just after the new roads are built. Therefore, landslides are increasing because of increased human activity. In every decade, due to the more growing development activities, more roads are required and these roads open up new areas, forest and other natural resources had heavily exploited. The combined result are an increase in landslides and erosion.

In numerous Himalayan towns, the problems of landslides and slope instability are now being noticed. Natural drainage often gets blocked or obstructed by human activity and surrounding hillsides get denuded because of increased human pressure. For construction purposes, dynamite is often used to blast the hills resulted further destabilisation of the slopes.

If afforestation was undertaken today, it might improve local conditions within a few years, but silt loads in the large streams will not decrease for decades (Hamilton, 1987). Forest undoubtedly play a significant role in inhibiting shallow landslides but had no effect on deep landslides. According to Ramsay (1986), forest cannot override the natural factors that cause landslides, especially in the Himalaya. A survey carried out in Kalimpong following the 1954 disastrous flood of North Bengal found that the primary cause of the landslides surveyed was susceptible geology. Human factors like deforestation, overgrazing and cultivation of slopes were subsidiary causes. Another study show that the unfavourable geology and heavy rainstorm caused massive landslide in most of the areas of the Himalaya though deforestation did play a role. The complex geological structure and certain susceptible rock type are more responsible for a large scale landslide in the Himalayan region.

Practically, landslides tend to come in cycles. After a large landslide, there will be a period of relative stability. Then a large number of landslides will occur almost simultaneously on otherwise long undisturbed slopes. In that case, large scale landsliding is followed by stability because time is required for the bedrock to weather and a layer of topsoil to accumulate before the next major slope failure. Forests will not be able to prevent major landslides, which occur more because of steep slopes and other natural factors like weak geology, seismicity and intense rainstorms. They can postpone the occurrence of major landslides but not stop them. Forests as compared to grazed and bare slopes, will get destabilished less frequently and small, shallow landslides will be prevented. But this will only facilitate the production of a deeper mantle—a thicker layer of top soil and ensure a larger landslide in the future. Therefore, the overall soil loss over a longer time scale will be the same with or without forests (Ives, et al., 1989).

Geological Effect

Higher a mountain rises, the steeper will be its slopes and the valleys will be more inclined. This will result in greater surface erosion, mass movement and erosion of stream valleys. When these steep slopes have to face heavy rainstorms and earthquakes, erosion will intensify further. So, keeping this view in mind we can conclude that some major geological reasons increase erosion in the Himalayan mountains. Human interference into the nature only accelerate the natural hazards. Any apparent neglect of the terraces must be stopped. Small landslides were sometimes deliberately triggered by water diversions in order to facilitate new terrace construction. Government agencies must take note of people's knowledge and of the circumstances which force people to take certain actions which they know may aggravate the hazards. They must work with the people for effective measures against impending hazards. Agricultural methods should be introduced in the Himalayan region.

References

Chandra, S. and Lahiri, V.K., 1985, *Forest Influences on Hydrological Parameters,* National Institute of Hydrology, Roorkee.

CSWCRTI, 1981, *Annual Report,* Dehradun.

Gregersen, H.M., Brooks, K.N., Dixon, J.A., and Hamilton, L.S., 1988, *Guidelines for Economic Appraisal of Watershed Management Projects,* FAO, Rome.

Hamilton, L.S., 1987, *What are the impacts of Himalayan Deforestation on the Ganges-Brahmaputra Lowlands and Deltas? Assumption and facts,* Mountain Research and Development, Vol. 7, No. 3.

Ives, Jack, D. and Messerli, Bruno, 1989, *The Himalayan Dilemma: Reconciling Development and Conservation,* United Nations University and Routledge, London and New York.

Mathur, H.N., Babu, R., Joshie, P. and Singh, B., 1976, *Effects of Clearfelling and Reforestation on Runoff and Peak Rates in Small Watersheds,* Indian Foresters, Dehradun, April.

Ramsay, H. and James, W., 1986, *Erosion problems in the Nepal Himalaya: An Overview,* in S.G. Joshi (ed.), Nepal Himalaya Geo-Ecological Perspectives, Himalayan Research Group, Nainital.

Sharma, C.K., 1987, *The Problem of Sediment Load in the Development of Water Resources in Nepal,* Mountain Research and Development, Vol. 7, No. 3.

Tejwani, K.G., 1987, *Sedimentation of reservoirs in the Himalayan region, India,* Mountain Research and Development, Vol. 7, No. 3.

Valdiya, K.S., 1985, *Accelerated Erosion and Landslide—prone zones in the Central Himalayan Region,* in J.S. Singh (ed.), Environmental Regeneration in Himalaya; Concepts and Strategies, The Central Himalayan Environment Association and Gyanodaya Prakashan, Nainital.

10

Effect of Environmental Problems in Himalaya

Introduction

This chapter deals with the major problem in the Himalayan region like—biotic pressure on forests—its reasons and effects. Secondly, how quarrying or open cast mining is not only causing environmental problems but also effecting the human, animal and aquatic life is also discussed here.

Another major problem in the Himalayan environment is landslide. It is causing loss of cross or rupees every year. In this chapter the readers will be acquainted with the names of some major landslides and the factors that are leading to the landslides. Important subjects like shifting cultivation and water polution are also hightlighted in this chapter.

A couple of most irritating environmental problems have erroped up in the world's highest mountain—the Himalaya. That has brought about a rapid degenerating change in that region during the last twenty five years. As a result, the well being of the entire Indian sub-continent as well as the lives of people dwelling there are at stake. Directly or indirectly human activities are responsible for most of the problems affecting the fragile environment, be it landslides, accelerated erosion, flash floods, lowering of the water table, scarcity of fuelwood and fodder or some other problem related to the environment.

Man's Interference

More than forty per cent of the total geographical area of the Himalaya is covered by forests. During the past fifty years, man has increasingly interfered with the forest ecosystem of the Himalaya as more and more wood is extracted for fuel, fodder for cattle and small timber for local use.

1. Cattle Rearing

The most important factor responsible for the ever increasing biotic pressure on Himalayan forests is perhaps the grazing by domestic animals. The rapid increase in cattle population has brought about a progressive degradation of the forest environment. The hill people rear animals for various purposes:

(a) *Milk production* — In each village they rear cows, goats and buffaloes to meet their milk requirements are met with locally. However, small settlements have to depend upon nomadic/semi-nomadic grazier communities for their milk supply. These grazier communities live in various parts of the Himalaya. They move upto the alpine meadows in summer and migrate to the lower hills in winter. Many graziers live near the major Himalayan urban settlements such as Srinagar, Dalhousie, Kulu, Dharamsala, Shimla, Dehradun, Mussoorie, Pauri, Joshimath, Almora, Ranikhet, Nainital, Pokhra, Kathmandu, Darjeeling, Gangtok and Thimpu. These people rear a large herd of milk producing animalsand sell milk in these towns. In the trans-Himalayan areas like Ladakh, Lahaul, Spiti etc. Yaks are the prime milk producers.

(b) *Farming*—Bullocks are used for ploughing the fields. These bullocks are short but sturdy. They are required to pull the plough on terraces that may be very small in size. On an average there is at least one pair of bullocks or yaks for every 7 or 8 families. Animals used for ploughing are usually stall-fed while those used for milk production are fed by grazing on forest/pasture lands.

(c) *Dung* — Modern fertilizers are used only in some of the foothill tracts. Generally animal dung is used for agriculture in the hills. Ashish (1981) wrote, "In order to maintain crop yields, each farmer likes to keep about 10 animals per hectare of cultivation. These animals are grazed on pastures and in forests during the day and only their night dung is collected. Cows require minimum attention as compared to stall-fed buffaloes."

Animals reared exclusively for their manure are allowed to graze on forests/pastures during the day and are kept in enclosures at night. The night dung is used for manuring the fields and the forest/pasture lands also benefit from the day dung of these animals as this is usually not collected by the villagers except in case of areas near the village.

Especially in tracts where the forests have been severely depleted. Animal dung is collected and used as a source of energy. Negi (1982b) states, "At present it is difficult to arrive at an exact figure about the cow dung consumed in the rural hearths but it can be assumed that this amount varies from 100 to 125 million tons per year".

(d) *Wool*—The famous rugs of Kashmir are made from the hair of domestic animals. Besides that it is also used for making woollen garments blankets and carpets. The people of Ladakh, Lahaul, and Spiti wear clothes made from the hair of yaks, sheep and goats. These animals are reared by many nomadic and semi-nomadic communities of the higher Himalaya. The tribes of upper Garhwal and Kumaun knit beautiful sweaters made from the hair of sheep and goats.

These animals are allowed to graze in the forest for several days at a stretch with shepherd dogs standing watch. Mainly because of the decrease in the activities of the predator species This practice has become prevalent in recent years.

Result of the overgrazing of forests and pasture are: (a) Gullies form in the tracks frequented by cattle. These are very common in the Siwalik hills where the formation of gullies is aided by the weak nature of the country rocks, zones of disturbance and highly erosive streams. (b) The grasses growing on the pasture are adversely affected. Overgrazing keeps them below the optimum height for metabolic activity. (c) Due to selective grazing of young and succulent grasses, only coarse and poor grasses are left. (b) Cattle hooves damage young regeneration, seedlings and even saplings. They render the soil more prone to erosion.

2. *Effect of Fires*

Forest fires in the Himalaya are mainly of two broad categories:

(a) *Intentional fires*—It includes forest fires that are set intentionally. It is called intentional fires. This may happen under the following situations:

1. When miscreants set fire to the forest with a specific intention of damaging the forest wealth.
2. The local people may set fire to the forest floor just before the onset of the monsoon season so as to induce a good growth of grass.

3. In Kashmir, the villagers set fire to the forest for inducing a rich growth of mushrooms during the rainy season.
4. To hide the stumps of illicit fellings.

(b) *Accidental fires*—Picnickers, trekkers and travellers may leave unextinguished cigarette butts, matches, camp fires etc. in the forest unintentionally this is called accidental fire.

These cause accelerated erosion, hamper regeneration, destroy valuable vegetative growth and wildlife.

3. Lack of Alternative Fuel

In the absence of coal, gas or other alternative sources of energy, the hill people have to depend upon scarce cattle dung and fuelwood obtained from the nearby forests for their energy requirements. In the Himalaya, it is common to see scores of womenfolk and children returning from the jungle with headloads of fuelwood or grass. Most villagers enjoy certain rights for the collection of dry stems, bark and leaves from the neighbouring forests to meet their energy requirements. But as the dry stems, bark and leaves are seldom enough to meet the energy needs of an average village having 30 to 40 families, the villagers are forced to remove more than what" is permitted under the rules. This is called 'overlopping' or 'illicit removal of forest produce'.

It causes the following damages:

1. Reduction in the forest cover, thereby leading to problems like accelerated erosion.
2. Decrease in the productivity of the forest.
3. Loss of regeneration.

Forest Conservation Act

Before the enactment of the Forest Conservation Act, 1980, there was a steady decline in the forest area of the Himalaya. This Act was enacted for the following reasons:

1. River valley projects entail the creation of large reservoirs. These lakes or reservoirs submerge vast areas of forest land, e.g., Bhakra Dam, Kalagarh Dam etc.
2. As the population of the hills increased the forest land was cleared for agriculture.

3. Roads are important links between people and places. The hundreds of kilometres of roads and canals that have been constructed in the Himalaya has led to a gradual reduction in the forest cover.
4. Cases of illicit felling have also reduced the forest area of the Himalaya.

Degradation of Forest Lands

1. Function of Tree Canopy

Accelerated erosion is caused by rapid degradation of forest lands. It is a well-known fact that leaves break the force of the falling rain water. Negi (1981) states, "the many tiered canopy of the Himalayan forests act as a brake on the force of the raindrops. The drops hit the crown of the topmost tree, then fall on the herbs and shrubs below. By the time the precipitation reaches the forest floor, the force is greatly reduced".

Ghosh and Rao (1979) have arrived at the following results regarding the interception of rainfall by various species:

Species	*Trees per hectare*	*Interception (%)*
Eucalyptus	1958	11.56
Pine	1156	22.10
Sal	1678	25.30
Acacia	0574	28.5

The areas with little or no vegetative cover receive the raindrops fall directly onto the forest floor. The raindrop causes splash erosion due to this kinetic energy. These drops combine together to form small channels which in turn give way to finger-like features known as rills. These ultimately lead to the formation of gullies. Such gullies are more likely to be formed in tectonically affected zones and in areas where the soil is more prone to erosion.

To keep the soil together, vegetation plays a vital role. In its absence accelerated erosion plays have. The root, with its network of countless root hairs, binds the soil together. Devoid of this, the top soil is quickly washed away by the rain water. Huge quantities of unprotected soil wash away with the force of the falling rain water and soil slumping results. Very often, the unprotected soil acts as a catalyst in the formation of massive landslides.

Himalayan forests also provide protection from snow erosion. The tree canopy reduces the quantity of snow falling onto the ground. Snow-melt water causes soil erosion. However, the thick layer of humus on the top soil provides protection against it.

Debris brought down from the uplands is deposited over cultivated fields, houses, roads, bridges and canals. The life of a number of river valley projects has thus been reduced.

2. *Effect of Sustained Biotic Pressure*

In many parts of Garhwal, Kumaun and Nepal a distinct change has been brought about in local vegetation. "Today large parts of Uttarakhand, once known for its dense forests of oak, burans, pine, deodar and other species, now lie denuded. Along the banks of the Himalayan rivers, hundreds of kilometres of land strips are now cluttered with cactus alone and the whole region is gradually turning into a desert. According to scientists, the broad-leaved trees which form rich humus were replaced by coniferous trees, which in turn are being replaced by cactus. If this trend continues for another 50 to 60 years, it will become almost impossible to save the Himalaya from total collapse". (Bhatt, 198). This statement is correct upto certain extent. There can be no doubt that a sustained biotic pressure on Himalayan forests has produced local vegetative changes in many tracts. At the same time it should be borne in mind that scrub lands and coniferous forests form an integral part of the ecosystem of the region.

The degradation of forests brings a drastic change in the local climate. The bare rocky slopes heat up the surrounding atmosphere by acting as a mirror, whereas dense vegetation acts as absorbent. Trees play a vital role in breaking the force of the prevailing winds which affect the mountain slopes.

3. *Role of Forest Floor Litter*

A forest cover helps the rainwater to seep into the soil, thus aiding in its storage as underground water. Negi (1981) states, "Water absorption power of a soil depends on its permeability. This is maintained by the roots (which act as a disintegrating agent), leaf litter, humus etc. In the steep Himalayan slopes, where the soil is shallow, the forest floor litter plays a major role. It absorbs about four times more water than ordinary soil and by easy decomposition, adds to the depth of the soil". Degradation of the forest cover has led to a fall in the local water table. Women and children have to walk further and further each day to fill their pots.

4. Degradation Leads to Floods

The thousands of cubic metres of water that pour down from the skies during the rainy season cannot be absorbed quickly if the mountain slopes are covered by degraded forests. This leads to flooding of the plains and the adjoining areas too. The unabsorbed rain water ends up in the nearby streams and rivers whose channels are not deep enough to accommodate this vast quantity of water. This phenomenon is known as flash floods. Ironically, with an increased biotic pressure on forests, flash floods are becoming more and more frequent in different parts of the Himalaya.

Even mild showers at many places in the western and central Himalaya lead to flash floods in streams and rivulets. In the Himalayan region, a severe flash flood occurred in the Alaknanda river in July, 1970. Unprecedented rains during the past 3-4 days before that brought a vast quantity of water into the river and its tributaries. The natural dam forming, the Gohna lake across one of the tributaries burst due to the huge volume of unabsorbed water that came down the mountain slopes. The ensuing flood brought misery to the thousands of people living along the river Alaknanda. Its effect was also felt in the plains of Uttar Pradesh was also effected by the flood.

5. Stability of Slopes

The roots of trees, shrubs and herbs bind the soil together and thus assist in the stability of slopes particularly in areas having a steep gradient. The vegetative cover plays a major role in the stability of slopes in the Himalayan region where the thin mountainous topsoil is susceptible to the slightest disturbance.

6. Enrichment of Soil

Surface run-off washes away the muddy component which is the important nutrient of the soil. The coarse and unproductive soil is left behind. Himalayan forest soils are very rich in organic matter, mainly due to a thick layer of humus over the topsoil. Huge torrents of surface run-off leach away this important constituent. But, if the same water penetrate through the top soil to the layers below carrying the important nutrients, the soil will be enriched.

7. Roots Help in Soil Formation

Himalayan mountain soils of today have formed over a period of million of years. The roots of plants play a very important role in soil formation. The roots of trees, shrubs and herbs are responsible for the disintegration of all types of rocks. They help in the physical breakdown of the underlying rocks. As they grow downwards, they penetrate into the small eracks due fissures, thus paving the way for water to form soil. In the absence of roots to hold it together, the soil can be washed away by surface run-off within no time.

Mineral Wealth

Since times immemorial the people living in the hills have been quarrying stones for building houses. Old workings of copper, iron and even gold are found in many parts of the Himalaya. Some of such important old workings have been found at Kiliarous near Dorus in Kashmir valley, Mathyana khal and Ghengar near Rudraprayag in Garhwal and at Dantagaon in Pauri district. However, regular quarrying for minerals started in the post-Independence era. From the Himalayan mines and quarries some important minerals are being extracted today. Some important ones away them are—copper, limestone, dolomite, lead, zinc, marble, magnesite, coal and iron. Building stones account for upto 80 per cent of the material extracted. In many tracts, limestone and marble are abundant.

Negi (1982a) states, "Limestone is a major mineral produce of the Himalayan region. In fact, the total number of limestone mines in the Himalayan region in 1977 was about 25% of the total number of limestone mines in the entire country. Most of these mines are privately owned and it is in these mines that improper exploitation and management is taking place."

Because of certain reasons not much attention has been paid to the environmental aspect of these quarries. Firstly, upto recent times, this problem was of a localised nature. Secondly, till recent times the emphasis was more on production. Thirdly, only recently the environmental impact of quarrying has begun to develop on the people. It was thought that only a small part of our society was affected. Fourthly, as compared to the mineral output of the entire country, the total mineral production in the Himalaya was not very substantial. Lastly, quarrying is carried out in remote and far-flung areas.

Environmental Problems

Quarrying or open cast mining is also causing some environmental problem. Let us see what are they:

1. Misuse of Precious Land

Through quarrying or open case mining we are practically misusing some valuable lands. Otherwise these lands could have very easily been put to use under agriculture, forestry or as a pasture. In other words, for all practical purposes, this land is lost forever so far as its use in agriculture or forestry is concerned. Reclamation techniques are yet to be adopted in the Himalaya.

2. Loss of Valuable Forests

Quarrying involves the removal of valuable forests. This is an immense loss which may never be recovered.

3. Valuable Topsoil is Lost

Topsoil is the base for the growth of all types of vegetation. To produce a few centimeters of topsoil nature takes thousands of years. When the overburden is removed in quarrying, this valuable topsoil is lost within a few hours.

4. Wild Animals Effected

Large variety of wild animals live in forests. The cutting down of vegetative growth disturbs their habitat. They are forced to move to other areas. Blasting and the sound of heavy machinery disturbthem very much.

5. Affecting Aquatic Life and Underground Water

(a) No infiltration of water into the soil is possible because of loss of valuable topsoil and vegetation. This leads to a fall in the level of the local water table.

(b) Devoid of the protective cover of the topsoil. The exposed rocks are subjected to the full action of the falling rain water due to lack of productive cover of the topsoil. Calcium the chemical present in the rocks, may be dissolved by the rain water, thus making it dangerous for human consumption, agricultural purposes or industrial uses. When this water containing calcium and some other chemicals reaches bigger streams and rivers, it may pollute them, thus affecting aquatic life. The run-off from the limestone quarries from Mussoorie contains large quantities of calcium carbonate.

(c) Trickling through the rocks this water containing a large amount of chemicals finds its way to the reservoirs of underground water. Thus, the underground water too gets contaminated. Drinking water in the localities where quarrying is done is adversely affected.

6. Road Construction

Construction of roads upto quarry to make it easily accessible causes damage to the vegetation and soil and results in numerous debris slides. It also affects the local drainage, agriculture and quality of the water that is available for human and animal consumption. Very often these roads are constructed violating of the established principles of engineering. In the hills, heavy rains during the monsoon season may cause landslides in these roads. This results in widespread havoc and it is a constant threat to life and property.

7. Hazardous Dust

Quarrying and related activities pollute the atmosphere also around the region let's see now:

(a) The dust is released at the quarry site may consist of pure calcium carbonate powder (in limestone quarries), silica dust (in quartzite quarries) and clayey dust (in slate quarries). This dust has an adverse effect on the health of the labourers involved in the extraction process.

(b) The roads constructed for transporting the quarried material are extremely dusty. Huge quantity of dust is blown high into the air whenever a vehicle passes. the drivers, labourers and the local people, this affects the surrounding vegetation also. Wildlife generally intends to stay away from such tracts.

(c) A large amount of carbon dioxide is released into the atmosphere in the process of conseting limestone into a more usable form.. This is injurious to health. Often strong winds spread this smog over almost the whole region.

Limestone quarries in Mussoorie were producing a large quantity of limestone in the seventies. Much of the slaking then was done in the furnaces spread all over Dehradun. The skies over this charming valley were filled with the smog produced by the furnaces. As a result, during that period several cases of *pneumenconosis* were reported there. Other diseases like silicosis and asbestosis are caused by this polluted air.

8. Release of Sediment

A vast quantity of sediment is released into the surface run-off by quarries. The base rocks are highly susceptible to the erosive action of the falling rain drops. Due to blasting country rocks become weak and more prone to the scouring action of the surface run-off. The erosive capacity of the sediment bearing surface run-off is immensely increased and it leads to the formation of rills and gullies in the fields lying below the quarries.

Not only that, even the run-off may deposit this unsorted sediment on forest land, agricultural land, roads. It may choke canals and stream channels, thus causing floods. Further, the water that contains high sediment is harmful not only for agriculture but for human consumption also. The large quantity of sediment in the stream water increases its erosive capacity resulting in accelerated erosion along stream and river banks.

9. Use of Dynamite

Use of dynamite is common for constructing roads, specially in hill areas. It weaken the host rocks so that they can then be easily extracted by hand tool or bulldozers. Blasting produces shock waves which weaken the country rocks.

According to the *Encyclopaedia Britannica,* the rate of detonation of standard dynamite is 3000 m/sec. Detonations are a special type of shock waves. They are elastic waves of large amplitude with a steep front which travel faster than acoustic waves. These shock waves have a constant amplitude and velocity. In some explosives, pressure of about 1,000,000 kg/sq cms and a temperature of upto 3000°C is produced. These shock waves may activate geological faults, lead to slope failure and weaken the country rocks, thus making them more susceptible to accelerated erosion.

10. Debris

The debris released during excavation operations is rolled down the hill slopes. This debris may cover valuable arable land, forest, pastures, roads, canals and even human settlements.

Landslides—Major Problem

Cases of disasters due to landslides are reported during the rainy year. It has been a major problem of the Himalayan environment since times immemorial. The slopes of Himalaya are the most unstable ones in the world. Many parts of the Himalayan chain are scarred by ugly looking repulsive wounds caused by landslides. Landslides are causing loss of crores of rupees every year.

Mathur (1980) has quantified the loss caused by landslides in the following manner: "A rough estimate indicates that the country suffers an annual loss of about Rs. 35 crore on account of man and vehicle hours, 50,000 man-hours on the country's 10,000 kms of hill @ Rs. 2 per hour and 5,000 vehicle hours @ Rs. 50 per vehicle. Loss of property settlements, houses, fields, orchards, forests, roads, goods delayed and damaged, cost of road maintenance and debris removal, if added might work out to be Rs. 1000 crore a year." Of these, bulk of the landslides occur in the Himalaya.

Some of the major landslides of the Himalaya are:

Jammu and Kashmir

1. Balinallah Landslide

While proceeding from Jammu along the road to Srinagar one comes across this major landslide for this landslide are: Firstly the undercutting by the Chenab river. Secondly, large scale see page of under ground water through the joint planes has made the rock and soil highly viscous. And lastly degradation of the vegetation.

2. Nashiri Landslide

At about 132 kms from Jammu on the Srinagar highway one comes across the Nashiri landslide.

It is the source of frequent and long road closures, traffic jams and immense hardships to travellers. Till 1913, the slope was fairly stable and was covered by a dense mixed forest till 1913. The construction of the Banihal cart road between 1914 and 1923 initiated the disturbance to the slope. However, no slide occurred in this zone till 1953. Kachroo and Hussain (1980) believe that the forest fire that broke out in the adjoining forest in 1953 finally caused slide. With the loss of the protective vegetal cover, the area downhill showed the first signs of slope failure. By about the same time, a small slip appeared in the bottom of the adjoining nallah.

In 1956, the slide took away parts of the road with it. A major slide destroyed the road at four places in 1967. In that year, the width of the slide zone was about 75 mts. It was about 1000 mts in 1973-74 and presently it is about 2000 mts. Debris continues to roll downhill in this landslide zone even today. A Nashiri by-pass road was constructed but a slide formed on this road too in 1967.

3. Acchar Landslide

While proceeding from Jammu to Srinagar this is the next major landslide zone. This landslide, called Acchar landslide, lies along a major tectonic disturbance that has shattered the rocks. A poor vegetative cover in the uplands and undercutting by the river Chenab has resulted in the whole mass sliding down

4. Khuni Nallah Landslide

Khuni Nallah landslide is located ar about 175 kms from Jammu on the road to Srinagar. This is another major landslide. The country rocks are highly fragile biotite schists that have been shattered by tectonic activity. During the rainy season this landslide becomes more active.

5. Malhori Landslide

On the Batote-Kistwar road is located the Malhori Landslide. It is about 39 kms from Batote. A massive slide occurred in this area in 1979. From time to time, tons of rock debris still roll down the hill slope.

Himachal Pradesh

1. Naina Devi Landslide

In a small hill near Bilaspur lies the Naina Devi landslide zone. At the top of it stands the famous Naina Devi temple. The main rock types are grey, highly fractured sandstones and purple claystones. Before the main slide took place small cracks were noticed near the bus stop for quite some time. Most people took these to be of local nature. However, after the unusual monsoons of 1978, numerous cracks were noticed on the floor of a temple store room. Cracks were also noticed at the bus stop of the Naina Devi township and in some nearby shops and houses. By mid-August, these cracks had considerably widened.

In August 1978, this small township experienced very heavy rain. On 25th August, 1978, when the people of Naina Devi township woke up the movement in the cracks was clearly seen for the first time. Slowly the mass crept down towards the valleys and within a few of hours it had slid down for a distance of about 2000 mts. In all about 145 structures were damaged.

Regarding this landslide Raju and Jalote (1980) state, "The area had been stable for a long time but the unprecedented rains had oversaturated the overburden material, thereby increasing the weight and pore pressure and reducing cohesion and the shear strength of the material specially at the interface of the claystone and the overburden".

2. Telangi Landslide

Between 2 and 3 kms stones along the Pangi-Kalpa road lies the Telangi landslide. It is another active landslide. Construction of this road began in 1965. The first slide occurred in 1973-74. After a major earthquake in Kinnaur in 1975 it became very active. Sharma et al. (1981) are of the opinion that the saturation of the overburden forming this slope is one of the major factors responsible for this landslide. Lack of vegetable cover in the uplands is another factor contributing to the formation of the Telangi landslide.

Garhwal and Kumaun

1. Bhagirathi Landslide

In the Bhagirathi river valley of Garhwal there are two small villages in the Mountain area—Gangnani and Dabrani. These are situated along the pilgrim route to Gangotri. These two places are important halting places for the thousands of pilgrims visiting the ancient shrines. The rock units of this tract are highly disturbed due to the main central thrust that separates the rocks of lesser and central Himalaya. Another factor responsible for this landslide is the poor vegetative cover in the uplands.

After continuous heavy showers for about a fortnight, a massive landslide occurred in the Kanoldiya Gad on 6th August, 1978. It brought down thousands of tons of debris. An area of about 4 kms long and 1 km wide was said to be covered by the debris. The debris blockaded the waters of the rivers Bhagirathi and Kanoldiya Gad, forming two temporary lakes. Later on when these lakes burst, floods occurred. As a result, many settlements were swept away and some others were severely damaged. These villages include Bhatwari, Gangnani, Gawana, Hina,

Joti, Jasyada, Dharkot, Kot, Maneri, Nantal and Sein. The Maneri-Bhali hydroelectric project was also severely damaged.

The flood that occurred due to the Bhagirathi landslide, also washed away upto 7 kms of motor roads in various sections. Itdamaged two motor bridges and eight small mule bridges too. Slide down of debris is still continuing.

2. *Kauntha Landslide*

Along the valley of the river Mandakini in Chamoli district, there is a small village, called Kauntha. It is the mountain hamlet of Chandrapuri. During the monsoons of 1979, thirty-nine people lost their lives when a massive landslide occurred in this area. Heavy rains coupled with weak and highly fractured rocks led to the formation of this landslide along the Kyonja Gad. It is a tributary of the river Mandakini. Approximately, an area of about 10 kms was affected damaging about 150 structures. Nearly a hundred cattlehead perished in the disaster. Though the rate of release of the debris has gone down considerably, this landslide is said to be active even today.

3. *Kalagad Landslide*

Along the Dehradun-Mussoorie road, to the north-west of Rajpur a landslide occurs along the Kalagad stream. This is called Kalagud landslide. In 1942, Auden published a report on this landslide. He observed that an active landslide along the Kalagad stream has partially affected the concrete bridge on the road. In his report, he has published a photograph of the bridge showing the pillars partially covered by the landslide debris. This landslide occurs along the zone of a major tectonic dislocation—the main central thrust.

Each year, particularly during the rainy season, a large quantity of debris comes sliding downhill. The shape of the landslide area resembles a leaf with its apex pointing towards downhill and covering an area of about 1.25 sq kms. Over the past fifty years, the level of the debris in the stream bed has risen. It has almost reached the level of the bridge at present.

The following factors are responsible for the formation of this slide:

(a) The main boundary thrust that has pulverised the country rocks. (b) An almost vertical slope in the long cliff which provides the major source material to this slide. (c) Rapid erosion by surface and subsurface run-off particularly during the monsoons. (d) Blasting for quarrying in the adjoining tract. (e) Overgrazing in the uplands.

4. Saknidhar Landslide

About 35 kms north of Rishikesh on the Rishikesh-Badrinath highway is located the Sakvidhar landslide. It lies along the zone of disturbance formed by the South Almora thrust as it cuts across the fragile lower Himalayan rocks. The construction of this road led to the formation of the landslide which is active even today. The poor vegetative cover immediately above the slope has contributed to the sliding of the debris. Another factor is the undercutting being caused by the river Alaknanda. It has caused excessive erosion in this zone.

5. Kaliasaur Landslide

Another major landslide is Kaliasaur landslide. It creates hardship to motorists going from Srinagar towards Badrinath. It lies in a tectonically active zone near Srinagar. Here the friable phyllitic rocks have been pulverised by an offshoot of the north Almora thrust. This landslide occurred soon after the motor road to Badrinath was constructed several decades back. The following factors are responsible for the formation of this slide: (a) Breaking of the country rocks by an offshoot of the north Almora thrust. (b) Poor vegetative cover in the uplands. (c) Undercutting by the river Alaknanda. (d) Seepage of surface and subsurface waters.

6. Nandparyag Landslide

Nandparyag is another landslide. It obstructs the free movement of traffic on the road to Badrinath. It is located near the town of Nandparyag, in the zone of the Alaknanda thrust (Negi, 1981). The following causes have been attributed for the formation of this landslide: (a) Dislocation of the country rocks by a deep-seated thrust. (b) Weak nature of the rocks. (c) Poor condition of the forests lying in the uplands. (d) Undercutting by the river Alaknanda. (e) Seepage of surface and subsurface waters that has made the debris highly viscous.

7. Belakuchi Landslide

Near Joshimath along the road to Badrinath is located the Belakuchi landslide. It is another problem area that causes immense hardships to travellers going to and from Badrinath. It is a giant landslide. Regarding this landslide Prasad (1977) states, "Because of the presence of a fracture zone nearby, the fractured and crushed rocks have fallen northwards into the valley to be accumulated as talus fans and stabilized at a fixed angle of repose. The toe of this was just above the slope. When the toe was cut down by the river in spate, an enormous debris avalanche followed".

8. Tawaghat Landslide

In the Tawaghat-Kalika Devi-Dharchula area of Pithoragarh district occur a number of landslides. It is in Kumaun district. Of these, the largest one is the Tawaghat landslide. This landslide is located near the town of Tawaghat. The main central thrust has produced a wide zone of weakness in this tract. On 15th August, 1977, after several days of continuing rain a giant landslide occurred in Tawaghat area. Thousands of tons of debris hurled down the mountain side on this day. Two eye witnesses, Gagan Singh and Jai Singh, the chowkidars of a rest-house at Tawaghat were, spell-bound. They witnessed gigantic rocks rolling down the mountain slopes at a great speed (Bahuguna, 1977). Life was disrupted in an area of about 50 sq kms due to this slide. More than 2 kms of road in various sections was affected. Forty-four persons died and a large number were injured. Over a hundred houses were damaged and many cattle perished.

This landslide occurred due to the factors noted below: (a) The main central thrust has weakened and pulverised the country rocks in a wide zone. (b) Overgrazing in the upper slopes. (c) Undercutting by the river. (d) High degree of slope. (e) The rocks became highly viscous due to seepage of water along the joints and bedding planes.

Nepal

Landslides severly affected Nepal too. These were brought about by a couple of inimical factors. They are:

(a) *Geological factors:* Firstly, disturbance due to weak nature of the country rocks. Secondly, presence of unconsolidated sediments.

(b) *Poor vegetative couer:* Poor condition of the vegetative cover in the uplands brought about by overgrazing; and unscientific fellings in the past.

(c) *Precipitation:* Very high levels of precipitation producing situations in which the total amount of rainfall received

Results of the Study Carried Out on the Landslides of Nepal by Leban (1979)

Ecological symbol	*Region name*	*No. of observations*	*Forest cover %*	*Forest landslides no./km forest*	*Non-forest landslides no./km non-forest*	*Landslides associated with streams and rivers no./km*	*Landslides associated with roads and trails no./km*	*No. of total landslides no./km*
II. A	Far Western Transitional Mountains**	5*	25	0.44		0.12	0.18	(1.57)
II. C	Western Transitional Mountains**	5*	40	0.15	0.72	0.16	0.02	(0.67)
II. D	Central Transitional Mountains	2*	45	0.84	5.13	0.88	0.04	(4.12)
III. E	Eastern Transitional Mountains	9	40	0.70	1.40	0.40	0.01	1.53
III. A&B	Far Western Middle Mountains	13	25	0.40	0.40	0.14	0.03	0.57
III. C	Piuthan Middle Mountains	8	20	0.60	0.61	0.19	0.02	0.82
III. D	Baglung Middle Mountains	9	35	0.23	1.43	0.35	0.03	1.39
III. E	Central Middle Mountains	14	25	0.32	0.49	0.35	0.04	0.84

Contd....

III. F	Kathmandu Middle Mountains	14	15	0.61	1.11	0.33	0.06	1.43
III.G	Eastern Middle Mountains	10	25	0.36	1.24	0.20	0.04	1.26
III. H	Far Eastern Mountains	9	20	0.40	0.73	0.21	0.97	0.94
	Average Middle Mountains	77	24	0.41	0.79	0.24	0.04	0.90
III. K&L	Eastern Mahabharat Lekh	13	40	1.28	1.67	0.36	0.05	1.92
III. M	Central Mahabharat Lekh	2*	40	0.43	4.78	0.48	0.09	(3.61)
III. N	Western Mahabharat Lekh	9	25	0.44	0.44	0.28	0.06	0.75
III. P	Far Western Mahabharat Lekh	10	75	0.17	0.60	0.31	0.06	0.65
	Average Mahabharat Lekh	32	43	0.70	1.01	0.32	0.06	1.21
IV. A	Western Siwaliks*	12	85	0.73	0.40	0.13	0.03	(0.84)
IV. B	Central Siwaliks**	9	75	0.83	0.28	0.19	0.06	(0.94)
IV. C	Eastern Siwaliks	14	80	1.15	2.75	0.53	0.02	2.02
Average of All		130	36	0.58	1.12	0.31	0.04	1.20

* This number of observations is considered to be too low. These figures were therefore not used to calculate averages.

** The figures for these ecological regions are not used for further calculations, except for the number of total landslides within the same ecological region.

within the span of a few hours is unable to quickly seep into the ground. Thus, this excess run-off produces landslides.

(d) Undercutting by streams and rivers.

(e) High degree of slope.

Based on satellite imageries Leban (1979) has carried out a detailed study of the landslides of Nepal. Observations reported by him for the different ecological regions of Nepal are tabulated on pages 480-81.

Sikkim

1. Siliguri-Gangtok National Highway

The Siliguri-Gangtok national highway is also not face of landslides. A number of major landslides occur there also. The largest slide is located at the ninth milestone. It has been a source of trouble since 1957. This slide is active even today. This landslide is the result of high degree of slope, weak country rocks and a degraded vegetative cover.

Some other landslides also occur on this highway. A combination of geological, climatic and vegetative causes from them. The greatest and most effective threats to the slope stability are cloud bursts and recurrent floods.

2. North Sikkim Highway

The North Sikkim Highway joins Gangtok and Chungthang. A major landslide zone is located in the rocks of the Daling series. Here landslides started about 30 years back. Between 1971 and 1978 the area subsided by about a metre. The causative factors of landslides in this tract are: Steep slopes, toe erosion, heavy rain and pulverised rocks.

Arunachal Pradesh

A large number of landslides in Arunachal Pradesh occur because of steep slopes and very high annual precipitation. The country rocks have been severely affected by the various incidence of tectonic activities. Another reason is—Arunachal Pradesh lies in an earthquake prone zone.

In this state an atrocious landslide has occurred at the 21 km point on the Kimin-zero motor road. This has been caused primarily by undercutting by the Bhareli river. The weak nature of the country rocks is the other reason. Another landslide zone is located between 43 km and 49 km point on the Tezu-Hayuliang road. Beside these two a landslide occurs on the road section between Baisakhi and Se La.

It affects nearly 20 kms of length thus creating problems for vehicular traffic.

Shifting Cultivation—Its Effect

The harmful practice of shifting cultivation in the Himalaya, is confined to the eastern Himalaya, especially in Arunachal Pradesh. The FAO report (1978) has defined this as, "Cultivation involving the removal and usually the burning of vegetation to create non-permanent clearings, which are fallowed to bush or forest for varying lengths of time; but also includes the temporary removal of vegetation for pasturage or other purposes of livelihood".

As per this method the cultivators do not cultivate a particular piece of land permanently. So, this system is called shifting cultivation. A patch of land is selected, all trees, shrubs and herbs are cut, left to dry and then set on fire. The price of land thus obtained is taken up for cultivation. Seeds are sown by dibbling small holes in the ground with the help of a wooden stick or metal piece. In this form of cultivation, no plough or animal power is used. All the operations are carried out by human labour. After a few years the cultivators move to another patch of land. The yield from these areas drops abruptly after the second year.

The yield start diminishing because the areas that have been under shifting cultivation there the soil nutrients are not replenished except by the ashes left behind after the burning operations are over. The cultivators then move to a fresh piece of land to return after an interval of more than 10 years. The time that passes between two series of cultivation on a particular patch of land. This is perhaps the most primitive form of cultivation. This form of cultivation has remained with the tribals inhabiting the tracts receiving heavy precipitation which creates conditions favourable for an abundant growth of vegetation.

Pattern of Operations

Though the land used under shifting cultivation usually belong to the tribe or community, when it is taken up for cultivation all rights over it are enjoyed by the cultivators. In the eastern Himalaya the shifting cultivators have to follow certain norms before starting the cultivation. A brief step by step detail will make the process clear.

To start the process a group of the village elders visit the area. After inspection they allot the piece of land for cultivation to a family or a group of families. After the *allotment* the recipient clears the area to be cultivated of all vegetation trees, shrubs, herbs and climbers. The vegetative rains are left to dry on the site itself.

By about a month or two the vegetal materials dry up. Then the cultivators return to the site. On a good sunny day the dried matter is burnt. The ash of the burnt material is the only form of replenishment available to the soil. So that is left on the site. This activity is carried on just before onset of the monsoon. Seeds of mustard, paddy, sorghum and some vegetables are soon as well as scattered all over the site. In case of maize, the seeds are placed in the small holes made in the soil. Then with a thin layer of soil the holes are covered. This is called *dibbling*. After the first few showers the young seedling begin to develop.

During the rainy season, a very dense growth of weeds comes up. They are removed by hand. This is called *weeding*. This process is done in the entire monsoon for three-four times. After full growth of the crop weedings may not be needed.

Now comes the stage when the cultivators have to protect the crop from any damage done by man, animal, bird, insect or disease. For the *protection* a group of persons stay at the site for 24 hours when the crop is about to be harvested. As soon as it is ripe and mature *harvesting* is done by using crude tools.

Then comes the process of *threshing and storing*. The harvest grain crop is kept in bundles. Threshing is done on a good sunny day by beating the stalks onto the ground, treading unless foot or by beating with a stick. In village big earthen pots are made to store the grains.

After retaining a certain quantity for the family, the rest of the grains and vegetables are taken for *marketing*. After the crop has been harvested, brought to the village and finally sold, the tribe or the community indulge themselves in *celebration.*

Adoption of Shifting Cultivation

This system of land use is limited to the medium or very high rainfall zone, about 500 cms and above remarks Negi (1982) after studying the areas in which shifting cultivation is common and the local climatic conditions. The density of population in these areas is relatively low. This system of land use generally prevail in the outer and lower hills of the eastern Himalaya, upto an elevation of about 1500 mts. Shifting cultivation is prevalent in Arunachal Pradesh.

The clouds loaded with clouds come directly in contact with the mountains in the western Himalaya. Here the moon-shaped hills face the vast open plains of Bangladesh, with Bay of Bengal in the south. This direct contact between cloud and mountain causes well distributed rainfall in southern and windward side of the hills.

The major factors that have led to the adoption of shifting cultivation in the eastern Himalaya are:

Pressure of Population—As stated above, the density of population in these areas is relatively low. So large patches of land are readily available to the tribals. However, with an increase in the pressure on the land, the interval of time between two successive series of cultivation on the same land has gone down to 10 years at present from 30 to 40 years in the past.

Feature and Weather— Steep slopes are not fit for settled cultivation. Land fit for terraced cultivation is insufficient. These conditions leave little choice for the people. Further, the undulating terrain and very humid climate with a dense vegetative growth has also led to the adoption of shifting cultivation.

Poverty and Impatience—The tribal communities prefer to continue with the age-old method of cultivation because construction of terraces on steep land is costly and the tribals are not rich enough to bear that cost. Moreover, to construct suitable terraces on steeply sloping land is a very time consuming. They do not have enough patience to go through this long procedure.

Absence of Storage Facility—Despite ample rainfall in this tract, the tribal people cannot retain and route this water wherever and whenever they need if for cultivation purpose.

Land Ownership —There are three patterns of land ownership in this tract, They are:

(a) Land is collectively owned by the villagers;

(b) Land is owned by individual families; or

(c) Land is owned by the tribal chief. He distributes the land amongst the members of their tribe. they, in turn, cultivate the land under shifting cultivation.

Dominant role in the lives of the shifting cultivators is played by the tribal or village chief. He may order his followers to move to new places. They do so at his command. These people still follow the primitive pattern of society. Socio-economic influences of modern civilization have not yet persuaded most of them.

Cultivators Basic Necessities—A small plot of land cannot meet the cultivators' basic needs of food, clothing, tobacco and other social expenses. They manage the extra necessary expenses by cultivating cash crops like tobacco, ginger, chillies, vegetables and mustard intermixed with paddy in the fields under shifting cultivation.

Disasters—Crop failures are the result of vagaries of nature, slope failure and climatic conditions. These are often attributed to evil influences and superstitions. This influenced the tribals to move from one land to the other.

Besides that the calamities like diseases, death and natural disasters like floods and earthquakes forced the tribal people to move to other places. Also, occurrences like elephant raids, epidemics of diseases and insect attacks force the tribals to move from one place to the other.

Diminishing Return—Cultivation on a patch of land continuously for more than two years leads to a sharp decline in crop yields. Lack of means to supplement the soil nutrients that have been consumed for plant growth has led to progressive diminishing yields. Under that situation the cultivators are forced to move from one patch of land to the other. The soil again fills up its nutrients in about 8 to 10 years. After that the cultivators grow crops on that patch of land once again.

Effect of Shifting Cultivation

A number of environmental problems crop up due to shifting cultivation. These are:

1. Cutting and burning of the vegetation reduces the forest cover. It leads to the loss of timber worth lakhs of rupees and also to the loss of other forest produce. Deforestation also results in the loss of a important source of raw material for paper and pulp, plywood industries. Thus the people of the area are deprived of a source of employment.

Thangam et al. (1978) have reported the results of their extensive studies on the material lost in a field due to burning during shifting cultivation operations was carried out by Thangam et al. (1978). In their study, they laid a 22 sq mts plot in a 3-year cycle in Siang district and a 10-year cycle plot in Subansiri district. The following results were obtained by them.

Shifting cultivation cycle (in year)	*Volume of wood burnt per hectare (in cu ft)*	*Details afforest growth*
3	10	Saplings of *Macaranga denticulata, Trema orientalis* and *grasses.*
6	40	Pole stage of *M. denticulata, Salmalia malabarica, T. orientalis* and grasses.
10	100	Small trees and pole crops of *M. denticulata, Dillenia* sp., *Duabanga* sp., *T. orientalis* and grasses.

In 1979, Thangam et al. estimated that on an average a yield of 2000 cm ft of commercial timber and 40 tons of firewood was produced by a forest adjoining the land under shifting cultivation. Thus, from an area of 2.30 sq kms the market value of the forest produce burnt was put to be Rs. 1.25 million.

2. Tribals working in the interior forest areas draw heavily upon the meagre forest resources and even encroach upon the adjoining forest lands.

3. Damage to the vegetative cover results in:

(a) Accelerated erosion;

(b) Silting up of canals and reservoirs;

(c) Depletion of the soil resources;

(d) Recurrent floods; and

(e) Local ecological changes.

4. An important part of shifting cultivation operations is burning of filled vegetal matter. But if, this fire is not properly supervised, then there is a danger of its spreading to the adjoining forest areas. In that case valuable forest produce of crores of rupees will be destroyed in

that fire. Small fires may destroy young trees upto the pole stage and shrubs, and may alter the status of the soil micro bio-ecology. Fires also disturb the local fauna by driving them away from the affected areas.

5. For a period of about 10 to 15 years hundreds of square kilometres of land is engaged in this form of cultivation. The land is left uncultivaqted for 10-15 years and thus is unable to produce anything. It is a great loss indeed. Had it been kept under regular cultivation, it could perhaps produce thousands of tons of foodgrains.

The fast increase in population has forced the shifting cultivators to reduce the period for which the land is kept fallow. The fallow period is drastically reduced. Now a stage has come in which the land is unable to regrow a forest crop. So, the nutrient the soil was getting in the form of ashes of the burnt vegetative matter no more getting that. This has reduced the fertility of the soil.

6. Due to repeated cutting down of certain trees, in many tracts of the eastern Himalaya, these species may become rare. The day is not far away they will even locally extinct. In many areas of Arunachal Pradesh, which are under shifting cultivation, some species that are becoming scarce are: *Acer hookeri—Betula alnoides, Ilicium griffithii—Michelia excelsa, Taxus* sp.—*Tetracentron sinensia.*

7. Ecological problems of a local nature are created due to repeated cut and burn of many parasites and epiphytes. These species are gradually eliminating from the area

8. Lower contents of total sesquioxides, aluminium, calcium, iron, potassium, phosphorus and a lower cation exchange capacity in soils of an area under shifting cultivation in comparison to the adjoining forests is reported by Jha (1979).

9. When a 8 to 10 years old forest is put under the process of sediment and water loss increases by 21 and 5 times respectively , reported Mishra and Ramakrishna (1980).

10. Floods have become a very serious problem in the Brahmaputra catchment. Much of the water coming into the Brahmaputra flows through the areas affected by shifting cultivation.

11. In eastern Himalaya a major obstacle in the path of development is the practice of shifting cultivation. Due to their nomadic habit,

modern facilities such as drinking water, health centres, schools etc. cannot be made available to them. For providing these infrastructures permanent settlements are must. Thus, the tribes practicing shifting cultivation can not enjoy the fruits of modern civilization as they keep on moving in order to remain near the land they may be cultivating at a particular point of time.

WATER POLLUTION

Another major problem in many Himalayan lakes is water pollution. Some of the worst affected ones are:

Dal Lake

Oen of the major tourist attractions of the Kashmir Himalaya. is the Dal Lake. It is a Srinagar valley with an area of about 11.45 sq kms. This water body occupies the central part of Srinagar valley. Ecological degradation is a major environmental problem in this lake. Factors responsible for this are:

1. *House-boats and Shikaras*—The degradation of the Dal Lake is indirectly caused by the heavy tourists to the beautiful Kashmir Valley. There are about 2000 house-boats and around 5000 shikaras in this lake. The seepage from these house-boats is released into the lake, shikaras cause water pollution.

2. *Sewage outwash*—Sewage outwash from the agricultural fields, vegetable gardens, residential areas and roads along the periphery of the lake pollutes the once crystal clear waters.

3. *Floating island*—Some small bodies of soil and vegetation floating on the lake is called floating island. Floating islands are used for growing vegetables. Degradation of Dal Lake is also caused by these floating islands.

4. *Sedimentary Elements*—Dr. V. Kaul of the Botany department, Srinagar University has estimated the annual deposition of silt into the Dal lake at 80 thousand tons (Usha Rai, 1982). About 20 years back a large amount of sediment was flowing into the lake from the Shankaracharya hill. This deposit of still into the lake from the adjoining areas leads to degradation. Of course this inflow has been checked upto certain extent by large scale afforestation.

Harmful Consequences

Some of the harmful consequences of eutrophication of the Dal lake are:

1. *Recession*—During the past 50 years the size of the lake has been reduced by about half. Negi (1982) has given the following figures to show this recession.

Year	*Area of the lake (in sq kms)*
1940	22.4
1950	21.0
1960	17.5
1970	14.8
1980	12.0
1985	11.45

2. *Chemicals*—The following are the significant changes in the level of various chemicals over the years:

(a) *Dissolved oxygen*—The level of dissolved oxygen during spring is about 7 mg/litre. In the peak tourist season this level falls to about 5 or even 4 mg/ litre.

(b) *Deficiency of Nitrogen*—The Dal lake is deficient in nitrogenous compounds. The NO_3-N level is about 7 mg/litre during the summer season and rises to about 15 mg/litre during the winter. The NH_2-N level fluctuates between 5 to 15 mg/litre in different seasons. The concentration of NO_2-N is moderate, varying between 120-440 mg/litre.

(c) *Some Others Chemical Levels*—The levels of some other chemical compounds have been shown in the following Table, produce by Kaul and Pandit (1979).

Chemical	*Level (in mg per litre)*
Calcium	25 to 45
Sodium	4.2 to 8.4
Magnesium	10 to 16
PO_4-P	214 to 450
Silica	1.9 to 8.0
Carbon dioxide	10 to 15
Sulphur dioxide	1.0 to 1.7

3. *Biological organisms*—Pollution has affected the habitat of a large variety of biological organisms that are found in the Dal Lake.

(a) Every year a large number of fish die in the Dal Lake The areas around the lake which forms the breeding ground for fish are highly affected. Moreover, unchecked sewage disposal and large inflow of silt into the lake havę adversly affected spawning and breeding grounds of carp fish. As a result in the recent past the number of fish in the Dal Lake has gone down.

(b) *Zooplanktons*—On the side towards Shankaracharya hill A number of zooplankton species are found. They are—*Keretalla* sp., *Pseudosida bidentata,* and *Sida crystallina.*

(c) *Phytoplanktons*—The habitat of phytoplanktons has also been adversely affected. Some of these are:

Astronella formosa	*CeratoOneis* sp.
Certium sp.	*Cymatopleura* sp.
Dinobryon *divergens*	*D. sertularia*
D. sociala	*Syndera acus*
Tabellaria sp.	

(d) *Weeds*—Unwanted weeds group in the lake in large quantity. Every year approximately 40 to 50 thousand tons of profuse weed graw in this lake. It is stated by Kaul (as quoted by Rai, 1978).

Lakes in UP Hills

To avoid the scorching summer of the plain, thousands of tourists go to Nainital area of the UP hills every year. There are seven lakes in Nainital area. This increases biotic pressure on the Nainital and Bhimtal lakes. The following factors are responsible for the ecological degradation of these lakes:

1. *Sewage disposal*—A considerable amount of sewage disposal from the surrounding hills is dumped into the serene waters of the Nainital and Bhimtal Lakes. This increases in the summer season.
2. *Sediment inflow*—Degradation of the vegetation on the surrounding hill sides increases the level of sediment inflow.

Every year a huge quantity of sediment flows into both these lakes.

3. *Boating*—A very popular game with the tourists is boating. The ecological balance of the lake waters is disturbed by the rising number of boats to accommodate the increasing number of tourists.

Environmental Problems

(a) *Low Oxygen Concentration*—During the spring season the level of dissolved oxygen in the Nainital Lake is less than 5 mg/litre. In Nainital lake, the level of dissolved oxygen during the spring season is less than 5 mg/litre. In summer, the peak tourist season, this falls to below 2.5 mg/litre (Negi 1982a). This condition is very dangerous for fresh water fishes (Ellis et al., 1946). They find it different to live there in that situation because they are very sensitive to low oxygen atmosphere.

(b) *Significance of Nitrogen and Phosphorus*—The biological activity of the lake system depends significantly on the level of nitrogen and phosphorus. Pant and Bisht (1981) have reported the following levels of these two elements in the Nainital and Bhimtal Lakes:

"In Nainital Lake, the NO_3-N reaches its peak value of 499 mg/litre during autumn and decreases to a lowest value of 76.25 mg/litre in spring. In this lake, the value of PO_4-P was found to be 35.3 mg/litre. In Bhimtal lake, the NO_3-N varies from 0.0 mg/litre in spring and late summer to 40 mg/litre in winter".

The records of the Water Works Division state that there is a rapid increase in the level of NO_3-N in the Nainital lake since 1960. Based on the records of nitrogen and phosphorus levels, it can be said that eutrophication has set in the Nainital Lake. The Bhimtal lake is in the pre-eutrophication stage.

(c) *Fishes Affected*—In these two lakes various categories of fishes are found. They are trout, mahaseer, mirror carp, asela, rohu, cattla, mirgal etc. Pollution of the lake waters increased in the mortality rate of fishes during the past 20 years or so. An estimate shows that about 40 to 50 kgs of fish die every year in the Nainital Lake (Das, 1978; Das, and Pande, 1978). Moreover, due to large inflow of sediments and sewage disposal the spawning and breeding grounds of carp fishes are destroyed.

(d) *Poisonous Chemicals*—Toxic substances such as ammonia, hydrogen sulphide and methane form at the bottom of the Nainital Lake due to the disposal of sewage. The physiology of biological organisms living in the lake are adversely affected by these poisonous chemicals.

(e) *Water molds*—Khulbe (1981) has reported that water molds occur in abundance in Nainital lake, particularly along its periphery and near sewage outlets. The following are the main water molds reported in the Nainital lake by Khulbe (1981): *Achlya apiculata Aphanomyces laevis Brevileginia linearis Protoachyla* sp. *S. diclina S. hypogyma Achlya flagellata Aphanomyces scaber Isoachlya anisosporva Saprolegnia crustosa S. ferax S. mixta*

Addition of organic matter contaminates water reducing its drinking quality, reports Khulbe. A number of water borne diseases in animal and human population being are caused by different pathagenic micro organisms which directly or indirectly are associated with the organic products.

(f) *Zooplanktons*—Large number of zooplanktons are found in all Lakes. But caphalodella sp. Sida sp. and Dakhnia sp. are found in Bhimtal Lake only.

(g) *Recession*—Once a part of the waterbody Mallital, on the northern end of National Lake, is at present a flat ground. Thus, what once was a part of the lake bottom is today a popular playground. The Nainital lake has receded. As a result of increased inflow of sediment into the waterbody the Nainital Lake has receded and this tract was exposed.

INCREASE IN CLIMBERS AND TREKKERS

A very popular sport in the Himalaya are mountaineer not and trekking. Most of these expeditions are spread out all over the Himalaya. The total number of mountaineering expeditions has gone up from about 20 each year in 1960 to over 200 per year now.

This has put an unprecedent pressure on the fragile environment of the Himalaya.

Effect on Mt. Everest

Around six to seven hundred people are involved in each expedition to Mt. Everest. It has *recee* and assault teams, support parties, supply groups and scores of other ancillary functionaries. These people need food, clothing and shelter for their 4 to 6 weeks long stay on the mountain. A majority of the expeditions follow the beaten path from Kathmandu to Namche bazar to reach Mt. Everest. All along this route they pollute the environment by leaving behind garbage and refuse.

In support of the above statement Capt. M.S. Kohli (1981) has said: "In June 1975, almost four years after the Himalaya was put on the world tourist map, I took a team to Mount Everest area in order to meet some old Sherpa friends. The Everest trail, once so clean and unpolluted had now turned into a 'garbage trail', full of food and juice tins, beer bottles, chocolate papers and all kinds of rubbish. I could not believe such a sight. The 'Shangrila land' was no more like the one I had seen in 1962 or even 1965. The 13,000 ft high Thyangboche Monastery in the shadow of Mount Everest which has been described as the most beautiful place on earth was full of garbage too. Two miles away was the Japanese hotel where rich tourists arrived by chartered flights and returned the next day after a quick look at Everest and Ama Dablam .. . The once uninterrupted sound of prayers which go on endlessly in the Thyangboche Monastery was now intercepted by occasional sounds of pilatus, porters and helicopters which landed in a nearby airstrip ...".

Black Peak Area

Black Peak a famous peak of Kashmir Himalaya. It is very popular with climbers. Today, nettle plants cover the mountain scopes. In this area sheep have been used for transporting climbing gear, food and clothing. They have eaten away all the wild flowers that grew along in this tract. There was a time when these flowers had provided additional beauty to the landscape.

Nanda Devi Area

Another popular trekking-cum-climbing trail in U.P. hills is Nanda Devi. This tract provided an excellent combination of adventure and beauty.It is easily successible. This nakes it more popular. It is hardly a day's drive from the plains of western Uttar Pradesh.

In the sixties only 1 to 3 expeditions were carried out into this region each year of which hardly one climbed any one of the three peaks of the area—viz., Nanda Devi, Nanda Devi east and Nanda Kot. In the seventies the number went up. At present, upto 10 climbing expeditions enter the Nanda Devi. This has changed the scenario of the once beautiful mountain trails by dumping garbage and refuse all over.

The leader of the Australian Changabang expedition 1980, is quoted by Kohli (1981) in the following words: "My party was fortunate to make the famous passage of the Rishi gorge and to climb within the fabled Nanda Devi sanctuary. This is an unforgettable experience. We were greatly disturbed, however, by the considerable amount of litter found at the camp sites along the gorge and at base camp within the sanctuary. We spent considerable time and efforts in trying to clean up the fouled areas and you can be assured that Australian parties respect your desire to keep these mountains clean".

EFFECT OF BUILDING ACTIVITIES

It is true that roads connect people and places; canals bring much needed water for irrigation, drinking and washing; and buildings provide shelter against wind, rain, snow and the sun. But these construction activities create a number of environmental problems. The total length of roads in the Himalayan region has increased during the past 40 years. There was a time when even the extreme end of Kumaun was not properly linked to the western extremity of Kashmir by roads. However, today it is possible to travel the Himalaya by road from one end to the other. With the launching of mammoth multipurpose projects a network of hundreds of kilometres of canals has been laid out in the Himalaya.

Multi-storeyed buildings have come up in many Himalayan towns. This has also upset the local environment. Blasting for breaking the rocks and release of debris during construction works cause environmental problems in the Himalaya. Let us see how—

Blasting

For breaking the hard country rock blasting is required. But it has some adverse effects. They are:

(a) Effect of explosion cannot be predicted. It may weaken more area than planned.

(b) Geological faults get reactivated due to blasting. This, in turn, may cause landslides and rockslides, thus endangering human life and property.

(c) During blasting operations headless persons may get hurt.

Release of Debris

The debris released during the construction cause serious environmental problems. Discussing this aspect of the problem in the Himalaya, Haigh (1981) wrote, "A recent survey of similar terrain in Tehri Garhwal and Dehradun districts (UP) indicated that on an average there are about ten large landslides (depositing an average of 500 cubic metres of debris) associated with each kilometre of road-bed. The author's study of a 6 kilometre stretch of road near Mussoorie in UP discovered 257 rock-falls resulting from the monsoon of 1978, and depositing the equivalent of a layer 0.3-0.4 metres deep on the road-bed itself. Further investigation suggested that these road-bed sediments alone contributed 7 per cent of the acceleration in the supply to the river bed".

Some major adverse impacts are produced by the debris slides on the Himalayan environment—

(a) Vegetative covers, are destroyed (b) It damages young regeneration. (c) Directly entering the streams or rivers that drain the area it adds to the problems of uccelerated erosion. (d) Fall in the local water table due to reduced infiltration. (e) Damage to Agricultural fields and human settlements are damaged.

DUMPING OF WASTE MATERIAL

In almost all the urban and semi-urban centres of the Himalaya sewage disposal has become a major problem. At Srinagar most of the sewage is dumped either into the Dal lake or in the river Jhelum. Bit at Udhampur (J&K) a small stream is the recipient of most of the sewage. on the outskirts of the city. These are the breeding grounds of various diseases. Sewage from Jammu town is dumped into the river Tawi. It makes the water unfit for human consumption. The waters of small nallahs and streams of Shimla get polluted when the sewage of Shimla town is released into them.

Whereas in Dehradun, most of the sewage finds its way either into the river Bindal which flows through the centre of the town or the river Rispana. Except in rainy season, there is virtually no natural flow of water in these two rivers. As a result, the sewage tends to stagnate and become the breeding ground for malaria and other diseases.

Kathmandu dumps its sewage into small streams and rivers thus polluting them. The sewage from Gangtok ends up in a large number of small nallahs and streams.

Aquatic and Wildlife Get Disturbed

Dumping of sewage materials in rivers and lakes has seriously affected the reptiles, fishes and aquatic birds living in them. These materials make the water unhygenic

Another threat is hunting. It was a major problem about 30 years back. However, this menace has considerably been checked with the enforcement of some strict laws.

Deer and antelope are the common prey species in the Himalayan region declining population of these species has given rise to revere competition among the exploiters. Disease also affect many species of wildlife. Quite a number of Hangul—Kashmir stag—died of Joh's disease in 1976-77.

Besides all these some human activities also disturb the wild animals. Those are—deforestation and visit to forest areas, sanctuaries and animal parks.

NATURAL CALAMITIES

Human activities are not the sole responsible factors that effect Himalayan environment. Natural process such as earthquakes, glaciers, avalanches and mountain streams too cause the degradation in this region.

Earthquakes

Seismically the Himalayan chain lies in a very active belt. There have been thousands of earthquake shocks some of them were very fatal worth mentioning among them were:

Kashmir earthquake (1885)—On May 30, 1885, a severe shock shook Srinagar and its surrounding areas. It caused extensive damage over an area of about 750 to 1000 sq kms. The shock was felt at places as far away as Shimla, Amritsar, Lahore, Rawalpindi, Jallandhar and Jammu. Around 3000 persons were killed during the earthquake said the scientists of the Geological Survey of India.

At Baramula, about 50 kms from Srinagar, huge cracks were formed in the earth's surface. At Patan, fissures as wise as 1 mt were reported. A massive soil slip occurred at Larrider near Baramula. At some places, Sand and water gushed out of these fissures at some places.

Assam earthquake (1897)—Due to the syntaxial bend in the most chain, the eastern Himalaya are more prone to reismic shocks. A very severe jolt shook the entire north-east India on 12th June, 1897, at about 5.00 p.m. It was on of the severest earthquake the world has ever known. Before the main shock, many small shocks were experienced for about a fortnight. It destroyed an area of about 200 thousand sq kms. The main shock lasted for about two and half minutes. The magnitude of the destruction was so severe that it seemed as if entire mountain sides had come sliding down.

Landslide damaged large areas of forests in the foothills of the eastern Himalaya. The vertical momentum was so large that huge boulders were thrown upwards. It completed destroyed many houses.

After a gap of about 50 years, another major shock was felt in this region on August 15, 1950. Its epicentre was near Rimalang the border of India, China and Burma. The shock development result of movement along the Mishmi thrust. This shock caused more damage than the previous one. Several areas were flooded. Huge landslides scarred the mountain slopes. These slides blocked the flow of the rivers Subansiri, Dibang and Tiding. When the debris blocking the Subansiri river broke after five days, the ensuing flash flooded an area of about eight hundred sq kms. Nearly 500 people lost their lives. River Brahmaputra and its tributaries changed their course at many places.

Kangra earthquake (1905)—In the Kangra valley of Himachal Pradesh violent shocks occurred in the early morning hours of April 4, 1905. The area affected ranged from parts of east Kashmir to Dehradun, Tehri and Mussoorie. Seismic instruments all over the world picked up this shock. As most of the people were asleep when this shock struck, the mortality rate rose very high. It has been estimated that about twenty thousand persons died. Thousands of cattle and wild animals were killed. Rivers and streams were temporarily blocked.

East Himalayan earthquake (1930)—This shock severely affected parts of the Darjeeling hills, Sikkim, Bhutan and Arunachal Pradesh. Many landslides were formed. Sand and water sprouted out from the fissures. It is also known as the hill earthquake.

Bihar-Nepal earthquake (1934)—This earthquake was on the severest shocks to have ever been experienced in India. It shook the hills of the lower Himalaya of Nepal and Kumaun. The main shock lasted for about 2.5 minutes. Many towns and settlements in the earthquake zone were in rubbles during this duration. There was widespread subsidence of land. Houses on ground sank deep into it. Often, only the roofs remained vissible. Large cracks developed on the earth's surface.

Extensive landslides occurred in the Himalayan foothills lying in the northern part of the affected area. Nearly ten thousand people died of which nearby three thousand died in Nepal. It is believed that this destructive earthquake was caused by the movement along a deep fault below the Indo-Ganga plain.

Kinnaur earthquake (1975)—A very high intensity earthquake struck parts of Kinnaur, Lahaul and Spiti in Himachal Pradesh at about 1.30 p.m. on January 19, 1975. It occurred due to movements along the Kaurik fault thereby causing widespread havoc and destruction. The township of Kaurik was the worst affected. Here nearly 50 people lost their lives and another 40 were injured. Nearly 300 houses were destroyed. It caused a massive landslide near Maling in Spiti valley. Another giant slide blocked the river Paro Chu near Sumdo.

Many other intstances of slope failure have been credited to this earthquake. Communications remained disrupted for many days. To bring relief to the worst affected areas helicopters had to be pressed into service.

Dharchula earthquake (1980)—In December 1980 an earthquake struck Dharchula and its surrounding tract in Pithoragarh district of Kumaun. Displacement along a deep seated fault is believed to have been the cause of this quake. A large part of Pithoragarh district was desolated.

Nepal earthquake (1988)—In 1988, a massive earthquake struck parts of Nepal and Bihar. Thousands of people were rendered homeless. Many lost their lives. Worst affected was the foothills zone of Nepal.

Uttarkashi earthquake (1991)—Uttarkashi and adjoining tracts of Garhwal were struck by a destructive earthquake on 20th October 1991. Hundreds of human lives were lost and there occurred widespread loss of property. To meet the needs of the suffering people, relief operations had to be carried out for a couple of months.

Glacier—Source of Erosion

In the snow bound areas of the higher and trans-Himalaya glaciers are the major source of erosion. In the Himalaya, the elevations to which glaciers may descend depends upon aspect, slope, precipitation and latitude. The Sarbal glacier of Kashmir descends to about 3700 mts and the Gangotri glacier descends to 4000 mts. The Khumbu glacier originates from Mount Everest. It terminates at 4900 mts. The Kanchanjunga glacier descends to an elevation of about 4200 mts.

Von Weissman (1959) (as cited by Vohra, 1981) has estimated that an area of about 33,200 sq kms where is approximately about 17 per cent of the total geographical area of the Himalaya, is under an ice cover. Studies carried out by different authors shows that several Himalayan glaciers have either slowed down or are receding.

Bhandari and Nijanpurkar (1981) state, "Several Himalayan glaciers have been studied. . . for example, the Nehnar in Kashmir, the Gara in Himachal Pradesh and the Changme-Khangpu in Sikkim. The studies found that most glaciers had a negative mass balance during the past decade, that is, more ice is melting than the snow that is accumulating. That glaciers are receding is also evident because several tributaries once connected with the mainbody of the glaciers are now disconnected—a consequence of the severe negative mass balance. If this trend continues, the consequences could be severe".

Unconsolidated debris are left behind by receding glaciers. This may become the source material for landslides. The huge areas vacated by glaciers remain barren for a considerable period of time. During this period excessive erosion of the morainic material takes place Moraines may block the passage of the melt water and form morainal lakes. Sometimes this debris dam burrts, releasing water with a great force. This causes flesh floods in the adjoining areas.

Snow Avalanches: The snow avalanches account for a large percentage of the earth's contemporary mass movement activity in terms of frequency of occurrence, mass energy and magnitude. In layman's language, the failure of snow-covered slopes is termed as *snow avalanches.* In areas where there is a rapid accumulation of a huge mass of snow, stress may increase by wind drafting, snowfall, blasting, seismic activity or biotic activity until the snow slope fails. Each year, thousands of avalanches occur in the high and trans-Himalaya, that individually involve masses between 10 and 10,000 tons and vertical displacements of over 1500 mts.

In the Himalaya, snow avalanches are of three types (modified and adopted after Perla, 1978).

1. *Loose slide*—It is an avalanche of loose snow in which the release of stress occurs in fresh snow, prior to the formation of intergranular bonds in the snow. Huge clouds of dry snow are formed as the avalanche rolls down the hill slope.
2. *Wet slide*—The snowslide, in this type of avalanche, moves down as a dense mass moving close to the surface at comparatively low speeds. Such avalanches do not normally produce dust clouds.
3. *Slab avalanche*—The snow breaks and slides down the hill slopes in the form of rough slabs. Avalanches tend to scour the hill slopes particularly when they are accompanied by rock fragments. This may result in the formation of gullies in weak zones. After the snow melts, these gullies are enlarged by accelerated erosion. Avalanches carry a vast amount of debris with them. When the snow melts, this material is left in the gullies. It is then washed away by running water and it finally ends up in the rivers.

Avalanches are a constant threat to forest land. Avalanches damage large chunks of forests. They destroy valuable forest produce. The excessive force of the on-rushing mass of snow, the trees at about breast height. Young trees and saplings are mostly damaged by avalanches. They can hardly offer the resistance to the mass of on-rushing snow. So, avalanches are a constant threat to forest land. Mostly the trees get damaged by avalanches to such an extent that they cannot be put to any other use except firewood.

Avalanches in Kashmir and Himachal Pradesh have been known to cover agricultural land and damage. The high altitude meadows or pastures that lie near the snowline are damaged by the avalanches in Kashmir and Himachal Pradesh. In 1968, in the Buga meadow in Garhwal, many sheep and goats were killed when an avalanche came down violently. It covered an area of about 300 sq mts within a very short time. The shepherds accompanying them barely managed to escape with their lives. Of course, this type of extreme cases happen in remote and inaccessible areas only.

Avalanches are not only a constant threat to the forest lands or meadows, but they are a threat for trekkers and villagers of the region too. About 10 years back a party of 15 trekkers who were on a training mission was buried alive in an avalanche near Gulmarg in Kashmir. The skeletons found at the lake of Rupkund in Garhwal are believed to be of either General Zorawar Singh's soldiers or those of pilgrims who were buried under a massive snow avalanche. A couple of centuries later their remains were discovered.

The village of Tunda in Ladakh was destroyed in a avalanche in 1838. Sixteen people died there in that incident. The famous Pin valley of Himachal Pradesh cut-off from the rest of the world when it was struck by a series of avalanches in 1978. For many weeks, air force helicopters had to be pressed into service to distribute relief materials to the affected people.

Erosion Caused by River and Stream

Intensity of movement of a river is proportionate to its volume of water and the rate of flow. In other words, the more the water a river carries the more is its erosive capacity. The geological work done by rivers depends upon its kinetic energy. The work done by rivers and streams in the Himalaya falls into three broad categories:

Erosion—In the Himalayan region, the flow of rivers result in erosion. It is the most distinguished characteristic of rivers there. By erosion of the rivers beds they make the river valley deeper. By erosion of the valley sides they broaden the river valley. Also the headward erosion increase the length of the river valley.

Transportation—The transportation of sediments may be carried out in the form of solution, suspension or siltation.

Drop in the speed of the water, changes in temperature and pressure conditions, and obstruction in the flow of the river result in deposition.

About 90 percent of the entire stream load eroded from the Indian sub-continent is carried by the drainage system of Himalaya. The denudation rate in India is 710 metric tons per year as calculated by Subramanium and Dalvi (1978). This is much higher than the world average. In comparison to that rate of denudation (in metric tons per year) for Asia is 342, North America 119, South America 84, Europe 69, Africa 41 and Australia 29. The Himalaya covers a very small proportion of the total surface area of the earth. Inspite of a very high rate of denudation it accounts for only about 10 per cent of the global denudation. The denudation carried out by any other river system in the world is far less than the denudation done by Himalayan rivers.

Subramanium and Dalvi (1978) have given the following table to show the erosion rates of the main river systems of the Himalaya.

River system	*Annual chemical load*	*Annual sediment load*	*Total annual load*	*Chemical erosion rate*	*Sediment erosion rate*	*Annual erosion rate*	*Physical/ chemical ratio*
	in unit of 10 kg/yr			*in unit of 10 km^2/yr*			
Indus	4.9	4.5	9.4	15.0	13.0	27.3	0.87
Yamuna	11.0	23.5	34.5	3.0	6.5	9.0	2.17
Ganga	13.8	86.0	99.8	0.13	0.82	9.4	6.32
Gandak	1.9	19.0	20.9	4.3	42.0	44.3	10.0
Brahmaputra	9.2	50.0	59.2	1.6	8.6	10.2	0.54

The following effects on the Himalayan environment has been adversely effected by the erosion caused by fast-flowing rivers and streams. Some of the major effects are: (a) The fast-flowing river water leads to undercutting of the valley sides thereby causing landslides. (b) Unsorted sediments are deposited over agricultural fields, settlements, roads and canals during flash floods. (c) Changes have been reported in the course of the rivers Brahmaputra, Kosi, Gandak, Teesta etc. This lays waste on a vast tract of land Law, 1968. (d) *Choe erosion*—A numbers of seasonal streams that originate in the Siwalik hills and drain into the Indo-Ganga plains is given this name. In Dehradun area choe streams are known as *raus*. In the foothills of Kumaun they are called *sots*. These streams are usually dry. Onlyin the rainy season a huge quantity of water gushes down, thereby causing high levels of erosion. The upper catchments of these streams consists of loose rocks that are easily eroded. In the monsoon season, the choe streams carry away this material and deposit them in the foothills. Valuable arable land, settlements, forests, roads and canals get submerged. Each year a vast tract of the region is desolated by the choe streams. The erosive capacity of these streams is intensified by: a) weak rocks in their catchments; b) the Siwalik hills have been affected by intense tectonic activity; c) heavy biotic pressure in the catchments of these streams; and d) high intensity of rainfall.

ELIMINATE POVERTY FOR DEVELOPMENT

To maintain a balance between development and conservation and to same the decaying Himalayan environment urgent measures are required to be taken. The prevailing critical situation in this region is effecting the economic status of the people there and poverty decreases the degree of success. So a sound conservation policy must take simultaneous focus on the conservation and protection of the Himalayan resources and take due account of the needs of those who depend on the resources for their livelihood.

Water Conservation

The Himalayan ecosystem has deteriorated to such an extent that today it is almost beyond recuperative stage. For regulating the water in the streams and rivers which originate from them some stringent measures are the utmost need of the day. From earliest times the aquafers in these mountains have supported life in the hills and also in the plains.

Water, today is in short supply in the mountains. The ever increasing population and the resultant increase in the demands of water for agricultural and domestic use and also degradation of the forest ecosystems has led to an alarming shortage of water which sustains all life forms and economic activities of human beings in the Himalayan region For regulating water and its harvesting in these frail mountains forest now hang to be primarily managed there. The aim must be to increase charging of the aquafers; flood control; soil conservation and regulating the water supply all round the year. Besides that the recurring floods and droughts experienced in different parts of the Himalaya, demand immediate steps for rehabilitation of the watersheds.

Measures to Protect Soil Conservation

Soil erosion is another major threat in the hills. The loss of soil is the foremost threat to the ecological safety and security of the inhabitants in the watersheds of various river systems that drain the degraded forests. The rapid deforestation and biotic pressure on the forests has increased the erosion and landslides.

To save the soil some important measures have already been initiated in both forest and non-forest lands. But enough is still not being done. In the affected areas conservation of the soil can be brought about by vegetative and mechanical measures. Improved methods of agriculture will also help to conserve the soil. Control of floods and landslides and proper land use planning on the basis of its capacity are also effective measures.

Importance of Forest

Barring the snowy wastes in the upper reaches, around 40 per cent of the land area of Himalayan mountain region is covered by forests. These forests provide both tangible and intangible benefits. People living in the rural areas of the Himalaya depend on forests for meeting their energy needs; for fodder; timber and many other products. So, for sustaining civilization in this mountain chain, efforts should be made to conserve the forests. Due to ban on green felling in the Indian Himalaya the process of deforestation has virtually stopped though degradation of the forest cover continues.Proper importance should be given to and increasing the productivity of the existing forest lands.

Wildlife Conservation

Himalayan wildlife is not out of danger. Some effective measures are required to be taken up for the conservation of wild life. With the setting up of the first national park in 1936, an effort was initiated to retain the rich heritage of wild animals in the Himalaya. Since then many protected areas and biosphere reserves have been constituted in the Himalayan region.

Inspite of these efforts during the past few years, Himalayan wildlife is still under damager. Some special efforts to protect the smaller animals are to be initiated immediately. For conservation of wildlife in areas other than national parks, sanctuaries and biosphere reserves; habitat management and improvement and stricter measures against poaching should be implemented.

Effect of Anthropogenic Disturbances

Severe damage has been done to plant and animal habitat by anthropogenic disturbances in the natural environment of Himalayan mountain chain. It is also effecting the ecosystem to varying degrees. Well-being of human life in the Himalayan region as well as a parts of south Asia may also be effected by the consequential loss of biodiversity. So, it is absolutely necessary to take some effective steps to check this cancerous problem in top priority.

Besides creating and preserving the biodiversity reserves, sharing of information regarding their importance will be an effective step.

Commercialization of Mineral Resources

Human activities like mining or quarrying have impetuously effected the Himalayan region. It has caused widespread destruction of the land all over this fact. Rich mineral resources are now being used for commercial benefits. This random mining is going on for a long time now, which is adversely affecting the vegetation, water resource, river, streams and finally topography.

For ecological well-being of the Himalayan mountain chain, an economical but effective technology is an urgent need. Mineral exploitation must be checked immediately otherwise the resultant events that may follow will be very horrible.

Water, Air and Noise Pollution

In Himalayan lakes and rivers water pollution is another major problem. It is severely effecting the aquatic life.

Moreover, blast of dynamite to construct roads, use of various technologies for mining are not only effecting the air, they are causing noise pollution too. Immediate steps are required to be taken to check of all these misance.

References

Ashish, M. (1981), The Effect of Overpopulation and Overgrazing in Kumaoun, *Sci. R.D. Mtns.*, pp. 25-26.

Bahuguna, S.L. (1979), The Himalaya: Towards a Programme of Reconstruction, *Him. For.*, pp. 141-51.

Bhandari and Nijanpukar (1981), Glaciers, *Sci. Today*, Dec. 1981, p. 60.

Bhatt, C.P., (1980), Ecosystem of the Central Himalayas and Chipko Movement, *DGS Sangh*, Chamoli.

Bordoloi, B.M., (1976), The Problem and Rehabilitation of the Jhumiya Families of the Hill Areas of Assam, N.E.I.C.S.S.R.

Borthakur, N.D., (1979), Improving Productivity of Jhum Cultivation with Special Reference to N.E. Hill Region, 63rd Sc. Cong.

Chaturvedi, M.D. and Uppal, B.N., (1953), A Study in Shifting Cultivation in Assam, ICAR Res. Series.

Cocklim, H.C., (1961), Study of Shifting Cultivation, *Curr. Anthropology.*

Das, S.M., (1978), Pollution in High Altitude Lakes, *Sci. Cult.,* 236-44(4).

Das, S.M. and Pande, J., (1978), Some Physico-Chemical and Biological Indicators of Pollution in Nainital Lake, Kumaon U.P. *Ind. Ecol,* 7-16, 5(1).

Didwal, R.S., (1980), Occurrence of Landslides in Jammu Province of J&K State and their Control, *Proc. Int. Symp.,* LS, 37-40, 1.

Haigh, M.J., (1981), Floods and Erosion in North India, *Him. Man. Nat.,* 4-8, 5 (2).

Hollaway, C.W., (1970), The Hangul in Dachigam, *Census Orix.,* 373-382, 10(8).

Kachroo, P.N. and Hussain, B., (1980), Genesis and Analysis of the Nashiri Slide, *Proc. Int. Symp.,* LS, 207-213, 1.

Kaul, V. and Pandit, A.K., (1979), Is the Dal Lake Dying? *Sci. Rep.,* 16 (11)

Khulbe, R.D., (1981), Frequency of Watermolds as an Indicator of Water Pollution, *Sci. R.D. Mtns.,*432-436.

Kohli, (1981), *Indian Mountaineer,* 38, Spring.

Law, B.C., (1986). Mountains and Rivers of India, 21st *Int. Geog. Cong.,* Calcutta.

Leban, P., (1979), Landslide Occurrence in Nepal, FAO/HMG/UNDP Project.

Mellor, N., (1978), Dynamics of Snow Avalanches, *Rock and Ava,* 735-792.

Mishra, B.K. and Ramakrishna, P.S., (1980), Shifting Cultivation and its Impact on Forested Ecosystems at Higher Elevations of the Himalaya, 67th Sci. Congress.

Negi, S.S., (1981), Forest and the Himalayan Environment: A Review, *Him. Man. Nat.,* 18-20, 4(10).

Negi, S.S., (1982a), Causes and Effects of the Deteriorating Himalayan Environment: A Preliminary Report, *Jour. Envi. Sci.,* 47-50, 25(2).

Negi, S.S., (1982b), Environmental Problems in the Himalaya; BSMPS Publ., Dehradun.

Negi, S.S., (1985), Managing the Himalayan Environment. PEBA Publ., Delhi.

Pant, M.C. and Bisht, J.S., (1981), Impact of the Changing Environment on the Lacustrine Fishes of Nainital, *Sci. R.D. Mtns.,* 437-448.

Pant, M.C. and Sharma, (1980)., Observations on Some Casual Winter Fish Mortality in Nainital Lake, *Proc. Nat. Sem. Res. & Devt.,* 248-254.

Perla, R.I., (1978), Failure of Snow Slopes, *Rockslides & Ava,* 731-752.

Rai, U., (1982), A Dying Lake, *Times of India,* 14-3-82.

Raju, K.C. and Jalote, P.M., (1980), Geotechnical Appraisal of the Naina Devi Landslide, HP., *Proc. Int. Symp.,* LS, I.

Sharma, et al. (1981), A Geological Report on Telangi Landslide, Distt. Kinnaur, HP, GSI Spl. Publ. No. 6.

Subramanium, V. and Dalvi, R.A., (1978), Some Aspects of Stream Erosion in the Himalaya, *Him. Geol...,* 822-834, 8 (II).

Thangam, E.S., (1977), Containing Shifting Cultivation, *Ind. Farm.,* (2).

Vohra, C.P., (1981), Himalayan Glaciers. *Him. Asp. Change,* 138-151.

11

Threat to Wildlife in Himalaya—A Regionwise Review

Introduction

Some of the rich wildlife species in the vast Himalayan chain are on the verge of extinction. It has become a major matter of concern not only for India but for the world. Due to large scale poaching and habitat destruction the Himalayan wildlife is under threat.

Besides that the land degradation process that is operating in the region has not only endangered the agricultural economy but has also threatened the survival of the fragile natural ecosystem.

This chapters mainly deals with the status of the wildlife in details with the number of national parks and sanctuaries created in different states to conserve the rich heritage of wildlife in the Himalayan chain. Implementation of Wildlife Protection Act, various prestigious wildlife conservation projects like Project Tiger etc. are also discussed in this chapter. The following zone-wise review of the problem will help us understand if more clearly. Here we will also know the steps taken by the government to check the menace.

Biographic Sub-region of Higher Himalaya

The biographic sub-region of higher Himalaya is situated above altitude of 2500-3000 m. Extending over the western Himalaya mountain ranges of Jammu and Kashmir, Himachal Pradesh and UP Himalaya, the region is characterized by different types of high altitude flora. Some of the prominent ones are—rhododendron, dwarf hill bamboo and birch forests. In between them are the alpine pastures and meadows. This range lies just above the belt of coniferous forest and below the snow line.

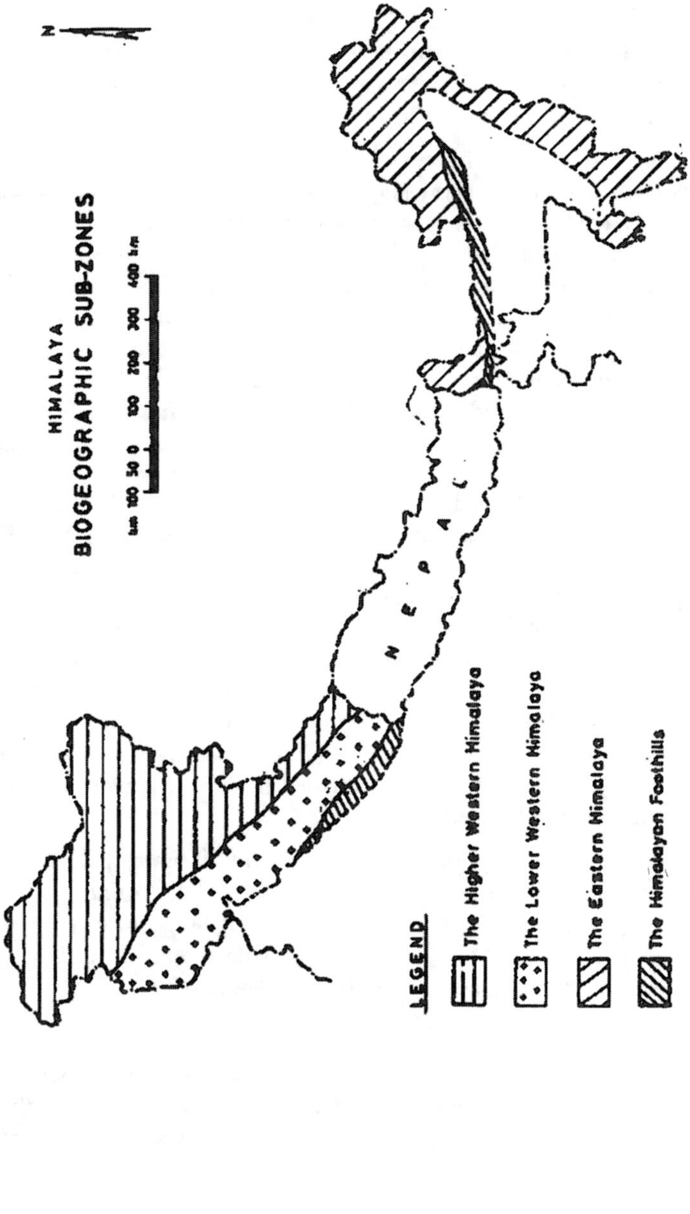
HIMALAYA
BIOGEOGRAPHIC SUB-ZONES
km 100 50 0 100 200 300 400 km
N
NEPAL
LEGEND
The Higher Western Himalaya
The Lower Western Himalaya
The Eastern Himalaya
The Himalayan Foothills

Fig. 11.1.

During summer the alpine pastures afford extension grazing grounds for wild animals. The entire alpine zone of the Himalaya is famous for rich variety of animal that are found here. In the winter season these animals move downwards into the coniferous zone. Variety of wildlife, a number of species of mammals and a large number of birds etc. are found here.

The mammals of the region include wild goats, sheeps, deers, marmots, mouse, hares, snow leopard. Markhor *(Capra falconeri)*, the finest of all wild goats, is very common in the region. It has thick and heavy cork screw shaped long horns. Kabul markhor *(Capra falconeri megaceros)* also occurs in the higher elevations in parts of the north-western Ladakh. At an elevation above 3500 m in this region Bharal (*Pseudois nayaur*), the principal sheep is found. It ascend to over 4800 m during the summer in search of food. The hangul or Kashmir stag *(Crevus elephus hangul)* and the musk deer *(Moschus moschiferus)* are the principal representative of deer family in the alpine zone of the western higher Himalaya. The hangul is the local representative of the European red deer. This deer is generally confined to the few pockets of Kashmir Himalaya. They are mostly found in the Dachigam Wildlife Sanctuary near Srinagar. The musk dear is found usually in the upper temperate, sub-alpine and alpine zones, particularly in the central and the western Himalaya. Musk glands of this deer found in males are used in the perfume making process. Both the hangul and musk deer move to high altitude pastures in search of food during summers. Due to large scale poaching and habitat destruction both these species have now become rare and threatened. In UP Himalaya for musk deer and for hangul in Kashmir Himalaya same remedial measures have been taunched to conserve them.

The most common animal in this region is Panthera pardus, the snow leopard. It is the most beautiful among predator mammals. But due to large scale hunting for its beautiful coat it has become very rare throughout the range. This animal oreys an wild sheep, goat and musk deer. With them it migrates in summer and winter.

Lower Western Himalaya

The lower western Himalaya includes the Himalayan mountain ranges of below ranges of below 2500 m elevation in Jammu and Kashmir, Himachal Pradesh and *Uttarakhand.* The main type of forests found in this region are sal forests, banj, Kharsu and moru oak forests and coniferous forests. These forests are spread over the altitudinal belt between 1500 m and nearly 2500 m. Leopard *(Panthera pardus)*, cheetal *(Axis axis)*, ghoral *(Nemorheadus goral)*, sambhar *(Cervus unicolor)*, tahr *(Hemitragus jemlahicus)*, ibex *(Capra ibex)*, jackal *(Canis aureus)*, brown bear *(Ursus arctos)*, black bear *(Selenarctos thibetanus)*, serotine *(Eptesicus serotinus)*, royle's vole *(Alticola roylei)*, mouse hare *(Ochotana roylei)*, serow *(Capricornus sumatraensis)* are found in this zone.

Harsh and cold climatic conditions and the changing topographic condition of this region suit the sheep and goat that are found here. Besides these, a large nmber of altitudinal migrant wildlife species are also seen in this biogeographic sub-zone. Otherwise they are generally found in the high altitude Himalayan alpine zone.

Himalayan Foothills

The Siwalik ranges and the *Bhabar* and *Tarai* belts in the extreme south constitute the foot hill zone of the Himalaya. The *Bhabar* belt extends from east to west to the south of the Siwaliks. It is characterized by the recent coarse deposits and the consequent low water table. The *Tarai*, with tall grassy field and savannah vegetation, is an outlying level belt corresponding with the thick deposition of fine silt to the south of the *Bhabar*. Because of the drainage is poor. Therefore, the water table is high. Peculiarity of this region is the natural monsoon forests. Sal, khair-sissoo and chir-pine at higher elevations are the most prevailing floral species of this region.

A number of various types of animals get rich and productive life support in the Himalayan foot hills mainly due to their location along the transition zone between the Himalaya to the north and the vast plains of north India to the south. As a result, as compared to other biogeographic sub-regions of the Himalaya, it has a vast variety of animals and in large number too. The foothills are the home of the typical big mammals of north India. The elephant *(Elephas maximus)*, tiger *(Panthera tigris)*, sambar *(Cervus unicolor)*, swamp deer *(Cervus duvauceli)*, cheetal *(Axis axis)*, hog deer *(Axis porcunis)*, barking deer *(Muntiacus muntjac)*, wild dog *(Cuon alpinus)*, jackal *(Canis aureus)*, hyaena *(Hyaena hyaena)*, sloth bear *(Melursus ursinus)*, black bear *(Selenarctos thibetanus)*, porcupine *(Hystrix indica)* are found in this region in abundance.

The Great Himalayan one-horned rhinoceros *(Rhinoceros unicornis)* which once inhabited the entire foothill zone is now considered a rare species confined in Assam foothills and in the Dudhwa National Park in the eastern *Tarai* region of UP. Another endangered animal is Cervus eldi, brow-antlered deer. Presently it is found only in few parts of the state of Manipur. It is considered to be most threatened deer variety in the world. This deer lives in a floating mass of vegetation deposits locally known as 'Phumdi'. Because of its floating habitat it is also named as 'dancing deer'. Some of the distinct wildlife species are now confined only to some particular parts of the Himalayan foothills, in the world. They are the gangetic gharial—*Garialis gangeticus)* hispid hare and pigmy hog—*Sus salvanius.*

Some beautiful birds are the speciality of this Himalayan region. *Lophophorus impejanus,* the Himalayan monal *Tragopan melanocephalus, (*western tragopan) *Lophura leucomelana* (Khalij) and *Catreus wallichii* (cheer pheasant) are the main species of birds found in this region.

Eastern Himalaya

Characterized by sal, evergreen and semi-green forests, the foothill zone of the eastern Himalaya extends upto the elevation of 1500 m. The states of Sikkim, Arunachal Pradesh, Nagaland, Manipur, Mizoram and hill areas of West Bengal and Assam form this sub-zone of the Himalaya. As you move up above this altitude upto 1800 m the type of natural vegetation changes into oak, magnolias, laurels and birches covered moss and ferns. As you move further above, from 2750 m to 4650 m you will come across the coniferous forests of pine, fir, yew and junipers, etc. These forest patches are marked with the undergrowth of scrubby rhododendrons and dwarf bamboos. The temperate belt in the region which extends at much higher altitude compared to the western Himalayan biogeographic sub-zone. It is characterized by the presence of Indo-Chinese wildlife species (Saharia, 1982). The typical ones are the leopard, *(Panthera pardus),* clouded leopard *(Neofelis nebulosa),* red panda *(Ailurus fulgens).* Three species of goat also inhabit the eastern Himalaya. In the forests of the region nearly at 2000 m lives the serow *(Capricornus sumatraensis).* From 1000 m to 3000 m you can come across the goral *(Nemorheadus goral),* a smaller goat antelope throughout the Himalayan range in the eastern Himalaya. Between the altitude of 2100m and 3000 m, the confined to the Mishimi hills and parts of Sikkim. It is said to be the largest member of goat-antilope family. Unfortunately now it has become rare. Besides, a great variety of birds and reptiles are also found in this eastern Himalaya region.

Status of Wildlife Conservation in the Himalayan States

A remarkable step to conserve the wildlife was passing of the Wildlife Protection Act in 1972. This Act empowers the state governments to constitute National Park or Wildlife Sanctuary for the purpose of protecting, propagating and developing wildlife and its environment in the areas of adequate ecological, faunal, floral, geomorphological, natural or zoological significance.

Wildlife Wings and Wildlife Advisory Boards were constituted to protect the wildlife. This Act made the state governments' effort to take necessary and suitable measures for the proper management of wildlife easier.

Before implementation of the Act there was only one National Park—the Corbett National Park. It is situated in the foothills of UP Himalaya. Besides that, in the entire Himalayan region of India there were nine wildlife sanctuaries including the Manas Wildlife Sanctuary in Assam. The Himalayan states of Uttar Pradesh, Himachal Pradesh and Assam took a great interest in the conservation and management of wildlife. They constituted national parks and wildlife sanctuaries under the provisions of the Wildlife Protection Act. Their effort increased the number of national parks to four and that of wildlife sanctuaries to forty-four in 1980. As a result in 1980, number of national parks and wildlife sanctuaries in the region respectively increased to as many as 4 and 44.

In the Himalayan mountains in 1995, there were 18 national parks (out of 61 in the country) 95 wildlife sanctuaries (out of 308 in the country) and 5 biosphere reserves (out of 14 in the country). This accounted for 5.21 per cent of the total geographical land surface of the region. National parks, wildlife sanctuaries and biosphere reserves respectively share 35.84 per cent, 50.84 per cent and 34.89 per cent of the total wildlife protected area of the region (Table 11.1). Table 11.2, 11.3, 11.4, 11.5, 11.6 and 11.7 present the statewise details of different categories of wildlife protected areas of the Himalaya.

The credit of having highest number of national parks (five) goes

TABLE 11.1

State/Regionwise Wildlife Protected Areas in the Himalaya

Sl. No.	*State/ region*	*Total protected area (km^2)*	*Percent of the total geo graphical area*	*Protected area percent of the total geographical area*	*Area under national parks (km^2)*	*Percent of the total protected area*	*Area under wildlife sanctuaries (km^2)*	*Percent of the total protected area*	*Area under biosphere reserves (km^2)*	*Percent of the total protected area*
1	*2*	*3*	*4*	*5*	*6*	*7*	*8*	*9*	*10*	*11*
1.	Jammu and Kashmir	1821	0.82	0.36	1141	62.66	680	37.34	—	—
2.	Himachal Pradesh	3929	7.06	3.56	1736	44.18	2193	55.82	—	—
3.	Uttar Pradesh	9984	19.53	8.12	3045	30.50	3084	30.90	4572	45.79
4.	Sikkim	988	13.92	11.60	850	86.03	138	13.97	—	—
5.	West Bengal (Darjeeling district)	1081	N.A.	N.A.	88	8.14	993	91.86	—	—
6.	Arunachal Pradesh	4393	5.24	N.A.	2507	57.07	1193	27.16	2500	56.91
7.	Assam	5237	6.68	1.57	430	8.21	4807	91.79	3270	62.44
8.	Nagaland	419	2.53	1.28	202	48.21	217	51.79	—	—
9.	Manipur	25	0.11	0.18	25	100.00	N.A.	N.A.	—	—
10.	Mizoram	600	2.85	0.80	600	100.00	600	100.00	—	—
11.	Tripura	1163	11.08	N.A.	—	—	1163	100.00	—	—
	Total	**29640**	**5.21**	**N.A.**	**10624**	**35.84**	**15068**	**50.84**	**10342**	**34.89**

N.A. = Not available.

Note: In some states some wildlife protected areas are being managed both as National Park and Wildlife Sanctuary/ National Park and Bioshpere Reserve/Wildlife Sanctuary and Biosphere Reserve.

to the Uttarakhand (UP Himalaya). The states of Jammu and Kashmir (3), has three and Himachal Pradesh and Arunachal Pradesh has two each.. In case of wildlife sanctuaries, Himachal Pradesh tops the list with as many as 29.

Whereas UP Himalaya has only three wildlife sanctuaries. But taking into account the coverage of protected area, the UP Himalaya has the highest proportion of 19.53 per cent of its total geographical land surface managed under wildlife protected area. So, it is clear that in comparison to the protected areas of Himachal Pradesh, the most of the protected areas of UP Himalaya encompass comparatively larger geographical land surface. Sikkim, Tripura and Himachal Pradesh have, respectively, accorded 13.02 per cent, 11.08 per cent and 7.06 per cent of their total geographical area, the status of protected areas.

TABLE 11.2

Wildlife Protected Areas in Jammu and Kashmir

Sl. No.	*Name of the wildlife protected area*	*Year of establishment*	*Area (km^2)*	*District/region in which falls*	*Principal wildlife species*
1	2	3	4	5	6
1.	Dachigam National Park	1981	141	Srinagar	Leopard, black bear, brown bear, serow, musk dear, hangul
2.	Hemis National Park	1981	600	Leh-Ladakh	Snow leopard, ibex, bharal, snow cock
3.	Kishtwar National Park	1981	400	Kishtwar	Leopard, civet cat, hangul, musk deer, markhor, ibex, grey langur
4.	Gulmarg Wildlife Sanctuary	NA	180	Gulmarg	Leopard, brown bear, ibex, markhor
5.	Jasrota Wildlife Sanctuary	NA	10	South-western part of J&K	Leopard, flying fox, cheeta, barking deer, wild boar, rhesus macaque
6.	Overa-Aru Wildlife	1981	400	Eastern Kashmir	Hangul, musk deer, serow, langur, leop-

Contd....

1	2	3	4	5	6
	Sanctuary			valley, Pahalgam	ard, brown bear
7.	Nandini Wildlife Sanctuary	—	34	Jammu region	Leopard, cheetal
8.	Ramnagar Wildlife Sanctuary	—	13	Jammu hills	Leopard, jungle cat, hyaena, sambhar, cheetal, wild beer, porcupine, pheasants, woodpeckers
9.	Shankaracharya hill Wildlife Sanctuary	NA	NA	Srinagar	Leopard, deer, black and brown beer, wild goat
10.	Surinsar Wildlife Sanctuary	NA	40	Udhampur	Leopard, wild boar, flying fox, pheasant, tits, jays, warblers
11.	Trikuta Wildlife Sanctuary	NA	3	Jammu hills	Hyaena, jungle cat, cheetal, wild boar, porcupine

NA = Not available.

Source: V.B. Saharia (1982), A.W. Rodgers and H.S. Panwar (1988), S.H. Prater (1972), Wildlife Institute of India, Dehradun.

Due to its various geoecological character, the Himalayan mountains, particularly the foothill belt of the Himalayan support a great variety of wildlife species. Besides, some of the rare faunal species are now only confined to the Himalayan region.For the conservation and protection of rare, endangered and threatened species some specific projects have been launched in different habitat types and ecological regions. Consequently, some of the national parks and wildlife sanctuaries have specifically been created.

Created in 1936, the Corbett National Park is the first national park of the country. It is situated in the Himalayan foothills of UP. It is famous for its beautiful natural landscapes and richness of wildlife. The park is one of the most ideal habitats of tiger in the Himalayan region and probably in the whole of the country. In view of this, the Corbett National Park was chosen as the venue for the inauguration of

TABLE 11.3

Wildlife Protected Areas in Himachal Pradesh

Sl. No.	*Name of the wildlife protected area*	*Year of establishment*	*Area (km^2)*	*District/region in which located*	*Principal wildlife species*
1	2	3	4	5	6
1.	Great Himalayan National Park	1984	1736	South-east of Kullu	Leopard, hyaena, civet cat, brown and black bear, musk dear, goral
2.	Pin Valley National Park	NA	NA	NE Himachal Pradesh	Snow leopard
3.	Bandli Wildlife Sanctuary	1974	41.4	Mandi	Leopard, jungle cat, black bear, goral, jungle fowl
4.	Chail Wildlife Sanctuary	1975	23.3	Solan	Goral, barking deer
5.	Churdhar Wildlife Sanctuary	NA	NA	Shimla and Sirmour	Leopard, musk deer, barking deer, wild boar, monal, khalij pheasant
6.	Darlaghat Wildlife Sanctuary	1974	44	Solan	Goral, barking deer, chakor, snow leopard, civet, fox flying fox, khalij, partridge, panther
7.	Daranghati Wildlife Sanctuary	1974	167	Kinnaur	Goral, black bear, serow, monal, khalij, musk deer, ibex, civet, marten, flying fox
8.	Gamul Siyahbehi Wildlife Sanctuary	1974	9	Chamba	Leopard, musk deer, bear, ghoral, serow

Contd....

1	2	3	4	5	6
9.	Govind Sagar Wildlife Sanctuary	1974	100	Bilaspur	Duck, teal, goose, crane
10.	Kais Wildlife Sanctuary	1954	13.7	Kullu	Leopard, black and brown bears, musk deer, goral, panther, civet, marten, flying fox
11.	Kelatop-Khajiar Wildlife Sanctuary	1958	47.3	Chamba	Jungle cat, civet, brown and black bears, goral, barking deer, serow, fox, flying fox
12.	Kanawar Wildlife Sanctuary	1954	54.3	Kullu	Leopard, tahr, serow, ghoral, bharal, musk deer, black bear, koklar, khalij, civet, marten, flying fox
13.	Khokhan Wildlife Sanctuary	1954	12.7	Kullu	Leopard, jungle cat, black bear
14.	Kugti Wildlife Sanctuary	1974	118.3	Chamba	Leopard, musk deer, barking deer, tahr, ibex, khalij monal, goral, serow, tragopan marten, civet, fox, flying fox
15.	Lippa Asrang Wildlife Sanctuary	1974	109.1	Kinnaur	Ibex, bharal, goral, monal, koklash, panther, fox, marten, flying fox, musk deer
16.	Majathal Wildlife Sanctuary	1974	92	Solan	Leopard, jungle cat, ghoral, barking deer, black bear, khalij, panther, marten, civet, fox, flying fox
17.	Manali Wildlife Sanctuary	1954	28.9	Kullu	Leopard, serow, barking deer, flying fox, tragopan, monal, khalij, tahr, goral, black bear, panther, civet, marten

Contd....

1	2	3	4	5	6
18.	Nargu and Winch Wildlife Sanctuary	1970	278.4	Mandi	Leopard, black and brown bears, musk deer, pheasants, jays, tits, goral, serow, monal, khalij, panther, marten, civet, fox, flying fox
19.	Naina Devi Wildlife Sanctuary	NA	163.5	Bilaspur	Leopard, wild boar, pheasants
20.	Pong Lake Wildlife Sanctuary	1983	NA	Una and Kangra	Aquatic fauna, migratory birds
21.	Raksham Chitkul Wildlife Sanctuary	1974	138.3	Kinnaur	Snow leopard, brown and black bears, monal, eagles, bharal, goral, koklash, snow cock, chakor, panther, fox, marten, flying fox
22.	Renuka Wildlife Sanctuary	1964	13. 5	Sirmour	Leopard, cheetal, wild boar, pheasant, jungle fowl, sambar, goral, barking deer, wild boar, khalij, peafowl, partridge, panther, fox, marten, civet, flying fox
23.	Rupi Bhaba Wildlife Sanctuary	NA	125	NE Himachal Pradesh	Leopard, mountain goats, sheeps, many species of birds
24.	Saichu-tan Nallah Wildlife Sanctuary	1974	103	Chamba	Leopard, musk deer, serow, snow cock, ibex, tahr, brown and black bears, goral, monal, tragopan, khalij, marten, fox, panther, chakor.
25.	Shikari Devi Wildlife Sanctuary	1974	213.5	Mandi	Leopard, hyaena, brown and black bears, goral, pheasants, barking deer, serow,

Contd....

1	2	3	4	5	6
					musk deer, monal, khalij, chakor, partridge, panther, marten civet, fox, flying fox
26.	Shilli Wildlife Sanctuary	1974	19.7	Solan	Jackal, goral, pheasant, jungle fowl, barking deer, chakor, khalij, partridge, panther, snow leopard, marten, civet, fox, flying fox
27.	Shimla catchment Wildlife Sanctuary	1958	10.3	Shimla	Leopard, black bear, goral, khalij, pheasant, monal, jungle fowl, barking deer, serow, koklash, panther, marten, civet, fox, flying fox
28.	Simbalwara Wildlife Sanctuary monal,	1974	55.4	Sirmour	Tiger, leopard, wolf, wild boar, cheetal, sambar, goral, barking deer, wild boar, marten, khalij
29.	Talra Wildlife Sanctuary	1974	76.2	Shimla	Leopard, wild boar, goral, jungle cat, pheasant, black bear, sambar, musk deer, khalij, koklash, monal, panther, civet, marten,
30.	Tirthan Sanctuary	1976	61.13	Kullu	Bear, deers, fox, goral, ibex, jackal, leopard, tahr, wolf and macaque
31.	Tundah Wildlife Sanctuary	1975	64.2	Chamba	Tahr, brown bear, black bear, musk deer, goral, serow, ibex, monal, tragopan, snow cock, khalij, leopard, panther, marten, civet

TABLE 11.4

Wildlife Protected Areas in U.P. Himalaya and Foothills

Sl. No.	*Name of the wildlife protected area*	*Year of establishment*	*Area (km^2)*	*District/region in which falls*	*Principal wildlife species*
1	2	3	4	5	6
1.	Corbett National Park (Tiger Reserve)	1936	1319	Nainital, Pauri Garhwal, Almora, Bijnor	Tiger, elephant, panther, sloth bear, nilgai, sambar, cheetal, hog deer, barking deer, muggar, crocodiles, wild boar, procupine, peafowl, red jungle fowl, partridge, goral
2.	Dudhwa National Park (Tiger Reserve)	1977	490	Lakhimpur Kheri	Tiger, rhinoceros, elephant, leopard, swamp deer, hog deer, panther, sambar, cheetal, wild boar, procupine, pea fowl, red jungle fowl, partridge
3.	Nanda Devi National Park	1982	630	Chamoli	Snow leopard, leopard, bharal, goral, serow, musk deer, snow cock, khalij, brown and black bear, tahr, sambar, wild boar, monal, pheasant, snow pigeon
4.	Rajaji National Park	NA	820	Dehradun, Haridwar, Pauri Garhwal	Elephant, tiger, panther, sloth bear, nilgai, sambar, cheetal, kakar, wild boar, procupine, peafowl, red jungle fowl, partridge
5.	Valley of Flowers National Park	1982	87	Chamoli	Leopard, civet, tahr, serow, musk deer, snow cock, khalij, vultures, eagles

Contd....

1	2	3	4	5	6
6.	Ascote Wildlife Sanctuary	1986	600	Pithoragarh	Musk deer, leopard, jungle cat, civet, goral, barking deer, brown bear
7.	Binsar Wildlife Sanctuary	1988	46	Almora	Leopard, civet, serow, goral, black and brown bears, khalij
8.	Govind Pashu Vihar Sanctuary	1955	953	Uttarkashi	Brown and black bear, snow leopard, bharal, musk deer, tahr, serow, goral, panther, sambar, wild boar, monal, pheasant, snow pigeon, civet
9.	Kedarnath Musk Deer Wildlife Sanctuary	1972	957	Chamoli	Musk deer, brown and black bears, snow leopard, bharal, tahr, serow, goral, civet, jungle cat, wolf, hyaena, barking deer, monal, khalij, pheasant, snow cock
10.	Kishanpur Wildlife Sanctuary	1972	227	Lakhimpur Kheri	Tiger, panther, sloth bear, sambar, swamp deer, cheetal, hog deer, barking deer, nilgai, pea fowl, jungle fowl, partridge
11.	Sonanadi Wildlife Sanctuary	1987	301	Pauri Garhwal	Tiger, leopard, elephant, cheetal, sambar, hog deer, bear
12.	Nanda Devi	NA	630	Chamoli	Leopard, jungle cat, civet, wolf brown bear, clawless otter, Himalayan weasel, bharal, ibex, Himalayan tahr, serow, goral, monal, pheasant, khalij, wood

Contd....

1	2	3	4	5	6
					cock, snow cock, hill pigeon, myna, Himalayan tree creeper, Himalayan woodpecker, Indian cuckoo, jungle fowl, bearded vulture, kite, fanthroated lizard, Asian house gecko, trinket snake, rat, snake, royal snake, king cobra
13.	Uttarkhand Biosphere Reserve	NA	3842	Tehri, Uttarkashi, Chamoli, Almora	Yalk, ibex, snow leopard, tahr, Himalayan weasel, brown bear, black bear, common other, Himalayam mouse hare, musk deer, red fox, panther, common mongoose, Indian wild dog, western trangopan, cheer pheasant, straited laughing thrush, black kite, goshawak, spotted owlet, osprey, king cobra, cobra, branded krait

NA = Not available.

Source: UP Forest Department, V.B. Saharia (1982), S.S. Negi (1995).

TABLE 11.5

Wildlife Protected Areas in Sikkim

Sl. No.	*Name of the wildlife protected area*	*Year of establishment*	*Area (km^2)*	*District/region in which falls*	*Principal wildlife species*
1.	Kanchanjunnga National Park	1977	850	Gangtok	Snow leopard, clouded leopard, marbled cat, black bear, red panda, tibetan wild ass, blue sheep, serow, goral, takin, musk deer, pheasant, partridge, binturong
2.	Fambong Lho Wildlife Sanctuary	NA	51	North-west of Gangtok	Leopard, hyaena, jackal, tahr, serow, flying fox
3.	Maenam Wildlife Sanctuary	NA	60	East of Gangtok	Leopard, bharal, serow, binturong, flying fox
4.	Kyongnosla Wildlife Santuary	NA	25	NA	Leopard, clouded leopard, civet, fox, bear, tahr, serow
5.	Singha Wildlife Sanctuary	NA	2	NA	Many interesting species of small animals and birds

NA = Not available.
Source: Saharia (1982), S.S. Negi (1995).

TABLE 11.6

Wildlife Protected Areas in the Himalayan and Sub-Himalayan Region of West Bengal

Sl. No.	*Name of the wildlife protected area*	*Year of establishment*	*Area (km^2)*	*District/region in which falls*	*Principal wildlife species*
1.	Neora National Park	NA	88	Jalpaiguri	Tiger, leopard, jungle cat, gaur, elephant, sambar, sloth bear
2.	Buxa Wildlife Sanctuary (Tiger Reserve)	NA	745	Alipurduar	Tiger, leopard, elephant, gaur, sambar, sloth bear, muntjac
3.	Mahananda Wildlife Sanctuary	1976	130	Darjeeling	Leopard, elephant, gaur, serow, tiger, sambar, swamp deer, hog deer, wild boar, gibbons, variety of birds
4.	Senchal Wildlife Sanctuary	NA	39	Darjeeling	Serow, Himalayan black bear, barking deer, goral, leopard, jungle cat, variety of hill birds
5.	Singalila Wild Sanctuary	NA	79	Darjeeling	Leopard, brown bear, deer, mountain goat, many species of birds

NA = Not available.

Source: V.B. Saharia (1982), S.S. Negi (1995).

TABLE 11.7

Wildlife Protected Areas in the North-Eastern Himalayan States

Name of the state	*Name of the wildlife protected area*	*Year of establishment*	*Area (km^2)*	*District(s)/ region in which falls*	*Principal wildlife species*
1	2	3	4	5	6
Arunachal Pradesh	1. Moiling National Park	NA	700	Subansiri-Siang	Tiger, leopard, goral, musk deer, takin, many varieties of birds
	2. Namdhapa National Park (Tiger Reserve)	1972	1807	Tirap	Tiger, panther, snow leopard, clouded leopard, golden cat, binturang, wild dog, wild buffalo, gaur, hog deer, parking deer, elephant, sloth bear, python, king cobra, Indian cobra monitor lizard, hornbills, pheasant
(P)NP	3. Dering Wildlife Sanctuary	NA	190	Subansiri-Siang	Wild buffalo, leopard, swamp deer, pheasant, migratory water fowl
	4. Itanagar Wildlife Sanctuary	NA	140.8	Subansiri-Siang	Elephant, gaur, tiger, panther, sambar, hog deer, barking deer, sloth bear
	5. Mahad Wildlife Sanctuary	1980	NA	Mishimi hills	Cloued leopard, leopard cat, jungle cat, hollock, gibbon, musk deer, spotted linsang, swamp deer
(P)NP	6. Pakhui Wildlife Sanctuary	1977	861.95	Kameng	Elephant, gaur, sambar, hog deer, barking deer, python, tiger leopard, sloth bear, slow loris

Contd....

1	2	3	4	5	6
	7. Namdhapa Biosphere Reserve	NA	2500	Lohit, Tirap	Tiger, leopard, marbled cat, hollock, monkey, giant heron, hodgson's hawk eagle, black eagle, shahin, hobby
Assam	1. Kaziranga National Park	1974	430	Jorhat	Great Indian one-horned rhinoceros, wild buffalo, elephant, gaur, leopard, wild boar, civet, otter, swamp deer, hog deer, sambar, tiger, python, pelican, partridge, florican
	2. Barnadi Wildlife Sanctuary	1980	26	East of Manas Sanctuary	Hispid hare, pigmy hog, elephant, tiger, leopard
	3. Deeper Beel Wildlife Sanctuary	NA	4	Kamrup	Rhinoceros elephant, gaur, tiger, leopard, mswamp deer, many terrestrial and aquatic birds
	4. Dibru Saikhowa Wildlife Sanctuary	NA	640	Tinsukia	Rhinoceros, elephant, gaur, tiger, leopard, swamp decr, birds
	5. Garampani Wildlife Sanctuary	1952	430	Diphu	Elephant, wild buffalo, leopard, common langur, hollock, tiger, swamp deer, pheasant
	6. Kachugaon Wildlife Sanctuary	NA	214	Kokrajhar	Tiger, elephant, gaur, rhinoceros, birds
	7. Laokhdwa Wildlife Sanctuary	1979	195	Nagaon	Rhinoceros, leopard, wild buffalo, swamp deer

Contd....

1	2	3	4	5	6
(P)NP	8. Manas Wildlife Sanctuary (Tiger Reserve)	1928	2840	Barpeta	Tiger, elephant, panther, gaur, wild buffalo, rhinoceros, golden langur, civet, otter, swamp deer, hog deer, sambar, pigmy hog, water monitor, wild boar
	9. Nameri Wildlife Sanctuary	NA	137	Sonitpur	Leopard, swamp deer, sambar, wild boar, birds
	10. Orang Wildlife Sanctuary	NA	65	Tezpur	Tiger, elephant, wild buffalo, rhinoceros, swamp deer, leopard, sambar, barking deer
	11. Pobha Wildlife Sanctuary	NA	45	North Lakhimpur	Wild buffalo, barking deer
	12. Pobitora Wildlife Sanctuary	NA	16	Marigaon	Rhinoceros, gaur, swamp deer, birds
	13. Sonai Rupai Wildlife Sanctuary	NA	195	Tezpur	Tiger, leopard, buffalo, swamp deer, hog deer, pigmy hog, hispid hare, rhinoceros, elephant, sambar, barking deer
	14. Kaziranga Biosphere Reserve	NA	NA	Kaziranga National Park and its adjoining tract in Assam	Rhinoceros, elephant, wild buffalo, tiger, leopard, wild boar, swamp deer, hog deer, capped langur, hollock, gibbon, Assamese macaque, variety of terrestrial and aquatic birds, reptiles
	15. Manas Biosphere Reserve	NA	NA	Manas Sanctuary and its adjoining area	Elephant rhinoceros, gaur, wild buffalo, swamp deer, hispid hare, pigmy hog, leopard, golden langur

Contd....

1	2	3	4	5	6
Manipur	1. Keibul Lamjao National Park	1977	25	Central	Brow antlered deer, wild goat, hog deer, wild boar, water birds, reptiles, fishes
	2. Sirohi National Park	1982	NA	Imphal	Tiger, leopard, wild boar, deer, antelopes, eagles, vultures, tits, warblers, owl, pheasant
Mizoram	1. Dampa National Park and Sanctuary (Tiger Reserve)	1976	600	Aizawal	Elephant, tiger, leopard, sambar, barking deer, Himalayan bear, wild boar, wild dog, wild cat, gaur, king cobra, python, hornbill, pheasant
Nagaland	1. Itanki National Park and Sanctuary	1975	202	Kohima	Elephant leopard, gibbon, gaur, bar king deer, wild bear, panther, tiger, pangolin, various kinds of birds and reptiles
	2. Fakin Wildlife Sanctuary	1983	6	Along the eastern border of the state	Tiger, leopard, jungle cat, gibbon, birds
	3. Pulicabadze Wildlife Sanctuary	1979	9.23	SE of Kohima	Jungle cat, gibbon, hair crested drongo, imperial pigeon, bush warbler
Tripura	1. Charilam Wildlife Sanctuary	NA	13	South of Agartala	Phayre's leaf monkey, serow, leopard, gaur, deer
P(NP)	2. Gumti Wildlife Sanctuary	NA	350	South Tripura	Leopard, gaur, serow, hornbill, pheasant
	3. Rowa Wildlife Sanctuary	NA	800	North Tripura	Elephant, gaur, tiger, serow, leopard, hornbill, pheasant

Contd....

1	2	3	4	5	6
	4. Sepamijala Wildlife Sanctuary	NA	18.53	West Tripura	Chasma badar, birds, primates
	5. Trishna Wildlife Sanctuary	NA	173	South Tripura	Tiger, elephant, gaur, serow, phayer's leaf monkey, hornbill, vultures, sparrow

NA = Not available.

(P)NP = Proposed National Parks.

Source: V.B. Saharia (1982), S.S. Negi (1995), A.W. Rodgers and H.S. Panwar (1988), S.H. Prater (1972), *Hindustan Times* (5 February 1996).

the launching of the Project Tiger, the country's most ambitious and prestigious wildlife conservation project, in 1974. The park was declared as the first 'Tiger Reserve' of India under Project Tiger. Now there are five tiger reserves (out of 23 in India in 1994) in different habitat types of the Himalayan region. These reserves are Corbett and Dudhwa in UP foothills, Buxa in sub-Himalayan West Bengal, Namdhapa in Arunachal Pradesh and Dampa in Mizoram. Besides, specific conservation schemes and projects are underway. Some schemes and projects for wildlife conservation are underway in Dachigam National Park in Jammu and Kashmir, Keibul Lamjo National Park in Manipur, Kaziranga National Park in Assam and Dudhwa National Park in UP foothills for hangul or Kashmir stag, brow-antlered deer or dancing deer, Indian rhinoceros and swamp deer or barasingha, respectively.

The Himalayan musk deer was declared threatened in 1973 due to large scale poaching and habitat destruction (Holloway 1973). The species was listed as 'vulnerable' in the Red Data Book of the International Union for the Conservation of Nature and Natural Resources (IUCN) in 1974. In collaboration with World Wildlife Fund (WWF) the Government of India launched a Himalayan musk deer conservation programme as a part of the 'Threatened Deer Programme' of IUCN in Kedarnath Wildlife Sanctuary in Uttarakhand. The main objective of this programme is to carry out an ecological study. In addition to this, measures are being taken for the breeding of musk deer in captivity with the objective of reintroducing the species into different natural habitats. It is reported that under the specific conservation schemes in different protected areas, the number of some animals particularly the tigers, rhinos and swamp deers has been increasing progressively. For their relocation in other areas some effective measures are also being taken on priority basis.

In the eastern Himalayan states tribal revolt and opposition from local people are the development of national parks and wildlife sanctuaries. Besides revolt, inadequate financial resources is also a major hurdle in the finalization and implementation of several wildlife conservation and development programmes in the region.

In Tripura, the proposals for the creation of Atharamura and Unokoti wildlife sanctuaries and the State Government's decision to convert the Gumti Wildlife Sanctuary into a national park covering a total area of 350 km^2 are being delayed due to mounting revolt and strong local opposition in the state (Hindustan Times, 5 February 1996). Fearing eviction local people are opposing the move to create national parks and wildlife sanctuaries. Besides, the insurgency operations are also hampering. The implementation of development and management plans in Sepahijala Wildlife Sanctuary covering are also being hampered for the same reason. The sanctuary is situated 33 km away from the state capital Agartala in the West Tripura district. It has a rich collection of wildlife, particularly birds and primates besides chasma badar which is reported to be found only in Tripura in the world.

During recent years, the population of deers, lions, blue bulls and tigers has increased considerably in this sanctuary. The state wildlife authorities are now planning to relocate them to Trishna and other wildlife protected areas of the state. Sepahijala is both conceived as a wildlife sanctuary as well as a research centre. Within the sanctuary there is a training institute too.

On the basis of the diversity of physiography, climate and natural vegetation the western and central Himalaya can be divided into seven east-west geo-environmental regions including the *Doons*. The salient geoecological characteristics of these regions have been presented in Table 11.8.

Land Resource

Land is the basic means of production. So of all the environmental resources it is the most valuable for mankind. In the development and growth of other biophysical resources of the earth land is the basic geographical element. All mankind's needs are fulfilled directly or indirectly by the land. Land is a very comprehensive term and includes the physical, biological and geological systems of the earth. With the increasing pressure of human activity on land, during the recent years, the man land ratio is gradually decreasing in the Himalayan region. Therefore, comprehensive and scientific study on the quantitative and qualitative variation of land resources and their reasonable exploitation and utilization to bring their potential into full play for meeting the needs of sustainable development, is one of the most important challenges to resource managers.

The tremendous increase in human population and the increased pace of techno-economic advancement in recent years have increased and diversified the demand of land and its products. As a result, the quality and quantity of land resources have continuously degraded and decreased. Under the impact of various bio-productive, techno-productive and non-productive landuse systems, the land and the whole environmental system of a geographical region changes positively or negatively.

The impact of some land transformation processes is limited to an area on which they are carried on, while the impact of others goes far beyond transforming the surrounding areas also. With the increase in the complexities of cultural ecology and advancements and development of human civilization, adverse impacts of landuse pattern have been increasing and the resulting process of land degradation has become more and more rapid, far reaching and often irreversible (Kostrowicki, 1983).

The most challenging and serious problem that the Himalayan region is facing today is the top soil erosion. Deforestation for meeting the demand of fuel and fodder and industrial raw material, expansion of agriculture on upslopes and marginal areas, construction of roads and settlements, overgrazing and other sort of unscientific land transformation systems have promoted the speedy process of land. Consequently, the survival of the natural ecosystem has been effected adversely.

12

Flora and Fauna of Kumaon Region

Introduction

The higher altitude of Kumaon region is covered by forests. The principal forest types in Kumaon consist of varieties of oak. Jim Corbett, who was a great naturalist and a shikari stayed here for sometime. During his stay over there he intensively traveled through the forests of this region and studied the large varieties of wild animals inhabited there. In his writings Jim Corbett has expressed his experience about the character and behaviour of these wild animals.

Forests of Kumaon being tropical forests there are a large varieties of reptiles too. Poaching and the reduction of natural habitat and tree cover have created a threatening situation for these species of wildlife.

For their protection some effective measures of government has naturally had some impact on the flora and fauna of the region. Various animal sanctuaries have been established. Famous Jim Corbett Park is one of them. It was named after the great naturalist.

A detailed description about the landscape of Kumaon region along with its forests and wildlife is depicted in this chapter.

Kumaon region is characterized by a wide varieties of wildlife. But many species are threatened with extinction due is large scale deforestation and poaching.

Jim Corbett in his stories have made some of the species of dur popular. Those are—sambar, spotted deer, hog deer barking deer (kakar). They are found to be roaming so freely in the sanctuary, that often a couple of them will come near your car and stare at you.

Besides these deers, a large number of wild elephants are found in the Corbett Park, especially in the spring and summer. But, wild animals like tigers and panthers are not so easily spotted there. It may be because either they are very cautions or their number is diminishing in spite of Project Tiger. Moreover, panthers are confined to the upper reaches and the park—above 1000 metres. Tigers are not found there. Other animals like wild bears, jackals, martens and foxes are found here in large numbers as we find varieties of monkeys, wild cat and leopard cat in plenty.

So far as aquatic life is concerned, the rivers are full of alligators, gharials as well as variety of fish. In Kumaon beautiful and variety of birds are also found in plenty.

All these together increase the beauty of the park. In other words, they make the park a paradise.

Jim Corbett was a great naturalist and a shikari. He lived in Kumaon for sometime. Famous Jim Corbett Park was named after him. In his writing he has beautifully expressed his vivid experience about his adventure while chasing tigers. Also we find description about the people animals of the area in these writings. In Jungle Lore he has written about his boyhood days in Kaladhungi in Tarai. He expressed his wanderings in the jungle in following words:

> "The jungles on both sides of the stream were teeming with game in the way of red jungle fowl, deer and pig that crossed the game track on their way to or from the fields."

But these jungles which were once frequently visited by tigers, panthers and pythous, unfortunately, the situation is not longer so. The deteriorating numbers of the wild animals, reptiles and birds are discussed in detail in the rest part of this chapter.

Wild Animals

Bearing the tiger, most of these animals, except for the tiger, are found higher up in the mountains too. The species of panther are generally seen in areas above 1,000 metres. But when it is unable to find its natural prey, it kills and carry away domestic cattle. This beautiful animal, once so large in number, is now rarely seen. Reduction of forest cover has caused this devastation. The Forest Department has classified it a vermin. Harassed villagers tend to eliminate it by poisoning the kill with pesticide. The panther dies when it returns to its kill, but it takes a long time to do so, and it dies in agony.

Found between 2,000 and 3,000 metres musk deer is at the mercy of hunters. The Japanese value its musk pods for their (unproven) aphrodisiac appeal, while the French want musk as a base for their perfumes. Only the male musk deer is in demand. The hunters while trying to trap it also indiscriminately utterly destroy the females and the young ones too. The musk pods are costlier than gold.

Another dangerous animal found at this altitude is Himalayan black bear. It is more dangerous than tiger. Jim Corbett's stories have some interesting descriptions of fights between bears and tigers, always started by a bear poaching a tiger's kill and, when challenged by the rightful owner, attacking the tiger, quite unafraid. Bear finds it easier to run uphill than dosn. So, if you meet a bear, run downhill, say the local villagers.

The brown bear is the largest carnivore in the Himalayas. It is well provided with fat to save itself from the severe cold in winter. It lives an mammals like mice, squirrets etc. Besides that it also cats domestic sheep and goats and carrion. It is devoted to its young ones which are born during its periods of seclusion. It devotes considerable time to the upbringing of its offspring. MK Ranjit Singhji describes how he once watched a mother bear trying to rest in the afternoon, while its cub ventured out on forays against its mother's wishes:

> Five times she had to bestir herself and bring back the errant cub. Calls and even cuffing was of no avail. Ultimately she half lay upon the cub, one huge arm pinning it to the ground, while it howled in protest. After a while she let it go, whereupon, thoroughly subdued, it curled up beside its mother and went to sleep.

Bear meat and skin are not in demand as they have no market value. Still bear is not safe from being hunt because it attacks the domestic cattle.

Snow leopard is found above the snow line. Its natural prey are varieties of sheep and goat families. They are—*thar, ghoral* and *bharal.* The shaggy that clings to the safer habitat of cliffs and rocky surfaces, but the *bharal* prefers open ground unless disturbed, when it leaps for the shelter of the crags. Both varieties of mountain goat are extremely curious. They easily fall prey to the poachers.

As a result their numbers are decreasing rapidly. The snow leopard is eavasive and rarely seen. Though its protective color is a reason behind it, but more so because it dwells in such an inapproachable place that the viewers can hardly reach there. Even if they reach its realm, they get too exhausted to notice it. Only those who have taken up residence in that inhospitable terrain have managed to see them. The Bedi brothers have a film on the snow leopard. They stayed in the region for several months to shoot the film.

The snow leopard is not white but varies from cream to grey, with flashes of rosettes. It is only about three feet long, but looks lengthy because of its bushy tail, which it can use as a wrap-around blanket. Its habitat ranges from 4,000 metres to 6,000 metres and *bharal,* which is its chief prey, is usually to be found up to an altitude of 5,000 metres. The hide of snow leopard is in great demand in the world of fashion. So its danger comes mainly from poachers in search of its hide, but also from the problems that threaten its food—wild goats, *thar, bharal* and *ghoral.* These have been reduced in numbers as they are being hunted for the cooking. Anothers reason is, domestic sheep and goats have been introduced into their exclusive habitat by an exploding population of sheep herders looking for fresh pastures. The *bhotia* sheep dog follows these domestic animals and disturbs the *bharal* female sheep, sometimes harrying them till they miscarry. Military air movements on the Himalayan border can cause the wild goats to stampede to their deaths. So being deprived of its natural food, the snow leopard faces extinction.

The Indian rock python is found mainly in the foothills. There it lives in caves, clefts in rocks, and abandoned burrows. It lives on warm-blooded animals. It can even swallow large animals like deer, jackal and wild boar. It feeds only once in a few days, and invariably becomes very lazy after a meal. Because of beautiful markings, it was widely hunted for its skin, which was used to make shoes and handbags, but trade in snake skin has now been banned.

Varieties of monkeys and in large number are found in these forests. The largest and commonest variety being the *rhesus macuaque.* It prefers to live around human settlements. The other variety found here is the common *langur* or *Hanuman monkey.* They inflict considerable damage on fruit and crops. To protect their crops from these monkeys, the villagers keep large bhotia dogs. Religious sentiment protects monkeys from being killed.

Reptiles

Since the forests of Kumaon are tropical forests, reptiles are found in plenty. Those reptiles are monitor lizards, pythons and many other varieties of snakes of which the *krait,* the *cobra* and the *Russels viper* are the poisonous ones. Of these, the most dangerous is the king cobra or *hamadryad,* another species mentioned often by Jim Corbett. In the monsoon season different types of tree and grass snakes are seen in this forest. Of course these species are harmless. An unusual snake is the Himalayan *pit viper* with bright green on top. It looks very much like the grass snakes. It is classified among the poisonous varieties.

Birds

230 species of bird are found in the western Himalayas out of 287 species of bird of the Indian hills as described by Salim. Ali. Like the flora, bird life varies with altitude and season. Perhaps the commonest of the birds, and the most beloved because of its cheerful, early morning song is the dark blue Himalayan whistling thrush. Jim Corbett has this to say about this songster:

> The privilege of catching the early worm is claimed by the Himalayan whistling thrush, better known as the whistling school boy. While walking through the Kumaon jungles in the half light between day and night, or night and day, a bird will flit by on silent wings pouring out a stream of golden song which once heard will never be forgotten.... Morning and evening he pours out his song while in flight, and during the day he sits for hours in a leafy tree whistling in a soft sweet minor key a song that has no beginning and no end.

Red billed blue magpie is one of the most striking birds in this forest. It is considered to be most destructive bird because it completes with monkeys in destroying the fruit trees. Partridge, wildfowl and *chukor* scramble up and down the hillside. Conservation laws have made them relatively safe from the cooking pot. Sunbirds and flycatchers are as colourful as the flowers. Flights of parakeets wheeling in formation at sunset are a marvellous sight. Raptors glide above the valleys and forests, occasionally coming out of a stately glide to swoop with swift ferocity on their unwary prey. The two most fascinating birds of the Himalayas are both raptors or birds of prey — the *golden eagle* and the *lammergeier.*

Yet another large Himalayan raptor is the bearded vulture. It is found between 200-4,000 metres and is sometimes seen sailing majestically even above 7,000 metres. These birds are not only magnificent specimens but are very rare. Bbecause they live at great heights, above 2,000 metres, little is known about their status and ecology. In frequently they desolate country. The lammergeier is, however, known to carry up huge bones like the femur of an ox. To feed itself on the marrow and fragments of these bones it drops the huge bones on the rocks from a great height and splinters them.

Game birds—jungle fowl, quail, pharianidae-pheasants and patridges—are considered to be the most distinctive bird family of Himalaya. Charming and fashionable feathors of some of there birds the attention of hunters and smugglers. As a result their numbers are gradually reducing. But as they breed in capitivity, their may incrrease again. Most popular Himalayan pheasants are the blood pheasant, the tragopan or horned pheasant, the monal or impeyan pheasant, the khaleej pheasant, the koklas and the chir pheasant. The mountain quail is an elusive and little-known member of the quail family. It is a small bird, slightly larger than the grey quail. It is said that this rare species was last seen in 1876. It is also believed that only about ten of them are available in the museums of UK and USA. All of these were collected from the areas around Mussoorie in Garhwal, and *Sher-ka-danda* peak near Nainital. To escape the harshness the central Asian platequ many species of birds come in from across the mighty mountain chain as do ducks and cranes. They follow the rivers galleys and come down into the Indian subcontinent. This is an interesting feature of bird life in the Himalayas.

The first Indian expedition to Mount Everest found three dead eagles at a height of 7,900 metres. Tenzing, one of the conquerors of Everest, also found a dead eagle on the South Col in 1952. A number of bird species breed in the higher Himalayas and move down in winter or even fly away to the Vindhyas and the Sahyadris. At about this time, the water bodies in the lower Himalayas attract many species of migratory birds especially from Siberia and the central Asian deserts.

Administrative Measures

Before British occupation there was a custom of planting *deodar* trees near temples. Jageshwar is a good example of this. Not much is known whether there existed any forest policy as such during that period. The forests were then extensive and rich and considered inexhaustible. The economy of the hill people depended entirely on the forests for fuel, animal fodder, medicinal herbs, leaf litter for manure, etc. The misuse of forest reserves was large because they were considered to be unfailing. The forest policy of the government has naturally had some impact on the flora and fauna of the area.

By 1824-25, these depredations were so extensive that Bishop Heber writes:

> Great devastations are generally made in these woods, partly by the increase in population, building and architecture, partly by the wasteful habits of travellers who cut down multitudes of young trees to make temporary huts and for fuel, while the cattle and goats which browse on the mountains prevent a great number of seedlings from rising. Unless some precautions are taken, the inhabited part of Kumaun will soon be wretchedly bare of wood, and the country already too arid will not only lose its beauty, but its small space of fertility.

Now, we find that Bishop Heber's predictions have come true to a large extent.

The formation of the 'Kham', the extension of railways and the consequent depletion of *sal* forests (*sal* wood is largely used for making railway sleepers) and the resettlement of refugees have destroyed the rich life of the tropical Tarai forest. Commissioner Traill attempted to reverse this process by his forest settlement of 1823. There was great public agitation against this reservation of forests. Subsequent legislation modified his settlement policy considerably. Finally, three main classes of forests have evolved in the Kumaon region—reserved forests, civil forests (protected forests under the administrative control of the Collector) and panchayat forests.

Huge landslips, especially in Nainital, where uninhibited and unrestricted construction work has weakened the mountain side are attributed to careless human activity. However, restrictions on quarrying and tree-felling have helped in the regeneration of forests. New forest projects and schemes now try to secure support and cooperation of the villagers. These measure were not only to protect the existing forests but also to educate the villagers in the fact that this resource which provides them with cattle fodder, fuel and fertiliser is not inexhaustible. Interestingly enough, villagers are no longer angry with forest officials who patrol the area to check the felling of trees. Rather their anger is directed at big contractors who fell trees for commercial purposes.

Besides attempting to save the existing forests from further depletion the new projects intend to bring more hill areas under forest cover. These projects involve the cooperation of NGOs. Moreover, the various animal sanctuaries have been established to protect special species. With their strict rules they also provide a safe shelter for other birds and animals. To educate tourists in the importance of protecting the animals which live in the area is their another major aim. The Corbett National Park near Ramnagar is the best known of these reserves. It is associated with Project Tiger, which has received considerable hype.

Apart from Corbett National Park, there are a few other smaller wildlife reserves in Kumaon. The Ascot Musk Deer Wildlife Sanctuary, located in the higher regions of Pithoragarh district, is meant to conserve the musk deer and its habitat. This rare species of deer is at risk because of the musk pod that the male of the species carries, and which is valued as an aphrodisiac and as a base for perfumes. It is not really necessary to kill the deer to extract the musk pod, but poachers are not interested in the preservation of the species, and also kill the females and the fawns in their senseless orgy of slaughter. This sanctuary covers an area of 284 sq km. The principal forest types found here consist of several varieties of *oak — banj, moru karshu, deodar, fir* and *spruce, blue pine* and, since it covers the higher altitudes, it is also home to *birch* and *fir rhodendron,* both tree and bush varieties, and *alpine scrub.* Apart from musk deer, this sanctuary houses leopard, jungle cat, civet cat, barking deer, serow, ghoral and brown bear.

The Binsar wildlife sanctuary is near Almora, and encompasses an area of about 50 sq km. This reserve too houses a similar variety of high altitude species, including the musk deer and the khalij pheasant.

It is not the lone hunter who is causing the damage, it is invariably big business that is doing the damage. The sellers of the musk pods, or the dealers in tiger bones and tiger and bear fat, who have no interest in trying to preserve a species. It is, therefore, not education or persuasion that will stop them. It is only heavy punishment. Till fairly recently, the man found with musk pods in his possession was not punished. He was punished only if caught in the act of poaching. Possession of banned articles must now become a serious offence.

Similarly, there are animals that are disappearing because their habitat is disappearing. The snow leopard, for instance, is being eliminated because its natural prey, the *tahr* and the *ghoral,* are being decimated by the sheep dogs that follow local shepherds into their habitat. It is essential that alien species are not introduced into national parks so that the balance of species is not upset.

Because of careless plucking and indiscriminately grazing many varieties of plants are becoming extinct. Goats poison the area around with their saliva while grazing. Some of these are of medicinal value, plants like *aconitum heterophullum* (*atees*), *rheum emodi (dolu) dioscorea deltoides (kin, gun, shingri mingri*). Some of these fast-disappearing plants are flowering plants of rare varieties — *podophyllum hexandrum, lilium wallichianum, gentiana kurooa.* How far this careless destruction can do damage to our world is incalculable. Also it is very difficult for the lay mind to comprehend. The only way to preserve our natural heritage seems to be to cordon off specific areas where the endangered animals and plants are to be found. Secondly, to enforce strictly the laws against grazing and poaching in these areas. Thirdly, to punish encroachers and law breakers in the strictest way. Either through indoctrination or through fear of punishment. The enforcers of the law—forest wardens, rangers, have o do their duty.

13

Nepal's Protected Areas and their Management

Introduction

Nepal is known for its natural beauty. It has the highest park on the earth. Quite a rapid progress has been made by Nepal in the sphere of conservation. Since the execution of the 1973 National Parks and Wildlife Conservation Act Noted has implemented a couple of conservation programmes successfully. Many wildlife species are recovering in these parks and reserves. In spite of that it is said chat the law has not been so effective as far trade is still going on in Kathmandu markets.

Moreover, protected areas are needed to be developed especially in the lower elevations of Himalaya.

This chapter mainly deals with the critical issues regarding wildlife protection throughout the Himalayan kingdom of Nepal along with some of the laws and rules are regulations in implmented to support the conservation of the wild species.

Nepali's Natural Beauty

For its commitment to the wildlife conservation and national parks, the His Majesty's Government of Nepal (HMG) has become the leader among the least developed countries (LDCs). Its natural beauty and topographical features which are ranging from subtropical lowland forest to the highest peak on the earth have made Nepal famous. Large population of the country, combined with high biological diversity inherent to regions in low latitudes with great geographic diversity, lead to particularly complex conservation problems and complex legislative and programmatic solutions for conserving native species while providing for rural development plans. Based on attitudinal propensity, Nepal can be divided roughly into five regions extending east-west, the protected areas of Nepal.

(i) the Outer terai—extending between approximately 75 to 200 m;

(ii) the Siwalik range—includes elevations approximately up to 750 m;

(iii) the Middle hills—the most densely-populated parts, with elevations ranging from 750 to 3,000 m;

(iv) the High mountains—with elevations ranging from 3,000 to 4,000 m; and

(v) the High Himalaya—includes elevations above 4,000 m.

The areas in the north of the main Himalayan peaks (rain shadow) have been referred to as the Trans-Himalayan Region (Heinen and Kattel 1992a). It is physiographically similar to the Tibetan plateau in that it is high altitude, cold desert. Comprising some of the land areas within two existing protected areas, this physiographic type is restricted to border areas in the central and western parts of the country.

In the rest portion of this chapter you will find information regarding the wildlife conservation programmes within the mountainous region of Nepal till 1994.

Reports of Heinen and Kattel (1992) and Heinen and Yonzon (1994) give us information about the animals of the region. Those reports have also discussed about the protected areas of lowland terai and inner terai regions. In their reports they have clearly expressed their opinion regarding the research and management programmes for conservation, this is followed by brief description of Nepal's mountain parks and reserves.

The 1973 National Park and Wildlife Conservation Act (NPWCA)

For the protection of areas and species in Nepal the Act has provided broad legislation (HMG 1973, 1977a; Heinen & Kattel 1992b). It has 34 sections. The purpose of this portion is to present main topics covered in the Act. Amendments and regulations are discussed later. Section 2 of the Act described four types of reserves, as defined in Nepali legislation:

(i) **National park:** This area set aside for conservation, management, and utilization of mammals, birds, vegetation, and landscape together with the natural environment.

(ii) **Controlled (strict) nature reserve:** This is an area of ecological significance or significant in other respects. It is set aside for scientific study.

(iii) **Wildlife reserve:** This area is reserved for the conservation and management of mammals, birds, and other resources and their habitat.

(iv) **Hunting reserve:** An area set aside for conservation and management of birds and mammals and other resources to provide hunting opportunities.

The first three categories correspond approximately to World Conservation Union (IUCN) Categories II, I and IV, respectively (IUCN, 1990 and 1993). Hunting reserves as defined above are not recognized by IUCN, but the broad goals of hunting reserves as managed in Nepal (e.g., Wegge, 1976a and b; Sharma, 1982) correspond approximately to IUCN Category VIII (Multiple Use Management Area).

The Act empowers The His Majesty's Government with the ability to create these four types of reserves and to transfer ownership for the purposes of declaring parks and reserves. Moreover, for the creation of a fifth type of reserve, a subsequent ammendment was allowed in the Act.

Nepal currently has 14 protected areas. Out of that 12 are under the jurisdiction of the Department of National Parks and Wildlife Conservation one, Shivapuri is under the jurisdiction of the Department of Soil Conservation, and one, Annapurna is under the jurisdiction of the King Mahendra Trust for Nature Conservation KMTCN, Table 13.1.

Entry into the wildlife reserves and national parks were permitted as per some sectors of the Act with an entry permit. Also prohibited activities like hunting, grazing and land clearing were allowed there again with a special permit. Subsequently these activities were allowed in wildlife reserves too.

TABLE 13.1

The National Parks and Reserves of Nepal. The Abbreviations used are: Wildlife Reserve (WR), National Park (NP), Hunting Reserve (HR), and Conservation Area (CA)

Name		*Location*	*Size (sq km)*	*Date of establishment*
Kosi Tappu	WR	Eastern terai	175	1976
Parsa	WR	Central terai	500	1984
Royal Chitwan	NP	Central terai	932	1973
Royal Bardia	NP	Mid-west terai	968	1976
Sukla Phanta	WR	Far-west terai	305	1976
Shivapuri	WR	Kathmandu valley	114	1985
Khaptad	NP	Far-west hills	225	1985
Makula Barun		Eastern Himalaya	2330	1991
Sagarmatha (Mt. Everest)	NP	Eastern Himalaya	1148	1976
Langtang	NP	Central Himalaya	1710	1976
Annapurna	CA	Western Himalaya	7000	1988
Dhor Patan	HR	Mid-west Himalaya	1325	1984
Rana	NP	Mid-west Himalaya	106	1976
Shey Phoksundo (Dolpo)	NP	Mid-west Himalaya	3555	1984

Other sections gave complete protection to Schedule I (protected) species Table 13.2, but stipulated that mad elephants or man-eating tigers, both protected species, may be destroyed. The use of a weapon against any wild animal, protected or otherwise, if it is in the process of attacking a person or domestic animal, was also permitted. The Act also permitted the also listed species which may be harvested for other purposes (e.g., trophy hunting), or collected for scientific purposes under Schedule Table 13.2.

Some section of the Act allowed the His Majesty's Government to fix quotas, grant licenses, impose closed seasons and restrict the acquisition, sale, or transfer of trophies of species which can be legally hunting (under Schedule II). Many sections of the 1973 Act dealt specifically with hunting legislation. Other sections empowered wildlife

TABLE 13.2

Protected (Schedule I) species in Nepal, as Stipulated in the 1973 National Parks and Wildlife Conservation Act (HMG) 1977) and Subsequent Amendments. The IUCN Listings are from IUCN (1990b) and the CITES Listings are from Favre (1989) and Jackson *et al.* (1990). IUCN Listings are: Endangered (E), Vulnerable (V), Rare (R), Indeterminate (I), Insufficiently Known (K), and Threatened (T). CITES Appendices are: I (Threatened with Extinction), II (Could become Endangered by Trade), and III (Protected by Individual Countries)

Species	*IUCN Listing*	*CITES Listing*
	Mammals	
Indian Rhinoceros	E	I
Wild Elephant	E	I
Wild Yak	E	I
Wild Buffalo	E	III (Nepal)
Gaur (Indian Bison)	V	I
Tibetan Antelope	—	I
Great Tibetan Sheep (Nayan)	I	I
Four-horned Antelope	—	III (Nepal)
Blackbuck	—	III (Nepal)
Swamp Deer	E	I
Musk Deer	V	I
Pigmy Hog	E	I
Hispid Hare	E	I
Gangetic Dolphin	V	I
Assamese Monkey	—	I
Linsang	—	I
Red (Lesser) Panda	K	II
Striped Hyena	E (N Africa)	—
Wolf	V	I (S Asia)
Brown Bear	—	II
Leopard Cat	—	(S & SE Asia)
Clouded Leopard	V	I
Lynx	—	II
Snow Leopard	E	I
Tiger	E	I

Contd....

Species	*IUCN Listing*	*CITES Listing*
	Birds	
Cheer Pheasant	V	I
Impeyan Pheasant	—	I
Crimson-horned Pheasant	—	III (Nepal)
White Stork	R	I
Black Stork	—	II
Bengal Florican	E	I
Lesser Florican	E	—
Sarus Crane	—	II
Giant Hornbill	—	I
	Reptiles	
Gharial (Gavial)	E	I
Indian Python	V	I
Yellow Monitor	I	I

officers to inspect and search, arrest without warrant, investigate, and confiscate illegally obtained wildlife products or weapons. Severe penalties were stipulated as per the Act. According to the Act offences are punishable by fines and/or jail sentences of upto 5 years depending on the severity of the offence. These sections empowered the wildlife officers to investigate, try, convict, and punish offenders of wildlife legislation. As per the Act the soldiers of the Royal Nepali Army were to protect the parks and reserves. Battalions are currently posted in most Nepali protected areas. With the enactment of the 1973 Act the Department of National Parks and Wildlife Conservation (DNPWC) was established. This was the first comprehensive legislation ever enacted with respect to wildlife in Nepal. It set in conservation of wildlife in Nepal.

Preliminary Regulations of the 1973 Act

Nepal's first protected area was the Royal Chitwan National Park. The first set of rules passed under the 1973 Act formalised the boundaries of the park and set the entry fees. Of course later the boundary was extended to cover more areas and the entry fee cunsent was also increased since then.

TABLE 13.3

Species which may be Harvested (Schedule 2) in Nepal. Fees are given in the Schedules (not Given Here) for Harvesting Individual Species for Hunting or Scientific Collection. Nepalese Citizens are Subject to Cheaper Hunting Fees, and no Fees for the Harvest of Non-game Species, Since Non-game Fees Cover only Scientific Collection

Large Game Mammals	
Himalayan Black Bear	Blue Bull (Nilgai)
Wild Boar	Serow
Blue Sheep (Naur, Bharal)	Hog Deer
Spotted Deer	Himalayan Tahr
Barking Deer	Goral
Small Game, Furbearers, and Non-game that may be collected for scientific purposes	
Mammals	**Birds (including eggs)**
Rhesus Monkey	All Eagles
Wild Dog	All Cranes
Wild Cats	Pheasants (except those on Schedule I)
Langur	
Jackals	All Vultures
Foxes	All Owls
Martin	Pied Hornbills
Marmots	All Geese
Hedge Hogs	All Hawks
Badgers	All Falcons
All Bats	All Ducks
All Squirrels	All Quail
Mongooses	All Passerines, Cuckoos, and Parakeets
Moles	
Voles	
Shrews	**Reptiles, Fish and Insects**
Rats and Mice	Marsh Crocodile and eggs
	All Turtles and eggs
	All Snakes (except python)
	All Lizards (except yellow monitor)
	All Fish
	All Insects

As per the regulations of the Act building of residences, cultivation, deforestation, grazing, making five and carrying weapons in the park was restricted. Even nets, poisons and baits, the items which could injure or kill the animals were also prohibited in the park. While the assistant wardens (gazetted third-class officers) were authorized to hear and decide cases of illegal removal of forest products, illegal fishing, entry without a permit, grazing of livestock, or camping in non-designated areas, the regulations of the Act empowered the Wardens (gazetted second-class officers) were given the power to hear cases regarding the destruction of boundary fences, poaching, and placing or spreading toxic materials. Whereas the cases of lesser offences were under the jurisdiction of the latter. As per rules published in 1979 the capture of wild was allowed. boar (HMG, 1979a). Later on it was announced that if a boar had come into agricultural areas it could be captured or killed by landowners without a license.

Amendments Passed

Subsequently, new rules for wildlife reserves were published (HMG, 1977b). They were almost similar to the ones published earlier for the Chitwan National Park. The only exception being the visitors were not allowed in the wildline reserves until an amendment was passed (HMG, 1982) and rules for the amendment was published (HMG 1985a). The entry fees were also increased to NRS 200 for Chitwan National Park (HMG, 1989a).

Poaching and catching of diseased or injured animals were allowed under the 1982 amendment, of course with the permission of the wildlife authorities. It further clarified that people might kill animals in self defense. The amendment also increased the fine and punishment for poaching tiger or musk deer and for various wildlife offences too. Section 16 of the 1973 Act stipulated that park or reserve authorities may allow for the removal of natural produce if it was considered important for the management of the area. This section also led to the legal removal of thatch grasses in the terai reserves for two weeks per year. Section 16 legalized the facility zones in Himalayan reserves which were to provide grazing, fuel wood, and timber to local residents.

The Himalayan National Park Regulations (HMG, 1979b) required visitors to be self-sufficient in fuel, but this has been frequently circumvented by individuals or small groups of trekkers who stay in locally-owned lodges because the regulations also allowed local residents use of firewood. The extent of this practice is not known and research is warranted, but the rules are enforced in the case of large, organized trekking parties or expeditions. The regulation also instructed that garbage is to be disposed of at the instructed places only. Facility areas for grazing and fuel wood collection by local residents might be specified by the warden.

Except three smallest parks—Shivapuri, Rara and Khaptad—rest of the mountain parks are inhabited. Although this is technically not legal, the letter of the law is met because villages and their surrounding agricultural holdings are excluded from (zones outside of) the park areas. Practically it was not feasible to remove the thousands of residents who live inside Langtang or Sagarmatha National Parks, for example. Unfortunately, deaths of some of the relocated residents occurred when they were removed from Rara. (Furer-Haimendrof, 1986).

A fifth type of protected area in Nepal (HMG, 1989b), was allowed to be created by the amendment of the 1973 Act. This would approximately correspond to IUCN Category V (Managed Landscapes and Seascapes; IUCN, 1990). This amendment was in response to a proposal for the Annapurna Conservation Area (Sherpa *et al.*, 1986; Rana, 1990). It represents another major advancement in Nepali conservation law. The Annapurna project allows for much more local participation and local utilization of resources than is the case of other types of reserves (Hough and Sherpa, 1989).

The Fourth Amendment was passed in 1992. It allows a part of tourist revenues to go directly into local community development for the purposes of managing buffer zones. The newest Amendment to the 1973 Act addressed issues related to the designation and management of buffer zones around national parks and reserves in Nepal. The Department of National Parks and Wildlife Conservation conducted several meetings with wardens of different management units to address site-specific issues of buffer zone management. This is an important provision due to the extensive nature of park-people conflicts in and around many of Nepal's protected areas (e.g., Heinen, 1993; Nepal and Weber, 1993; Sherpa, 1993). Based on the population density and needs, the buffer zone management is expected to vary. greatly in different areas. Research is needed on the implementation of this potentially innovative and progressive programme.

Hunting Licenses

The Third Amendment Rules (HMG, 1985b) revised the hunting rules which were published soon after the implimentation of the 1973 Act. (HMG, 1974b). Hunting is allowed in Nepal from mid-October to mid-April, though closed seasons may be imposed locally. Four types of hunting licenses can be issued in Nepal: district, general, bird, and supplementary. District licenses are for Nepali citizens only. The are sold at District Forest Offices and are valid for 10 days. They allow hunters to shoot a maximum of three mammals and ten birds, with quotas given for individual species. Several small species— rabbits, pheasants, doves, peafowl, duck and a number of larger mammals, wild boar, spotted deer may be hunted with district licenses.

General licenses are issued by the Department of National Parks and Wildlife Conservation. These licenses are valid for 15 days in the terai and 21 days in the mountains. Hunters are permitted to hunt up to four mammals and ten birds. They are allowed to hunt only in the areas that are specified in the licenses, with quotas given for individual species. A few mammal species may be hunted with general licenses in addition to those permitted under district licenses. Bird licenses may be purchased for the bird species and quotas permitted under the general license.

Along with a general license anyone can purchase a supplementary lincense. It is valid for 9 days in addition to the time stipulated on the general license. In excess of the general license quota any mammal species listed on the general license may be hunted with a supplementary license by paying the appropriate fee. Several other large game species such as wild sheep and goats may only be hunted with a supplementary license. Registration fee is required to reserve a block of forested land for the use of the hunter. It is in addition to the regular fes listed per animal. All trophies are to must be submitted to DNPWC for inspection. A certificate is then issued to the hunter indicating the animal was legally obtained.

International Convention

Nepal was among the first country to sign CITES (the Convention on International Trade in Endangered Species of Wild Fauna and Flora, Favre, 1989) in 1975. DNPWC is the management authority for Nepal. It has served twice on the standing committee representing the Asia region. The country has participated in all General Assembly meetings and standing committee meetings. Despite being abide by the CITES legislationl Heinen and Leisure (1993) found during the period of their study that there were 76 shops in Kathmandu selling illegal wildlife products to foreign tourists. It was a violation of the law. This issue was raised at a July 1994 meeting sponsored by the Nepal Forum for Environmental Journalists and World Wildlife Fund-Nepal and has received quite a bit of recent press in the country (Anonymous 1994).

His Majesty's Government has no bilateral migratory bird treaty although about 30 per cent of Nepali's bird species are migratory. However, Nepal and other countries in the Indian subcontinent agreed to take part in the international waterfowl census sponsored by the International Waterfowl Research Bureau (IWRB), based in Great Britain (IWRB, 1987). Nepal has no formal agreements for the management of international reserves though several exist. To begin with late Rajiv Gandhi, then a Member of Parliament, was requested to establish the Valmiki Sctuary. A 460 sq km reserve this sanctuary was located on the Indo-Nepalese border in the Indian State of Bihar, adjacent of Chitwan National Park. As part of the Heart of the Himalaya Conservation Programme recently the Chinese Government has recently created a new reserve in the Autonomous Region of Tibet. It is sponsored by the USA-based Woodlands Mountain Institute. This reserve borders Sagarmatha, Langtang, and Makalu-Barun National Parks in Nepal (Jackson *et al.,* 1990).

The secretariat of the South Asia Association for Regional Cooperation (AARC) is located in Kathmandu. Besides Nepal and India five other South Asian countries are members of this organization. They are—Bangladesh, Bhutan, the Maldives, Paleistan and Sri Lanka. Though primarily Concerned with economic and political matters, a number of meetings of this organization have focussed on the conservation problems including water and soil erosion (Verghese, 1985). Pradhan (1989) suggested that SAARC should also deal with major ecological issues like forestry, parks wildlife conservation etc.

Nepal is a State Member of the World Conservation Union (IUCN). It regularly sends participants to IUCN General Assemblies. Besides that, Nepal is also host to the eight nation-member International Center for Integrated Mountain Development (ICIMOD), which begun in 1983. ICIMOD has had many accomplishments to date, including co-sponsoring a seminar on Himalayan protected areas (KMTNC/ICIMOD, 1985). In 1978 Nepal became a party to the World Heritage Convention (Hales, 1984). Chitwan and Sagarmatha National Parks are the two listed Nepali World Heritage Natural Sites. For manpower development in Sagarmatha National Park and alternative energy projects World Heritage Trust funds were used under this agreement (Jeffries 1984).

Parks and Reserves of the Nepal Himalaya

The majority of Nepal's land area comprise the Middle hills. Still only two small protected areas are located entirely within these areas. However, it should also be pointed out that all of the Himalayan parks contain some areas below the tree line, especially along river valleys, and therefore afford some protection to flora and fauna associated with the Middle hills region. Much of the land area within Nepal's protected area system is at very high elevations, hence is relatively unproductive from a biological point of view. Some consequences of this are discussed later. In this section, each park and reserve found in mountainous regions of the country is profiled on a case by case basis. The first two listed are reserves that fall mostly in the Middle hills zone, and the next seven fall primarily in the High Mountain and High Himalayan zones as described earlier.

Shivapuri Wildlife Reserve

The Shivapuri Wildlife Reserve covers an area of 114 sq km., Located along the northern rim of Kathmandu Valley this reserve is managed by the Department of Soil Conservation. Several watershed projects had been underway in Shivapuri since 1976 as the area supplies the bulk of the drinking water for Kathmandu Valley (Carson, 1985). There is a monastery located within Shivapuri. In 1976 the area was declared a wildlife reserve but management has not yet changed to DNPWC. Due to deforestation and overgrazing in the area, there was concern that the quality of the city's water supply was in damages. Of course, firewood collection and grazing have been controlled, and forests are regenerating after the creation of Shivapuri reserve. Wildlife recorded in the area include leopard *(Panthera pardus)*, Himalayan black bear *(Selenarctos thibetanus)*, wild boar *(Sus scrofa)*, barking deer *(Muntiacus muntjak)*, and many species of birds. As the forests continue to regenerate it is expected that avian diversity and abundance should increase.

As the reserve is near to Nepal's largest population centre, it has great potential for attracting both foreign and national visitors, especially for the purposes of conservation education. A visitor's centre aimed at Nepali nationals would be a great asset in the reserve because it is possible for school groups to visit the area on a daily basis from any part of Kathmandu Valley.

Khaptad National Park

Established in 1985, Khaptad is located. In the Middle hills of the Far Western Development Region is located the Khaptad National Park. Established in 1985 it covers 225 sq km area with head quarter at Khaptad Lekh an elevation of about 3,000 m. Side slopes around the Lekh and within the park descend to elevation of about 1,850 m. Very few foreign tourists visit Khaptad. The most important feature of the area is the Ashram of the Khaptad Baba. Believed to be a Hindu Saint, the Baba has lived in the area for over 40 years. It is said that this park was created at his suggestion. A small lake is also located there. It is considered to be a sacred one. Forests of the region are protected in large part by religious devotion to the area. The main management problem reported in Khaptad is its religious significance which effectively protects it from other forms of encroachment such as poaching or wood cutting. Many medicinal herbs are found in the park, and a herb farm was formerly located on Khaptad Lekh.

Wildlife recorded in Khaptad include leopard, Himalayan black bear, wild dog *(Cuon alpinus),* wild boar, musk deer *(Moschus moschiferus),* goral *(Nemorhaedus goral),* Himalayan tahr *(Hemitragus jemlahicus),* and as well as a host of bird species. Inhabitants of the area graze livestock on Khaptad Lekh; this practice has not been studied, and there are plans to restrict it. A research programme began in 1994 in which uses of biotic resources in and around Khaptad for local markets are being surveyed and monitored.

Makalu-Barun National Park/Conservation Area

One of the unexploited regions in the mountains of Nepal is Baruun Valley. It is uninhabited too. Covering an area approximately 400 sq km, Barun Valley is located at the east of Sagarmatha National Park. It is biologically important area. A preliminary wildlife survey listed 131 species of birds including two new to Nepal. One of which had not been recorded in the country since 1846. Fourteen other rare species of birds were recorded, as were 25 species of mammals. Among the latter were five protected species: Assamese monkey *(Maccaca assamensis),* red panda *(Aillurus fulgens),* musk deer, great Tibetan sheep *(Ovis ammon hodgsoni),* and snow leopard *(Panthera uncia).*

The report (Taylor-Ide, 1985) proposed complete protection for the area, as a National Park (Taylor-Ide and Shrestha, 1985). A seminar on the region was also held that year (KMTNC/ICIMOD, 1985). The Barun Valley contains some of the last remaining cloud forests in Nepal. The most recent proposal called for the creation of an integrated National Park and Conservation Area; it allows for the designation of facility zones in which many human uses are permitted, and other more strictly protected zones.

The Chinese have recently established Qomolangma Nature Preserve, adjacent to both Sagarmatha and Langtang National Parks, in the Tibetan Autonomous Region. It spreads over an area of approximately 25,000 sq km. The total protected land area within the region (including Makalu-Barun) is more than 30,000 sq km. Legislation to designate Makalu-Barun was passed in late 1991, and implementation began in 1992.

Sagarmatha (Mt. Everest) National Park

The Sagarmatha National Park was established in the year 1976. In 1980 it was declared a World Heritage Natural Site (IUCN, 1990). The park is 1,148 sq km in area with the headquarters located in Namche Bazaar at an height of 3,500 m in Solu Khumbu District. From STOL airstrip at Lukla it takes almost a day to reach there. The ticket office of this park is located in Jorsale, a four-hour walk from Lukla. Much of the land are within Sagarmatha is above the tree line. Many sites of religious importance are found in Sagarmatha. There are monasteries too at Tengboche and Thame. It is the most visited mountain national park in Nepal. More than 10,000 foreign visitor visited Sagarmatha in recent years. This park is inhabited by about 2,500 people, majority being Sherpa. They are a group of Tibetan origin which lived for centuries as shepherds and long distance traders (Jeffries and Clarbrough, 1986). Theirfucl requirements and that of the foreign tourists have adversely effected the existing forests there (Sherpa, 1993).

Important wildlife include Himalayan tahr and black bear, musk deer, red panda, wolves *(Canis lupus)*, and wild dogs. Snow leopards may occur in the park, but there are no recent sightings. About 120 species of birds have been recorded. Recent surveys suggest that populations of both Himalayan tahr and musk deer have increased since the area has been protected (e.g., Kattel, 1987, Kattel and Alldredge, 1991). Several projects have been designed to improve the environmental situation in Sagarmatha and the lives of its resident population under the patronage of the Himalayan Trust, the Government of New Zealand, and the World Heritage Trust (Coburn, 1982, 1985; Jeffries and Clarbrough, 1986).

Langtang National Park

Langtang National Park was established in 1976 with an area of 1,710 sq m. Langtang was recognized as an important protected area by Caughly (1969). The lower slopes are as low as 450 m. It is dominated by subtropical sal forests, and several peaks are over 7,900 m tall. Approximately 25 per cent of the area is forested, and slightly over 30 per cent is under permanent ice and snow. The rest is composed mainly of alpine grass and scrub (Borradaile *et al.*, 1977). It is the only mountain park with direct road access from Kathmandu. This accessibility makes Langtang more important for tourism. There is great ethnic diversity within this region. The important religious sites in Langtang include the monasteries at Sing Gompa and Kyanjin, and the sacred lake at Gosaikund situated at a height of 4,350 m. To exclude village areas from the park and provide facility areas for local residents for grazing and fuel wood, like Sagarmatha Langtang is also zoned Permits are issued to the local residents to cut timber. His Majesty's Government operates two cheese factories in Langtang (Carter, 1987). As a result localized deforestation is caused which effects and disturb to wildlife, especially the endangered red panda (Yonzon and Hunter, 1991). Major wildlife species include leopard, musk deer, Himalayan tahr and black bear, wild dog, red panda, barking deer, goral, serow *(Capricornis sumatraensis)*, wild boar, and several hundred species of birds. The endangered snow leopard and clouded leopard *(Neofelis nebulosa)* were also recorded in the original survey (Borradaile *et al.*, 1977). Five strict nature reserves, all in inaccessible areas, have been declared inside of Langtang.

Annapurna Conservation Area

The Annapurna region contains the major peaks of Annapurna I,

II, III, and IV, Annapurna South, the Nilgiri Himalaya, Tilicho, Lamjung, and Machapuchre. It is considered to be one of the most important areas for tourist trekking in the world. Encircling this mountain system there is a 320 km trekking route. That route passes through six districts in Nepal. Over 50,000 trekking permits have been issued annually in recent years. The trekking route passes along the south side of the Himalaya in one of the highest rainfall areas in Nepal, and along the north side in a high altitude desert. This climatic diversity has caused a great deal of cultural and biotic diversity. Many of the peaks located within the circuit are over 7,600 m in elevation. Muktinath, which is considered to be a sacred place for both Buddhists and Hindus, is situated at 3,650 m in the north-west part of the Annapurna region.

Besides a national park (Bunting and Wright, 1985), the Annapurna region has beem proposed as a Recreation Area (Sakya, 1982), and the plan under implementation is a Conservation Area (Sherpa *et al.*, 1986; Rana *et al.*, 1990), approximately corresponding to IUCN Category V (IUCN, 1990). The entire area of 7,000 sq km, is zoned into the following categories: wilderness zone, special management zone, protected forest, seasonal grazing zone, intensive use zone, and biotic/ anthropological zone. The management plan calls for maximum participation and management of the area by local people (Hough and Sherpa, 1989). It is estimated that revenues from tourism will completely support management activities. Under the Annapurna Conservation Area Project (ACAP) the reserve is managed by KMTNC. ACAP is an integrated project and it concentrates on five different areas: forestry and wildlife, alternative energy development, community health, conservation education, and tourism (Jain, 1994).

Dhor Patan Hunting Reserve

The only hunting reserve in Nepal is Dhor Patan Hunting Reserve with an area 1,325 km. The area was initially surveyed by Wegge (1976a), and the reserve was established in 1984. The headquarters is located in Dhor Patan Valley at 2,850 m in height and extends north to a height more than 5,500 m. Few tourists visit the area. It is most famous for blue sheep *(Pseudois nayaur,* Upreti, 1977). Goral, serow, Himalayan tahr, barking deer, and wild boar are also hunted here. Dhor Patan is operated on a block system which contains a total of 7 locks which are registered to two guides who specialize in blue sheep hunting and have mostly foreign clients. Hunters are subject to quotas on individual animal species. Not more than 20 licenses are issued for the entire area to any one.

The field staff of the Department of National Parks and Wildlife Conservation in Dhor Patan do not check bag limits that is a major problem with the regulations. Licensing is conducted at the main office in Kathmandu. After a party has done hunting, the guide fills in a standard form indicating how many of each species were harvested. This system has been subject to abuse. It is unlikely that any large game populations are currently over-exploited in Dhor Patan, however, and some could probably tolerate heavier hunting pressures (Heinen and Kattel, 1992a). Other species recorded in Wegge (1976a) were lynx *(Felis lynx)*, red panda, musk deer, wild dogs, and snow leopard. The major management problem is grazing by domestic animals during the rainy season. The management is unable to control it even today. As prevailing in some other Himalayan protected areas, here also, with due prior permission, residents may collect dry fuelwood and cut timber for their personal use.

Rara National Park

Covering an area of 106 sq km the Rara National Park is the smallest reserve in Nepal. It was established in 1976 and can be reached in a two to three day walk from Jumla. It is connected with Kathmandu and Nepalgunj with regular air service. Rara is important for its scenery. It contains lake Rara, which at 10 sq km in area, is Nepal's largest lake. The lake is situated at a height of 3,050 m, and the park extends to a height over 3,950 m. Two villages were previously located in Rara. Total population was 650 people. The remainings of these villages are still found in Rara. This was considered inconsistent with the goals of the park (Bolton, 1976b), and the residents were relocated in Bardia District in the terai. (Furer Haimendorf, 1986). Though Government has provided a guest house, not many tourists visit Rara.

Mammals recorded in the park include Leopard, black bear, wild dog, musk deer, goral and serow are found in Rara National Park (Bolton, 1976). Many species of waterfowl visit the lake during spring and fall migrations. The vegetation includes high elevational forests of scrub and alpine meadows.

Shey Phoksundo National Park

Shey Phoksundo National Park is located in Dolpo and Mugu

Districts in the Midwestern Development Region of Nepal. Proposed as a park in 1980 (Kattel, 1980) and finally established in 1984, this park conversion area of 3,555 sq km. The headquarters is located near Phoksundo Lake, about a two-day walk from the village of Shey, one of the field sites used by Schaller (1977) in his classic Himalayan wildlife study. The park is located mostly within the trans-Himalayan region of the country. On the south it is bordered by Daulagiri Himalaya. Due to the rain shadow effect of the high peaks, it is physiographically more similar to the Tibetan Plateau (Sakya, 1978). The region is sparsely populated by ethnic groups of Tibetan origin. Because Dolpo is a food-short area, and because of historical political disputes, most of the region was restricted for foreigners until 1989. Several important Buddhist religious sites are located within the park. There are monasteries at Shey, Yangir, and Samling, as well as Shey Mountain, sacred to Dolpo Buddhists.

The western side of the park was used as a field site for Jackson and Ahlborn's (1986) study of snow leopard, the first long term field study on the species. Shey Phoksundo also has many other species characteristic of the Tibetan Plateau including wild yak *(Bos grunniens)*, great Tibetan sheep, Tibetan antelope *(Pantholops hodgsoni)*, and brown bear *(Ursus arctos)*. Of course the presence of these species has not been confirmed in recent years. But much of the area has not been systematically explored. Because of the known and possible presence of many rare species, Shey Phoksundo is an extremely important protected area. The other important wildlife in the area include blue sheep, Himalayan tahr, black bear, goral, serow, leopard, wolf, musk deer, wild dog, wild boar, and snow leopard. General wildlife surveys are urgently needed. The preliminary one was completed by Yonzon (1990). But due to its restricted status at that time Jackson and Ahlborn (1986) were not granted permission to conduct a regional survey in the area.

New Hunting Reserves in Nepal Himalaya Proposed

Within the vicinity of the existing wildlife reserves and parks five more hunting reserves, in addition to the Dhor Patan, were proposed in 1970s and 1980s to harvest the surplus animals without putting the existing ones in other reserves in risk. But unfortunately the proposal did not materialize for various reasons (UNDP/FAO 1973 and 1980).

After the survey of Langtang National Park, first by Caughly

(1969) and later by Wegge (1976a) it was stated that seven game mammals were recorded there. They are: goral leopard, black bear, tahr, serow, barking deer, and wild boar. The endangered musk deer was also recorded. Besides all these, evidence of the presence of snow leopard was found. However, the most popular species in the Nepal Himalaya is blue sheep. It is common in Dhor Patan but absent in Rasuwa. All districts containing national parks and wildlife reserves were later declared closed to hunting, and Rasuwa Hunting Reserve was never established.

At the time of this writing, another proposal has been circulated for the establishment of a protected area in the Kanchanjunga region of eastern Nepal (D. Miller *pers. comm.*, USAID, Kathmandu).

Kanchanjunga forms part of the border between Nepal and Sikkim. It is the worlds third highest peak. The area has great possibility for attracting foreign tourists. It was proposed to establish a protected area in the Kanchanjunga region of eastern Nepal (D Miller Pers Comm., USAD, Kathmandu). No action has yet been taken on this proposal. But at a time when tourist entries are increasing into the country, it is essential to give importance to the high mountain environment in other Himalayan reserves. By expending trekking opportunities into other areas such pressure on heavily visited site such as Sagarmatha and Annapurna can be reduced.

Discussion: Future Needs of Research and Management

Ives and Messerli (1989) identified several important. Heinen and Yonzon (1994) suggested that upto a great degree the problems regarding research on watersheds in the Himalaya.are connected to biodiversity conservation research in Nepal. Ives and Messerli (1989) also identified many important problems. For example, research and management project to date have been generally narrowly defined, limited in scope, and limited in duration. Most wildlife studies have been done in one National Park. Thus most effort so far have focused on single species, especially large mammals, for less than three years' duration. For designing long-term biodiversity conservation plans the limitations of such programmes may have serious impacts. These observations suggest several priorities for biodiversity conservation in Nepal within the existing legislative and institutional framework. They were first presented by Heinen and Yonzon (1994) and are only briefly outlined here.

1. Developing a comprehensive biological inventory and monitoring programme for all lands under the conrtrol of the Department of National Parks and Wildlife Conservation and the Department of Forestry with the intent of developing a national data base. The first priority for the conservation of biodiversity in Nepal. Though various branches of His Majesty's Government have at times addressed inventories of selected species in certain areas, the information generated thus so far is insufficient for long-term conservation needs for several reasons. There is little detailed information on most areas, and essentially no long-term surveys judge the general features of species diversity by any measure. In view of the current rapid human population growth (HMG, 1987a and b and 1991), the dependence of rural people on natural products for foods, building materials and medicines (e.g., Lehmkuhl *et al.*, 1988), and the growth of tourism (Kayastha, 1985), the need for such work is most necessary.
2. The second priority for biodiversity conservation in Nepal is a critical review and reconsideration of all species listed on Schedule II of the National Park and Wildlife Conservation Act of 1973. Heinen and Yonzon (1994) suggested that many Schedule II species should be fully protected according to what is known about their status and distribution in the country and in the region as a whole.
3. To identify and conserve particular important and unique habitats, some of which may be rather small in area is the third priority. The Department of National Parks and Wildlife Conservation has traditionally concentrated on large mammals and relatively large areas and important features only. It has not paid proper attention to protect some important unique habitats that have high biological diversity of groups other than large mammals (e.g., remnant broad-leaved forests in the eastern Himalaya, Inskipp, 1989). A good deal of the land area within Himalayan reserves in Nepal is above the snowline (e.g., 30 per cent of Langtang), and is thus biologically unproductive. There is a great need to develop more protected areas at intermediate elevations (Hunter and Yonzon, 1993). The Eastern Himalaya are recognized as one of the conservation 'hot spots' of the

world (Myers, 1988 and 1990; Dinerstein and Wikramanayake, 1993), which are areas of high biodiversity and high rates of habitat destruction.

4. Suitable study to evaluate reintroduction possibilities for particular species is the another priority for biodiversity conservation. As Heinen and Yonzon (1994) showed there have been several successful reintroductions in the terai region of the country, but many more could be planned. There may be less need in the mountainous parts of Nepal, although many wildlife populations are known to be depleted, there are few actual extirpations. Thus, habitat protection (number 3, above) may be more important in the context of mountainous regions of the country.

With His Majesty's Government of Nepal the Government of the Netherlands has recently begun a project that will develop a biodiversity classification system for the country. It will test the representativeness of the protected area system for its geographic and physiographic coverage, (UNDP/HMG, 1993). This approach is referred to as gap analysis (Scott *et al.* 1991). This will allow an objective assessment of where additional reserves may be needed from a biological standpoint. This is the first nationwide attempt to do so in Nepal, and clearly meets priority 3, above.

For Makalu-Barun National Park/Conservation area the United Nations Development Programme (UNDP/HMG, 1993) has proposed a conservation development project. This project concentrates mainly on local development in the context of reserve management. The reserve includes areas ranging from heights of 435 m to several peaks above 8,000 m, and includes some of the last remaining cloud forest in Nepal (Anonymous, 1990), thus it addresses priority 3. The proposal is also preliminary, and there are plans to address further priorities 1 and 3 for the eastern region of the country, by biological consultants over the 5 year duration of the project. Biodiversity inventories are underway in Makalu. Barun and Annapurna (Dr. P.R. Sakya, *per. comm.* Woodlands Mountain Institute). The final results are awaited.

Besides the large projects, there are many smaller actions initiatives begun or proposed that address some aspects of the listed priorities. Yonzon (1993) has begun the most comprehensive biological inventory and monitoring scheme in Nepal to date. The project focuses on Shivapuri Wildlife Reserve. It includes many of the staff of the Natural History Museum. Researchers have inventorised species diversity for many groups of vertebrates, invertebrates, and plants, and have done extensive interviews with local people about their uses of various forest products. Heinen (1994) proposed to list out the sacred forests of Nepal to develop a national data base and rapid ecological monitoring scheme for such sites that could be used throughout the country. Gadgil (1991) and others have discussed the potential importance of sacred forests for conservation in Asia, but they have not been systematically studied in Nepal to date and these sacred forests may be important in maintaining populations of some plants and animals. These two initiatives, although currently narrow in scope, could be expanded and clearly address aspects of priorities 1 and 3.

Heinen and Yonzon (1994) discussed many species that should be moved from Schedule II to Schedule I. Other work (e.g., Sharma, 1982; Heinen and Leisure, 1993) points to gaps in law enforcement with regard to certain species harvested for hunting and/or sold by the fur industry, outside of the existing system of parks and reserves. Progress has been made on part of this issue in the July 1994 meeting on Illegal Wildlife Trade in Nepal. This meeting was sponsored by the Nepal Forum for Environmental Journalists (Anonymous, 1994). Besides focusing on the illegal fur trade in tourist markets in Kathmandu, the meeting also discussed illegal trade in tiger *(Panthera tigris)* and rhino *(Rhinoceros unicornis)* parts. The larger issue to address is how many species that are currently harvestable under Schedule II should be completely protected under Schedule I. There has been no broad discussion of this issue to date. Trophy hunting is no longer permitted for several Schedule II species like leopard, Himalayan black bear and sloth bear through these could be harvested under license until the late 1980s.

Conclusions

Since the passage of the 1973 National Parks and Wildlife Conservation. Nepal has made relatively rapid progress in the conservation sector. As a result protected areas have been created that now cover over 15 per cent of the land area of the country. This has also heeped in the recovery of numerous species within parks and reserves. Several concerns are also evident that there are park-people conflicts in most management units in Nepal. The status of many species outside protected areas is also uncertain. Much of the land area within the current reserve system being above snowline, it is biologically unproductive. Major needs for the future include the extension of biodiversity inventory and monitoring programmes for large areas of the country, and the creation of additional reserves in physiographic areas that are currently under-represented in the protected area system. In addition, a great deal of information suggests that conservation law must be implemented expecially in cases in which external markets threaten some species. Besides that it should also be implemented in areas outside of national parks and reserves. There has been some progress made on most of these issues in recent years. The new buffer zone legislation is particularly encouraging in this context. It may help to reduce some pressure from parks and reserves. At the same time it provides benefits and incentives to local residents to maintain protected areas, thus improving park/people relations. In addition, the recent (July 1994) attention paid to issues of illegal wildlife trade outside the protected areas is also encouraging. It led to greater awareness of the issue both inside and outside His Majesty's Government. As Nepal enters the 21st century with rising affluence, rising populations, and rising numbers of foreign tourists. Continual research and monitoring of progress in the conservation methods and increasing awareness of the issue are the needs of the day.

References

Anonymous, (1990), *The Makalu-Barun National Park and Conservation Area Management Plan,* DNPWC/Woodlands Mountain Institute, Kathmandu.

Anonymous, (1994), *Wildlife Conservation in Nepal* (Nepali text). Report of the Wildlife Watch.

Blower, J.H., (1971), *Proposed National Park in Khumbu District.* Unpubl. Report to the Secretary of Forests, Kathmandu Group of Nepal Forum for Environmental Journalists, Kathmandu.

Bolton, M., (1976), *Lake Rara National Park Management Plan.* FAO/UNDP National Parks and Wildlife Conservation Project, Nepal, Working Document No. 3, Rome.

Borradaile, L.M. Greene, Moon, L., Robinson, P. & Taite, A., (1977), *Langtang National Parks Management Plan.* FAO/UNDP National Parks and Wildlife Conservation Working Document No. 7, Rome.

Carson, B., (1985), *Erosion and Sedimentation Processes in the Nepalese Himalaya.* ICIMOD Occasional Paper No. 1, Kathmandu.

Carter, J., (1987), *Organizations concerned with forestry in Nepal.* ICIMOD Occasional Paper No. 3, Kathmandu.

Caughly, G., (1969), *Wildlife and Recreation in the Trisuli Watershed.* Trisuli Watershed Development Project Report No. 6, Kathmandu.

Coburn, B, (1982), Alternative energy sources for Sagarmatha National Park. *Parks* 7(1): 16-18.

Coburn, B., (1985), Energy alternatives for Sagarmatha National Park. In: *People and Protected Areas in the Hindu Kush Himalaya, ed.,* Anonymous, KMTNC/ICIMOD, Kathmandu, pp, 71-72.

Dinerstein, E. & Wikramanayake, E. (1993). Beyond 'hot spots': How to prioritize investments to conserve biodiversity in the Indo-Pacific Region. *Cons. Biol.* 7(1): 53-65.

Favre, D.S., (1989), *International Trade in Endangered Species.* Martinus Nijoff. Boston, MA.

Furer-Haimendorf, C. von, (1986), Foreward: Development with a human face. In: *Bikas/Binas: Development and Destruction, eds.,* Tuting, L. & Dixit, P., Geobuch, Munich, pp. 11-14.

Gadgil, M. (1991), Conserving India's biodiversity: the societal context. *Evolutionary Trends in Plants 5* (1): 3-8.

Hales, D.F., (1984), The World Heritage Convention. In: *National Parks, Conservation, and Development, eds.,* McNeely, J. & Miller, K.R., Smithsonian Institution Press, Washington, DC. pp. 67-74.

Heinen, J.T., (1993), Park-people relations in Kosi Tappu Wildlife Reserve, Nepal: A socio-economic analysis. *Environmental Conservation* 20(1): 25:34.

Heinen, J.T., (1994), Agenda for Sacred Forests. *Habitat Himalaya Newsfile,* Kathmandu. 2(1): 2-4.

Heinen, J.T. & Kattel, B., (1992a), Parks, people, and conservation: A review of management issues in Nepal's protected areas. *Population and Environment* 14(1): 49-84.

Heinen, J.T. & Kattel, B., (1992b), A review of conservation legislation in Nepal: past progress and future needs. *Environmental Management* 16(6): 723-733.

Heinen, J.T. & Leisure, B., (1993), A new look at the Himalayan fur trade. *Oryx* 27(4): 231-238.

Heinen, J.T. & Yonzon, P.B., (1994), A review of conservation issues and programmes in Nepal: from a single species focus toward biodiversity protection. *Mountain Research and Development.* 14(1): 61-76.

HMG, (1973), National Parks and Wildlife Conservation Act 2029. *Nepal Gazette* 2029: 11-28 B.S.

HMG, (1974a), Royal Chitwan National Park Regulations 2030. *Nepal Gazette* 23 (46): 1-11.

HMG, (1974b), National Parks and Wildlife Conservation Rules 2030. *Nepal Gazette* 23(47): 1-27.

HMG, (1977a), *National Park and Wildlife Conservation Act 2029 (1973).* Official English Translation. Ministry of Law and Justice, Kathmandu.

HMG, (1977b), Wildlife Reserve Regulations 2034. *Nepal Gazette.* 35(26): 1-16.

HMG, (1979a), National Park and Wildlife Conservation Regulations 2030, Second Amendment. *Nepal Gazette* 28(40) : 1-3.

HMG, (1979b), The Himalayan National Parks Regulations 2036. *Nepal Gazette* 29(21): 1-14.

HMG, (1982), National Park and Wildlife Conservation Act 2029, Second Amendment. *Nepal Gazette* 2039: 9-8, B.S.

HMG, (1985a), Wildlife Reserve Regulations 2034, First Amendment. *Nepal Gazette* 35(26): 1-16.

HMG, (1985b), National Park and Wildlife Conservation Regulation 2030, Third Amendment. *Nepal Gazette* 35: 1-16.

HMG, (1987a), *Statistical Yearbook of Nepal.* National Planning Commission, Bureau of Statistics, Kathmandu.

HMG, (1987b), Nepal Tourist Statisties, Tourism Kathmandu.

HMG, (1988), Himalayan National Park Regulations 2036, First Amendment. *Nepal Gazette* 38(31).

HMG, (1989a), Royal Chitwan National Park Regulations, First Amendment. *Nepal Gazette* 39(7): 1-5.

HMG, (1989b), National Park and Wildlife Conservation Act 2029, Third Amendment. *Nepal Gazette* 24: 1-13.

HMG, (1991), *Statistical Yearbook of Nepal.* National Planning Commission, Central Bureau of Statistics, Kathmandu.

Hough, J.L. & Sherpa, M.N., (1989), Bottom up versus basic needs: integrated conservation and development in the Annapurna and Michiru Mountain Conservation Areas in Nepal and Malawi. *Ambio* 19(8): 434-441.

Hunter, M.L. & Yonzon, P.B., (1991), Altitudinal distributions of birds, mammals, people, forests and parks in Nepal. *Cons. Biol.* 7 (2): 420-423.

Inskipp, C., (1989), *Nepal's Forest Birds: Their Status and Conservation.* International Council for Bird Preservation Monograph No. 4. Cambridge.

IUCN, (1990), *United Nations List of National Parks and Protected Areas 1990.* IUCN Publications, Gland, Switzerland.

IUCN, (1993), *United Nations List of Parks and Protected Areas.* IUCN Publications, Gland, Switzerland.

Ives, J.D & Messerli, B., (1989), *The Himalayan Dilemma: Reconciling Development and Conservation.* Routledge, London.

IWRB, (1987), *International Waterfowl Census, Midwinter Counts, Asia 1987.* IWRB Report, Slimbridge, England.

Jackson, R. & Alborn, G., (1986), *Himalayan Snow Leopard Project. Final Progress Report* Unpubl. Report to the National Geographic Society, Washington, DC.

Jackson, R., Nepali, H.S. & Sherpa, A.R., (1990), *Aspects of Wildlife Protection and Utilization in the Makalu-Barun Conservation Area.* Makalu-Barun Conservation Project Working Paper Publication Series Report 11. DNPWC/ Woodlands Mountain Institute, Kathmandu.

Jain, N., (1994), *Trekking Tourism in Annapurna Conservation Area, Nepal.* Unpublished Ph.D., Dissertation, University of Michigan, Ann Arbor.

Jeffries, B.E., (1984), The Sherpas of Sagarmatha. In: *National Parks, Conservation, and Development, eds.,* McNeely, J. & Miller, K.R., Smithsonian Institution Press Washington, DC. pp. 127-132.

Jeffries, M. & Clarbrough, M., (1986), *The Story of Mt. Everest National Park.* Cobb/Horwood Publications, Auckland, New Zealand.

Kattel, B., (1980), *Preliminary ecological survey of Shey-Phoksundo area and boundary recommendations for the establishment of Shey-Phoksundo National Park.* Unpubl. Report to DNPWC, Kathmandu.

Kattel, B., (1987), *The Himalayan Musk Deer Ecology Project.* Unpubl. Annual Report to DNPWC, Kathmandu.

Kattel, B. & Alldredge, A.W., (1991), Capturing and handling of the Himalayan musk deer. *Wildl. Soc. Bull.* 19(4): 397-399.

Kayastha, N., (1985), *Development of Tourism in Nepal for South Asia Regional Cooperation.* Center for Economic Development and Administration. Tribhuvan University, Kirtipur, Kathmandu.

KMTNC/ICIMOD, (1985), *People and Protected Areas in the Hindu-Kush Himalaya.* Proceedings of an International Workshop on National Parks, Kathmandu.

Lehmkuhl, J., Upreti, R.K. & Sharma, U., (1988), National Parks and local development: grass and people in Royal Chitwan National Park, Nepal. *Environmental Conservation* 15(2) 143-148

Myers, N., (1988), Threatened biotas: "Hot Spot' in tropical forest. The *Environmentalist* 8(3): 187-208.

Myers, N., (1990), The biodiversity challenge: expended hot spots analysis. The *Environmentalist* 10(4): 243-256.

Nepal, S.K & Weber, K.E., (1993), *Struggle for Existence: Park-people Conflict in Royal Chitwan National Park.* HSD Monograph, Asian Institute of Technology, Bangkok.

Pradhan, B., (1989), *SAARC and its Future.* Malla Press, Kathmandu.

Rana. D.S., (1990), *(ed.) Annapurna Conservation Area Project: Three Year Retrospective Progress Report,* March 1986-December 1989. KMTNC, Kathmandu.

Rana, S.J.B., Pandey, P.D. Mishra, H.R., (1986), *Conservation for Development.* Unpubl Report to KMTNC and DNPWC, Kathmandu.

Sakya, K. (1978). *Dolpo: The World Behind the Himalayas.* Kathmandu: Jore Ganesh Press. Sakya, K. (1982). *Annapurna National Recreation Area.* Unpub, Report to DNPWC, Kathmandu.

Schaller, G.G., (1977), *Mountain Monarchs.* Univ. Chicago Press, Chicago, IL.

Scott, J.M., Csuti, B., Smith, K., Estes, J.E. & Caicco, S., (1991), Gap analysis of species richness and vegetation cover: An integrated biodiversity conservation strategy. In: *Balancing on the Brink of Extinction, ed.,* Kohm, K., Island Press, Washington, D.C. pp. 282-297.

Sharma, U.R., (1982), *Proposal for the Creation of Bara Hunting Reserve.* Unpubl. Report to DNPWC, Kathmandu.

Sherpa, M.N., (1993), *Grass Roots in a Himalayan Kingdom.* In: *The Law of the Mother: Protecting Indigenous Peoples in Protected Areas, ed.,* Kemf, E., Sierra Club Books, San Francisco, CA, USA. pp. 45-52.

Sherpa, M.N., Coburn, B. & Gurung, C.P., (1986), *Annapurna Conservation Area. Nepal: Operational Plan.* Unpubl. Report to KMTNC, Kathmandu.

Shrestha, T.B., Sherpa, L.N., Banskota, K. & Nepali, R.K., (1990), *The Makalu-Barun National Park and Conservation area Management Plan.* Unpubl. Report to DNPWC and Woodlands Mountain Institute, Kathmandu.

Taylor-Ide, D., (1985), *The Barun Valley Report.* Unpubl. Report to DNPWC, Kathmandu.

Taylor-Ide, D. & Shrestha, T.B., (1985), The Makalu-Barun Park-A Proposal. In: *People and Protected Areas in the Hindu-Kush Himalaya, ed.,* Anonymous, KMTNC/ICIMOD, Kathmandu. pp. 129-132.

UNDP/FAO, (1973), *National Parks and Wildlife Conservation-Nepal.* Unpubl. UNDP/FAO Project Document, Rome.

UNDP/FAO, (1980), *National Parks and Wildlife Conservation-Nepal.* Project Findings. Unpubl. UNDP/FAO Project Document, Rome.

UNDP/HMG, (1993), *Makalu-Barun National Park and Conservation Area Project Document.* NEP/92 G31/A/01/99. Global Environmental Facility, UNDP, New York.

Verghese, B.G., (1985), River waters: Doubts hamper agreements. In: *South Asian Regional Cooperation, ed.,* Dharmdasani, M.D., Shalimar Publishing House, Varanasi, India, pp. 155-163.

Wegge, P., (1976a), *Himalayan Shikar Reserves.* FAO/UNDP National Park and Wildlife Conservation-Nepal. Document No. 5, Rome.

Wegge, P., (1976b), *Terai Shikar Reserves.* FAO/UNDP National Parks and Wildlife Conservation-Nepal. Document No. 4, Rome.

Yonzon, P. B., (1990), *The 1990 Wildlife Survey of Shey-Phoksundo National Park, Dolpo, West Nepal.* Unpublished Report to KMTNC, Kathmandu.

Yonzon, P.B., (1993), *A Biological Survey of Shivapuri Watershed / Wildlife Reserve.* Unpubl. Report to DNPWC, Kathmandu.

Yonzon, P.B. & Hunter, Jr, M.L., (1991), Cheese, tourists, and red panda in the Nepal Himalayas. *Cons. Biol.* 5(2): 196-202.

14

Rich Flora and Fauna of Arunachal Pradesh

Introduction

The north-eastern region comprises of seven sister states. If is one of the biological 'hot sports' of India. It encompasses 2,55,037 sq km area. In other words it covers over 7.7 per cent of Indian land area. ?Among those seven states Arunachal Pradesh is the largest one. This state is situated on the easternmost tip of the 2500 km long Himalayan belt. It is surrounded by China on the north and north-east, Indian states of Nagaland and Assam on the east and east-south and Bhutan on west.

The forests of Arunachal Pradesh are suitable for the invasion and origin of different types of biological components and life support systems. Due to differences in topographical climatic and geological conditions of the region various types of plants and animals are found there. The vegetation cover of Arunachal Pradesh extends from foot hills to the snow-covered peaks. The region can be divided in eight zones. Details about the zones with special stress on wildlife, mammals, birds, reptiles and aquatic life are to be found in this chapter.

Arunachal Pradesh—A Synopsis

Arunachal Pradesh is perhaps the most suitable state for invasion and origin of varieties of biological components and life support system. It possesses the richest flora and fauna in comparison to any other area of this size not only in Himalayan region but in the world.

It is the only Indian state which possesses major cat species like tiger, leopard, clouded leopard and snow leopard, as well as some lesser cats like golden cat, pallas's cat and marbled cat. Some non-human primates seen here are: capped langur, slow loris, Assamese macaque, stump tailed macaque, rhesus macaque, pigtailed macaque and hoolock gibbon. All the Indian goat antilopes—serow, goral and taken—are found here.

Among the high altitude species musk deer, Himalayan black bear, bharal, red panda occur here. Endangered animals like herbivores gour, wild buffalo, elephant are found here. Besides that some mammals like rodents, livets, liusang, bats, mongoose and shrews are quite popular here. Above all. Arunachal Pradesh has a rich variety of birds too.

Although a number of positive measures have been taken to save the wildlife and the habitats and about 10.7 per cent area of the state has been declared protected area and certain protective measures have been implemented, it seems that there are not sufficient. Wildlife parks and sanctuaries should be managed properly, especially from the ecological point of view. Above all, some stricter laws should be enforced to check the destructive human activities.

Geographic and Climatic Features of Arunachal Pradesh

At the north-east of the country is the Arunachal Pradesh Covering an area of 83,740 sq km it lies between 26°28' to 29°30' N latitudes and 91° 30' to 97° 30' E longitudes. Bounded by Bhutan, China and Indian States of Nagaland and Assam on the west North and east respectively, this State enjoys natural frontiers of mountains. This territory is further contigously connected with the main plain land of Assam.

The eccentric chain of hills and deep valleys around dominate the topography and features of this state. Latitudinal locations and altitudes influence its climate. December and January are the coldest months. During that period temperature fluctuates between 5°C and 20°C in the altitudes below 900 m. In altitudes above 3,000 m the mean temperature is below the freezing point. July and August are the warmest months when average daily temperature below 900 m remains 27°C or so. The region receives heavy rain fall almost everywhere. Annual rainfall in different parts of the state varies between 100 and 400 cm.

The variable differences in the geological climatic and topographical conditions of Arunachal Pradesh result in an extensive diversity of flora and fauna. Its rich vegetation covers are real treasure houses of wonderful varieties of plants and animals. About 61.5 per cent of land of Arunachal Pradesh is under forest cover.

Eight Habitat Zones

Varied ecological situations and bioclimatic conditions of Arunachal Pradesh are caused by its wet ecological conditions, tropical to moderate temperature and its mountain soils. Based on these conditions the whole region can be classified in eight broad habitat zones. They are: (i) evergreen and semi-evergreen forests, (ii) subtropical broad-leaved forests, (iii) subtropical pine forests, (iv) temperate forests, (v) bamboo forests, (vi) degraded and mixed forests, (vii) grasslands and (viii) aquatic domain. Characteristic feature and salient flora (when necessary) of different bioclimatic zones are as follows:

(i) *Evergreen and semi-evergreen forests (up to 600 m):* Along the hill slopes, moist valleys and catchment areas occurs the evergreen forests whereas the semi-green forests are foundalong the foot hills and river banks. The average feature of the jungle is thick, hygrophilous and dense. Majority of the plants are woody with great dimensions. The climax forest has very little undergrowth. The overhead is so dense that the light intensity below is as low as 1 per cent of that of above. Most of the climbing plants and some of the epiphytes are also woody. The rate of turnover of mutrients is very high due to rapid decomposition of dead biomass. In semi-evergreen forest the deciduous trees are also making up dominance. There is well differentiated middle storey below the upper storey. The evergreen predominates in the lower canopy. The undergrowth is heavy and grow abundantly lianas and climbers grow here.

Prominent plants of this forest are: *Dipterocarpus macrocarpus, Shorea assamica, Altingia excelsa, Tetrameles nudiflora, Ailanthes grandis, Terminalia myriocarpa, Mesua ferrea, Cinnamomum glaucescens, Echinocarpus assamicus, Quercus milroyi, Syzygium formosum, Mallotus nepalensis, Croton joufra, Polyalthia jenkinsii, Sterospermum chelonoides. Elaeocarpus aristatus, Artocarpus lakoocha, Canarium strictum, Pterygota alata, Firmiana colorata, Sterculia villosa, Acrocarpus fraxinifolius, Duabanga grandiflora, Bombax ceiba etc.*

(ii) *Subtropical broad-leaved forests (800-1900 m):* The strong influence of human interference is quite prominent in this landscape. In same places of former natural vegetation are seen. The characteristic feature of the forest is warm-temperate owing to its development in mid-altitudinal situation and moderate climate. The forest is quite rich in flora. This area is of marked seasonal changes which is more related to precipitation than to temperature. This area has semi dry winters and hot less-wet summers. Winters are not intense cold with no month averaging below 5°C, while summers are hot with temperature averages over 20°C.

Prominent plant of this forest are: *Schima wallichii, S. khasiana, Castanopsis indica, Quercus lamellosa, Q. spicata, Michelia oblonga, Ostodes paniculata, Ulmus lancifolium* etc.

(iii) *Subtropical pine forests (1000-1800 m):* In Arunachal Pradesh this type of forest extends both in the subtropical and temperate belts. Pines are characteristically adapted to the conditions of prolonged winters, short growing seasons, shortage of moisture and high winds. It is commonly seen in rain shadow areas. The flora is not very rich here. Pine community areas are generally of immature sandy soils, or low-lying marshy soils.

Prominent Plants of this Forest are: *Finns roxburghii, P. wallichiana, P. merkusii* etc.

(iv) *Temperate forests (1800-3500 m):* This type of forest is found in all districts almost as a continuous belt. In addition to trees, the temperate forest possess many shrubs ferns, herbs and mossess at ground level. Therefore the forest if fairly productive. Though plentiful but usually gentle rainfall is well distributed seasonally in the temperate forest zone. Strong winds, plentiful clouds and frequent fog and mist are characteristic of the winter half of the year. Winters are cold with snowfalls. Summers are rather warm and humid. The natural vegetation is deciduous type. The trees have abundant foliage which they shed in autumn. This abundance of leaf litter. This gives plenty of humus for the soil by decomposition.

Prominent plants of this forest are: *Quercus kamroopii, Q. glauca, Populus ciliata, P. gambleii, Magnolia campbellii, Rhododendron* spp., *Abies delavayi, Taxus baccata* etc.

(v) *Bamboo forests (up to 2000 m):* Large expanse of pure or dominant bamboo vegetation are hardly seen and it tends to occur mostly in scattered clumps. Usually bamboos appear in abundoned jhum field or they are present as undergrowth in tropical evergreen forest. The amount of ground vegetation varies. Rotting leaf litter at ground and the rate of turnover of nutrients are not so high.

Prominent varieties are: *Bambusa tulda, B. pallida, Dendrocalamus hamiltonii, D. sikkimensis, Pseudostachyum polymorphum* etc.

(vi) *Degraded forests (up to 3000 m):* Various adverse anthropic forces have been responsible for the growth of degraded or modified forests by destructing the primary natural vegetation. These forests are in different stage of succession and are far from reaching a climax condition. These forests are poor having less floral diversity, inferior quality trees, weeds and shrubs.

Varieties of plant found here are: *Macaranga denticulata, Mallotus tetracoccus, Callicarpa arboreal Glochidion* sp., *Mikania micrantha, Eupatorium odoratum* etc.

(vii) *Grasslands:* Grasslands constitute an important feature of vegetation in riverine plains and at high altitudes. Transition from forest to grassland may be gradual or abrupt. The vegetation is dominated by grasses though sedges form part of flora. Present state and distribution of grasslands are man-made to a great extent as a result of prevailing landuse pattern.

Different varieties are: *Saccharum spontaneum, S. arundinaceum, Neyraudia arundinacea Thysanolaena maxima, Imperata cylindrica* etc.

(viii) *Aquatic domain:* The simple flow of water coming out of hill slopes is noticed very often. The landscape in Arunachal Pradesh is dissected by network of streams and river courses. The six major rivers in the State—Kameng, Subansiri, Siang, Dibang, Lohit and Tirap—finally join the river Brahmaputra in Assam. Excepting some small tributaries the rivers in Arunachal Pradesh flow round the year. Their beds are full of stones and pebbles. There are also a few natural lakes and stagnant water bodies. The lotic and lentic water bodies in the State measure about 620.46 sq km and 29.97 sq km respectively. There are two large wetlands measuring 20,200 hectare (Biswas and Trisal, 1993).

Animals in Habitat Zones: A Review

(i) *Evergreen and semi-evergreen forests:* In evergreen forest amphibians and reptiles are seen in plenty. There life at ground level is almost suppressed by a thick cover of rotting humus. In that condition the ground animals, other than amphibians and reptiles, find their movement difficult. Arboreal forms and flying creatures are found in abundant in this forest zone. Birds here are having well developed climbing feet and short stout legs. They are mostly arboreal. Large animals in this zone live in open areas.

On the other hand in semi evergreen forests many ground living forms like rodents, deer and a numnber of carnivores are found here. In this forest zones the canopy is less dense than evergreen forests. So the forest floor gets emparatively more sunlight.

Mammals in habiting there are: *Elephas maximus* (Elephant), *Cervus unicolor* (Sambar), *Muntiacus muntjak* (Barking deer), *Bos gaurus* (Gaur), *Bubalus bubalis* (Wild buffalo), *Funambulus pennanti* (Stripped squirrel), *Ratufa bicolor* (Malayan giant squirrel), *Rattus blanfordi* (White-tailed rat), *Caprolagus hispidus* (Hispid hare), *Nycticebus coucang* (Slow loris), *Macaca mulatta* (Rhesus macaque) *Macaca* (Tiger), *Panthera pardus* (Leopard), *Felis chaus* (jungle cat), *Felis viverrina* (Fishing cat), *Viberra zibetha* (Large Indian civet), *Prionodon pardicolor* (Spotted linsang), *Canis aureus* (Jackal), *Vulpes bengalensis* (Indian fox) etc.

Some common birds seen there are: *Pardicula manipurensis* (Painted bush quail), *Arborophila torqueola* (Hill partridge), *Bambusicola fytchii* (Assam bamboo partridge), *Pellorneum rufipes* (Spotted babbler), *Trichastoma tickellii* (Tickell's babbler), *Pomatorhinus schisticeps* (Scimitar babber), *Muscicapa pava* (Red-breasted flycatcher), *Rhipidura albicollis* (Fantail flyceatcher), *Tesia olivea* (Ground warbler), *Cettia flavolivacea* (Bush warbler), *Anthreptes singhalensis* (Ruby cheek), *Aethopyga satura* (Black breasted sunbird) etc.

Lauge varieties snakes and a number of lizards are also found here. The important ones are being: *Python molurus* (Rock python), *Bungarus fasciatus* (Banded krait), *Naja naja* (Common cobra), *Ophiophagus hannah* (King cobra), *Psamodynastes pulverulentus* (Pit viper), *Ptyas mucosus* (Rat snake), *Elaphae radiata* (Copper head), *Typhlops diardi* (Blind snake), *Oligodon cinereus* (Kukri snake), *Dendrelaphis cyanochloris* (Tree snake), *Ophites jara* (Wolf snake), *Paers monticola* (Snail eater) *etc., and Varanus bengalensis* (Common Indian monitor), *Cyrtodactylus khasiensis* (Gecko), *Cosimbotus platyurus* (House gecko), *Ptyctolaemus gularis* (Blue-throated lizard) etc.

Many amphibian forms like frogs and toads are found here. The important ones are: *Amolops afghanus, Rana kuhlii, R. limnocharis, Philautus cherrapunjiae, P. namdaphaensis, Rhacophorus leucomastyx, R. maculatus, R. maximus, R. reinwardtii, R. namdaphaensis, Bufo himalayanus* etc.

(ii) *Subtropical broad-leaved forests:* The ecosystem in this forest zone less productive than the evergreen and semi-evergreen forests. This has adverse effects on the productive animal life. Most of the animals are not habitual to this situation; it comprises life forms from neighbouring forest zones, both upper and lower. Number of animals in this forest zone is not much. That too is also gradually reducing.

Some commonly found mammals are: *Presbytis pileatus* (Capped langur) *Macaca mulatta, Macaca assamensis, Presbytis entellus* (Common langur), *Hylobates hoolock, Callosciurus macclellandi* (Himalayan stripped squirrel), *Funambulus pennanti, Petaurista petaurista* (Giant flying squirrel), *Budorcas taxicolor* (Takin), *Bos gaurus, Cervus unicolor, Muntiacus muntijak, Felis chaus, Felis bengalensis* (Leopard cat), *Felis manul* (Pallas's cat), *Mustela* sp. (Weasel), *Cuon alpinus* (Wild dog), *Selenarctos thibetanus* (Himalayan black bear), *Ailurus fulgens* (Red panda),

Some of the birds are: *Lophura leucomelana* (Kaleej pheasant), *Polyplectron bicalcaratum* (Peacock pheasant), *Gallus gallus* (Red jungle fowl), *Grus nigricollis* (Blacknecked crane), *Arborophila torqueola* (Hill partridge), *Turdus naumanni* (Dusky thrush), *Turdus boulboul* (Grey winged black bird), *Zoothera citrina* (Ground thrush), *Streptopelia tranquebarica* (Spotted dove), *Columba pulchricollis* (Ashy wool pigeon), *Megalaima franklinii* (Golden-throated barbet), *Copsychus saularis* (Magpie robin), *Rhycticeros undulatus* (Wreathed hornbill), *Pycnonotus caefer* (Red vented bulbul), *Dicrurus paradiseus* (Racket tailed drongo) etc.

Reptiles are less common in this forest zone. Some important ones are: *Bungarus niger* (Black krait), *Naja naja, Trimeresurus monticola* (Blotched pit viper), *Python molurus, Boiga ochracea, Sphenomorphus maculatum, Cyrtodactylus khasiensis* etc.

(iii) *Subtropical pine forests:* Prominent predators in thus forest are wolf, jungle cat, pallas's cat etc. Here animals like deer and fur-bearing animals are active throughout the year. Here birds tend to migrate. There are a number of rats, squirrels and flying suirrels too in this forest zone. The animals stringently live on plant-difined diets are supported by leaves, shoots and seeds of the trees and also by herb-like plants of this forest.

Mammals seen there are: *Cervus unicolor, Muntiacus muntjak, Ratufa bicolor, Petaurista magniflcus* (Hodgson's flying squirrel), *Hystrix indica* (Indian porcupine), *Bandicota bengalensis* (Mole rat), *Canis lupus* (Wolf), *Felis manul, Selenarctos thibetanus* etc.

Some common birds are: *Eudynamys scolopacea* (Indian koel), *Streptopelia tranqueberica* (Red turtle dove), *Corvus macrorhynchus* (Jungle crow), *Otus spilocephalus* (Eastern spotted scope owl), *Polyplectron bicalcaratum, Venellus* (Lapwing), *Ceryle lugubris* (East Himalayan pied kingfisher), *Dicrurus paradiseus, Sturnus malabaricus* (Grey headed myna), *Monticola rufiventris* (Chestnut bellied rock thrush) etc. Except geckoes *(Cyrtodactylus khasiensis)* reptiles are infrequently seen.

(iv) *Temperate forests:* Due to high productive atmosphere of temperate forest zone varieties of forms are found here and in large numbers. But abundance and variety reduce to a great extent beyond a height of 3000 m. Both herbivores and carmivores mammals are seen here, that too in large numbers. Rich species of birds like woodpeckers, titimices, thrushes, warblers, finches and tragopans are also found in this forest zone. But they generally leave the forest during the lean season. Reptiles are not seen here.

Some important mammals are: *Hystrix indica, Moschus moschiferus* (Musk deer), *Ailurus fulgens, Capricornis sumatraensis* (Serow), *Nemorhaedus goral* (Goral), *Pseudois nayaur* (Bharal), *Bos grunniens* (yak), *Marmota bobak* (Himalayan marmot), *Selenarctos thibetanus, Panthera pardus, Canis lupus, Mustela* sp., *Panthera uncia* (Snow leopard) etc.

Some common birds are: *Ithaginis cruentus* (Blood pheasant), *Columba leuconota* (Snow pigeon), *Macropygia unchall* (Bar tailed cuckoo dove), *Lanius tephronotus* (Grey backed shrike), *Dicrurus leucophaeus* (Assam grey drongo), *Corvus macrorhynchos, Hypsipetes madagascariensis* (Black bulbul), *Dinophieum benghalense* (Golden backed woodpecker), *Carduelis spinioides* (Himalayan green finch), *Myiophoenus caeruleus* (Whistling thrush), *Phylloscopus maculipensis* (Grey faced leaf warbler), *Parus monticolus* (Grey backed tit), *P. rubidiventris* (black tit), *Tragopan satyra* (Satyr tragopan) etc.

(v) *Bamboo forests:* No large animals are usually noticed in bamboo forests. Of course for bamboo vegetation elephants prefers to traced this forest. Some amphibians are reptitles are accustomed to west part of the forest floor. In search of rodent prey some species of cat family visit bamboo forest from neighbouring forests. During the blooming of bamboo plants Vandeleuria oleracea (long tailed rat) and Cannomys badius (bamboo rat) are seen in this forest.

Bamboo forests are usually dense thus not easy to enter. Rotting leaf litter at grounds is not so high.

Either for food or for nesting number of birds visit bamboo forests. Common among them are: *Bambusicola fytchii* (Assam bamboo partridge), *Francolinus francolinus* (Assam black partridge), *Paradoxornis rufipes* (Assam redhead parrot bill), *P. gularis* (Assam greyhead parrot bill), *Gracula religiosa* (Eastern hill myna), *Dendrocitta vagabunda* (Northeastern tree pie), *Pycnonotus jacosus* (Assam red whiskered bulbul), *Leiothrix argentauris* (Himalayan silvereared mesia), *Alcippe nipalensis* (Nepal quaker babbler), *Loriculus vernalis* (Indian lorikeet) etc.

(vi) *Degraded forests:* In degraded forests animals are not plenty in number. It develops animal community depending upon richness, growth and size of the forest. A thinned state of fauna of the neighbouring natural forest is seen here. *Gollunda ellioti* (Bush rat) is especially associated with bush and scrub jungles. Here it builds densely woven nest of stalk fibres and grasses. It consumes seeds of *Lantana* that grows profusely in degraded forest.

(vii) *Grasslands:* It is a wide open space. Before interference of man the grasslands were homeland of rich animals. Here most animals are of deceptive colour. For reptiles and amphibians grassland is a suitable habitat. Birds find their food and prey here. Some small herbivores are seen here. They include squirrels rats, hares etc.

The common mammals are: *Bubalus bubalis, Sus scrofa, Funambulus pennanti, Caprolagus hispidus, Bandicota bengalensis* (Mole rat), *Cuon alpinus Canis aureus, Vulpes bengalensis* etc.

Some of the birds are: *Calandrella raytal* (Ganges land lark), *Eremophila alpestris* (Elwee's homed lark), *Hirundo rustica* (Eastern swallow), *Limosa limosa* (Eastern black tailed godwit), *Accipiter nisus* (Sparrow hawk), *Falco severus* (Hobby).

A few available reptiles are: *Typhlops* sp., *Elaphae taeniura, Ophites jara, Ptyas mucosus* etc. Among the amphibians, *Rana kuhlii, Rhacophorus maculatus* etc. are seen.

(viii) *Aquatic domain:* The fishes are represented by both cold and warm water inhabiting forms. The important species are: *Schizothorax richardsonii, Danio aequipinnatus, D. rerio, Aspidopariya jaya, Barilius barna, B. bendelisis, B. bola, Semiplotus semiplotus, Puntius chola, Acrossocheilus hexagonolepis, Chagunius chagunio, Tor tor, T. putitora, Grossocheilus latius, Garra lissorhynchus, Badis badis, Amblyceps mangois, Bagarius bagarius, Xenentodon cancila, Mastacembalus armatus, Labeo dero, L. pangusia* etc.

Existence of Certain Animals in A Partiecular Zone

As we have seen in previous paras various vertebrate species have occupied all major habitat zone of Arunachal Pradesh (Pal, 1993b). In comparison to less vegetated, cooler and dries situations of higher altitudes, wild animals prefer to like in moderately high altitude zones approximately of about 1800 m and in semi-tropieal, undulating terrain, wide grasslands, marshy lands, dense jungles and rivers. While some species are distributed in more than one zone, some others are restricted to a particular zone under a specialize situation. Existence of animals in a particular place depends on certain factor like food, shelter, water, prevailing temperature soil condition etc. Less member of animals are found in or near zones which are covered with snow almost throughout the year. In upper temperate zones blue sheep, snow leopard, musk deer etc. are found because they can tolerate the cold condition of the zone. On the other hand bharal Himalaya black bear, red panda serow and goral are seen in large numbers in upper temperate zone.

Factor Determine the Distribution of Vertibrates

The soil characteristics determine the distribution of vertibrates. Of course, edaptive factors influence the distribution of various vertibrates in this region is not clear. It has been observed that the burrows of various field rodents are distributed on hill slopes and edges of table lands not flooded by rise of water level. They collect around shrub in drier season. This sort of shifting may be due to remaining in close proximity to the soil in which moisture regime is higher due to presence of root systems. Species like, pangolin occurs in rock crevices and burrows. The gravelly plains are preferred by many reptiles viz., garden lizard, Indian monitor, banded krait, rat snake etc. They often take shelter in rock crevices.

Distribution of Vertebrate and Mammal Species

Vegetal covers, vegetation type and vertebrate species are clearly associated. Elephants feed on a variety of plants. Due their specialized trunk and thick tongue they can feed even on thorny plants. Their major food items are banana plants and grasses. They can feed on a great deal on bamboos also. Wild boars are in the habit of eating both animal and vegetable. They consume different types of roots, underground steems as well as dead and decaying flesh. Only wild boars and elephants are capable of pulling out or digging the underground stems and rootsGour and Sambar prefer green grasses. In absence of that they consume fruits and leaves.

Major mammals mainly living in trees are: common langur, capped langur, rhesus macaque, slow loris, hoolock gibbon stripped squirrel and flying squirrel. The stripped squirrels build nest of twigs and leaves on big trees. They feed on variety of fruits, barks leaves and seeds. The flying squirrels are mainly active in night. They glide between canopy of tall trees in deep jungles and are mainly frugivoruous.

The langurs, macaques and hoolock gibbons live in highly social groups in evergreen and broad-leaved forests. They appear to feed on trees like *Dalbergia, Artocarpus* and many others. In rain forests slow loris is found. It generally does not come to ground. It is a very seclusive nocturnal animal that feeds on tender shoots, leaves, fruits, insects, birds' eggs etc. The growing mammals like mice, squirrels etc prefer to live in areas where there are plenty of foods and shelters than the areas with poor vegetation.

Endangered Habitat and Wildlife

The territory of wild animals are very much disturbed by the destruction of forests. The trend is so prominent here that a number of species, once distributed over large areas of the State, are protected/ endangered or vulnerable (Tikader, 1983). Golden cat, leopard cat, tiger slow loris, hoolock gibbon, wild buffalo etc. all these animals have been included under the Schedule I of Wildlife (Protection) Act, 1972. Animals under other Schedules are the capped langur, flying squirrel, Chinese pangolin, gaur, elephant, serow, wild dog, leopard, barking deer, sambar, Himalayan black bear, rock python etc. They also fave the same problem in relation to their habitat loss. Of the avi-fauna some of the endangered species are: eastern white stork, peacock pheasant, Bengal florican, partridges and different species of hornbills. The white winged wood duck is a vulnerable species.

Because of their popular, the major share of conservation attention has been drawn by birds and mammals. On the other hand, though invertebrates are about 75 per cent of all animals there, they get less attention because they are not so eminent.

In comparison to Arunachal Pradesh in Eastern Himalaya, hardly any other area of that size has such a huge varieties of flora and fauna. This state is quite rich in reptiles, amphibia and fresh water fishes. But, unfortunately the rising pressure on forest and forest produce caused by rising population and their different destructive activities have endangered the survival of wild life there. To check the situation some effective measures and their strict implementation are the need of the day that too on priority basis.

References

Ali, S. and Ripley, S.D., (1983), *Handbook of the Birds of India and Pakistan* (Compact ed.). Oxford Univ. Press: Delhi, Oxford, New York; xiii + 737 pp., 131 pl.

Annandale, N., (1906), Notes on the fresh water fauna of India—VIII. Some Himalayan Tadpoles. *J. Asiat. Soc. Bengal* 2: 289-292.

Badola, S.P., (1975), Fish fauna of the Garhwal hills—Part II. Pauri-Garhwal, U.P. *Indian J. Zool.* 16(1): 57-70.

Bhatnagar, R.K., (1972), Reptilian fauna of Dehradun district (Uttar Pradesh)—Part I. *Cheetal* 15(4): 15-23.

Bhatnagar, R.K. and Mishra, P., (1972), Revised synopsis of birds of Dehradun and adjacent hills—Part II (Columbiformes to Piciformes). *Cheetal* 14(4): 40-53.

Bor, N.L., (1938), A sketch of the vegetation of Aka Hills, Assam. *Ind. For. Rec.* (n.s.) *Bot.* 1(14): 103-221.

Briggs, F.S., (1931), Birds observed in the neighbourhood of Ranikhet. *J. Bombay Nat. Hist. Soc.* 34: 1072-1079.

Boulenger, G.A., (1920), A monograph of South Asian, Papuan, Melanesian and suctalian frogs of the genus *Rana Rec. Indian Mus.* 20: 84-221.

Chandola-Saklani, A., Bist, M., Kar, A., Seikhar, C. and Lakhera, P., (1987), Wildlife in Garhwal Himalaya: Breeding and behavioural biology. In: *Western Himalaya: Environment, Problem and Development,* eds., Pangtey, Y.P.S. and Joshi, S.C., Gyanodaya Prakashan, Nainital. 1: 242-262.

Chatterjee, A.K., (1989), Survey of snow leopard and associated animals in Mouling National Park, East Siang District, Arunachal Pradesh. *Arunachal Forest News* 7(1-2): 10-15.

15

Beautiful But Little Known Species: Red Panda on the Brink of Extinction

Introduction

Generally found at an altitude of 2200-4800 m in the temperate forests of the Himalaya Red Panda, a fresh-eating animal, is at the verge of extinction. It is said to be the sole representative of the family Ailuridae. Increasing human intrusion into its already diminishing habitats has threaten its existence. It has some regional names also—Sankam (Lepcha), updonga (Bhutia) and Kundo (Nepali). Red Panda prefer to live in forest with plenty of bamboos.

Presently it is found only in some protected areas in India and Nepal. The Calcutta Zoological Garden was the first to house it. Its existence can be secured only by rising public awareness about this animal.

A Carnivore Under the Family Ailuridae

In spite of the fact that flora and fauna beside maintaining the systemic balance of the mountain region also play vital ecological roles, they have not received due attention in the Himalayan region. This has put the exceptionally rich species in danger. So much so, some of them are on the verge of extinction.

Red Panda is one of the many faunal species found in this region. It is a flesh eating animal under the family Ailuridae. It is also unique in many respects. *Ailurus fulgens fulgens* (the Indian Red Panda), and *Ailurus fulgens styni* (the Red Panda of China) are the two subspecies of Red Panda. The split between these two species appears to have occurred some three million years ago, when torrential river flows cut increasingly into the rising mountain ranges during the formation of the Himalaya, creating the deep and wide valley of the Brahmaputra. After it was brought to the attention of the Western world in 1821 the small carnivore received its common name of Panda. Although the exact origin of the word 'Panda, is not yet known, perhaps the Red or Lesser Panda received this name after the Giant Panda had already been discovered. Its taxonomic classification as *Ailurus fulgens* in 1825 is attributed to the French zoologist Frederic Cuvier *(Ailurus* meaning 'cat like' and *fulgens* meaning 'brilliantly coloured').

The Red Panda also has several regional names. It is also called *sankam [Lepcha]; kundo [Nepali]* and *updongka* [*Bhutia*]. Taxonomy of the red panda has been the subject of debate. Since its discovery, it has been alternately placed at different points of time in the Procyonidae, Ursidae, Ailuropoda and in Ailuropodidae. The difficulties of determining whether certain characteristics of Ailurus are phylogenetically conservative or converge with those of other species found in similar ecological habitats caused the disputes in this connection. Evidence based on fossil records, serology, karyology, behavioural biology, anatomy, physiology and reproduction reflects closer affinities of the red panda with Procyonidae than with Ursidae. Since its ecological and digestive specialisations are quite distinct from other species endemic in the region, it has been placed in the separate family of Ailuridae which derives from Procyonidae.

Red Panda occupies a highly specialised region where many zoogeographical areas meet. This is proved from the fact that this animal is found at altitudes between 2200-4800 m in the Himalaya extending from Eastern Nepal through West Bengal, Sikkim, Bhutan and Arunachal Pradesh. The Red Panda habitat is characterised by preserve of mixed deciduous and coniferous forests. The animal is also found in high mountains in northern Burma, and in Sichuan and Yunan in western China. It prefers to live in temperate forests with bamboo thickets. At present the animal is highly endangered. Its survival is at stake. The IUCN has rightly put red panda in its critical category list. The fact that it is the single representative of the family Ailuridae enhances its importance form the biological point of view.

In the red panda habitat several other mammalian and bird types are also found. Some these are listed below:

TABLE 15.1

List of Mammalian and Bird Species Sharing the Red Panda Habitat

MAMMALIAN SPECIES	
Binturong	*Arctictis binturong*
Yak	*Bos grunniens*
Takin	*Budorcas taxicolor*
Jackal	*Canis aureus*
Tibetan wolf	*Canis lupas chanco*
Serow	*Capricornis sumatraensis*
Leopard cat	*Felis bengalensis*
Jungle cat	*Felis chaus*
Marbled cat	*Felis marmorata*
Golden cat	*Felis temmincki*
Himalayan tahr	*Hemitragus felmahicus*
Indian pangolin	*Manis crassicaudata*
Chinese pangolin	*Manis pentadactyla*
Himalayan marmot	*Marmota babak*
Yellow-throated marten	*Martes flavigula*
Stone marten	*Mastes foina*
Musk deer	*Moschus moschiferus*
Barking deer	*Muntiacus nuntjak*
Yellow-bellied weasel	*Mustela kathiah*
Himalayan weasel	*Mustela sibirica*
Goral	*Nemorhaedus goral*
Clouded leopard	*Neofelis nebulosa*
Great Tibetan sheep	*Ovis ammon hodgsoni*
Bharal	*Ovis nahura*
Himalayan palm civet	*Paguma larvata*
Tibetan antelope	*Panthelops hodgsonii*
Common palm civet	*Paradoxurus hermaphroditus*
Hodgson's Flying squirrel	*Petaurista magnificus*

Contd....

Tiger civet	*Prionnodon pardicolor*
Lesser Giant squirrel	*Ratufa bicolor*
Himalayan Black bear	*Selenarctos thibetanus*
Brown bear	*Ursus arctos*
Large Indian civet	*Viverra zibetha*
Hill fox	*Vulpes montana*
BIRD SPECIES	
North Indian crested goshawk	*Accipiter trivirgatus indicus*
East Himalayan bersa sparrow hawk	*Accipiter virgatus affinis*
Roufous-necked hornbill	*Aceros nepalensis*
Eastern hoary-throated barwing	*Actinodura nepalensis vinctura*
Common iora	*Aegithina tiphia tiphia*
Indian pied hornbill	*Anthracoceros malabaricus*
Himalayan golden eagle	*Aquila chrysaetos daphnea*
Little spider-hunter	*Arachnothera longirostris longirostris*
Hill partridge	*Arporophila torqueola*
Blyth's baza	*Aviceda ferdoni*
Indian black-breasted baza	*Aviceda leuphotes leuphotes*
Buzzard	*Buteo buteo japonicus*
Long-legged buzzard	*Buteo rufinus rufinus*
Indian large-tailed nightjar	*Caprimulgus macrurus albonotatus*
Common Indian rosefinch	*Carpodacus erythrinus*
Nepal dark rosefinch	*Carpodacus nipalensis*
Chir pheasant	*Catreus wallichii*
Greater coucal	*Centropus sinensis sinensis*
Rusty-flanked tree creeper	*Certhia nipalensis nipalensis*
East Himalayan pied kingfisher	*Ceryle lugubris guttulata*
Orange-bellied leafbird	*Chloropsis hardwickii*
Hen-harrier	*Circus cyaneus*
Yellow-headed fantail warbler	*Cisticola exilis tytleri*
Great parrotbill	*Conostoma aemodium*

Contd....

White-rumped shama	*Copsychus malabaricus indicus*
Large cuckoo-shrike	*Coracina novaehollandiae*
Yellow-breasted greenfinch	*Curduelis spinoides*
Nepal cutia	*Cutia nipalensis nipalensis*
North-eastern tree pie	*Dendrocitta vagabunda vagabunda*
Fire-breasted flowerpecker	*Dicaeum ignipectus*
White-bellied drongo	*Dicrurus caerulescens*
Little forktail	*Enicurus scouleri scouleri*
Indian blue robin	*Erithacus brunneus*
Eastern ruby-throat	*Erithacus pectoralis confusus*
Asian koel	*Eudynamys scolopacea*
Indian red jungle fowl	*Gallus gallus murghi*
White-spotted laughing thrush	*Garrulax ocellatus*
Rufous-chinned laughing thrush	*Garrulax rufogularis*
Striated laughing thrush	*Garrulax striatus*
Hill mynah	*Gracula religiosa*
Himalayan griffon vulture	*Gyps himalayensis*
Eastern rufous sibia	*Heterophasia capistrata bayleyi*
Eastern swallow	*Hirundo rustica*
Blood pheasant	*Ithaginis cruentus tibetanus*
White-backed munia	*Lanchura strata*
Rufous-backed shrike	*Lanius schach*
Black-headed shrike	*Lanius schach tricolor*
Monal pheasant	*Laphophorus impefanus*
Himalayan silver-reared mesia	*Leiothrix argentauris argentauris*
Eastern red-billed leiothrix	*Leiothrix lutea calipyga*
Eastern plain-coloured mountain finch	*Leucosticte nemoricola*
Kalij pheasant	*Lophura leucomelana melanota*
Bartailed cuckoo-dove	*Macropygia unchall tusalia*
Himalayan Great barbet	*Megalaima virens marshallorum*
Long-billed calandra lark	*Melonocorypha maxima maxima*
Crested bunting	*Melophus lathami*
Red-tailed minla	*Minla ignotincta*

Contd....

Blue-capped rock hrush	*Monticola cinclorhynchus*
Blue rock thrush	*Monticola solitarius*
Verditer flycatcher	*Muscicapa thalassina*
Indian maroon oriole	*Oriolus traılli traılli*
North Indian scops owl	*Otus scops sunia*
Green-backed tit	*Parus monticolus monticolus*
Yellow-cheeked tit	*Parus xanthogenys*
Scarlet minivet	*Pericrocotus flammeus*
Himalayan crimson breasted woodpecker	*Picodes cathpharius cathpharius*
Northern ruddy-breasted crane	*Porzana fusca fusca*
Long-tailed broadbill	*Psarisomus dalhousiae dalhousiae*
Indian red-breasted parakeet	*Psittacula alexandri fasciata*
White-cheeked bulbul	*Pycnonotus leucogenys*
Green-billed malkoha	*Rhopodytes tristis tristis*
Spot-winged starling	*Saroglossa spiloptera*
Grey bushchat	*Saxicola ferrea*
Fire-fronted serin	*Serinus pusillus*
Himalayan whitetailed nuthatch	*Sitta himalayensis*
Crested serpent eagle	*Spilornis cheela cheela*
Mountain hawk-eagle	*Spizaetus nipalensis nipalensis*
Rufous turtle dove	*Streptopelia orientalis orientalis*
Himalayan wood owl	*Strix aluco nivicola*
Himalayan brown wood owl	*Strix leptogrammica newarensis*
Tibetan sandgrouse	*Syrrhaptes tibetanus*
Paradise flycatcher	*Terpsiphone paradisi*
Satyr tragopan	*Tragopan satyra*
Grey-winged blackbird	*Turdus boulboul*
Northern Bustard quail	*Turnix suscitator plumbipes*
Eastern whiskered yuhina	*Yuhina flavicollis flavicollis*
White-throated ground thrush	*Zoothera citrina*

Features and Food Habits of Red Panda

Red Pandars are generally active in night. and they prefer to rest of trees. Their breeding season starts as winter sets in. They are found in pairs during that period. Mating is generally recorded in the months of January, February and mid-March. After gestation of between 120-160 days, births usually occur during the monsoon in the months of June, July and August. They give birth an average of two cubs. At birth, the new borns weigh 110-130 gm and have a growth increment rate of 7-20 gm per day. They begin to eat solid food at the age of approximately 90 days. The eyes and ears of the newborn cubs are closed at birth. The head, body and tail are covered with thick and woolly, gray-buff fur. The planter surfaces of the feet remain unfurred since birth and the skin is pink. For the first 7-10 days after they are born, the panda cubs can not move. During those days the mother stays curled around them most of the time, except when she goes out for food. During the absence of their mother, the cubs sleep in a curled or semi-curled position. At the age of 18 days, their eyes open and the young become mobile.

Like members of the cat family, the red panda also has the habit of shifting its cubs from one shelter to another, apparently for nursing. 3 or 4 such shiftings of the cubs are noted in captivity. Mother kills the cubs if she is disturbed by them or if they are born weak. No information is available on the dispersal pattern of juveniles and the rearing of neonates in the wild. In captivity, the cubs may be reared uptil the age of 4 months by their mother. During that period she teaches them how to climb and feed, and perform other related activities. The cubs become independent at the age of 3 months.

The red panda is a nocturnal, crepuscular and arboreal animal. It is most active at dawn, dusk and night. For the remainder of the day, it rests sleeping on trees or other elevated surfaces in a curled position. They may be found in resting posture in the trees even during heavy rain and windy and chilly weather. Even snowfall does not affect this activity. Activity patterns however change over the year according to the variation in climate. Though red panda can climb they generally prefer to wander on ground. These small animals have low basal metabolic rates, approximately one-third of that of the similar-sized cats and comparable to that of tree sloths. The red panda can further reduce the metabolic rate at low ambient temperatures, without concomitant reduction in body temperature.

Though it is flesh-eating animal, the red panda is accustomed to bamboo feeding also. That is why red panda is found in forests with plenty of bamboo. The gut of this animal is not capable of breaking down the chemically fibrous constituent materials of bamboo which mainly comprise cellulose, still a major part of diet of red panda consists of bamboo, which is, of course of law energy value. The arboreal nature of this animal and its typical resting posture may have developed as an every-saving strategy. Most herbivores have developed some means of digesting the cellulose that constitutes the structural material of plant cell walls. Mammals like cattle, sheep, goat and antelopes have a modified stomach where plant material is fermented by bacteria. In the rumen of these animals, cellulose is converted to simple sugar and assimilation then takes place. In non-ruminants like the rabbit, the elephant, the horse and the rhinoceros, plant material is partially digested in the stomach and small intestine before being passed into the caecum, where final breakdown of cellulose takes place.

How the red panda has adapted to bamboo feeding despite the fact that there is no organ to digest and assimilate this diet and its digestive system is that of a carnivore is defficedt to understand. It is also remarkable that the animal has a dental formula typical of a carnivore, except for the missing fourth upper premolars. However the teeth themselves bear closer resemblance to those of herbivores. The cheek teeth are unusually large, broad and flat with a complex pattern of ridges specialised for bamboo feeding. The animal can digest only 20 per cent of the total amount of bamboo leaves it consumes. The low metabolic rate for the species is consistent with low fertility and low rates of growth. In addition to bamboo, the red panda in the wild takes small mammals, birds' eggs, blossoms and berries. The animal in captivity was found to hunt birds.

In comparison to other fissipeds the feeding behaviour of red panda is quite different. The smaller foods like blossoms, berries and small leaves are compressed with the incisors. To eat bamboo leaves, while either in sitting or standing position, this animal holds the leaves firmly in the palm of a single fore paw and then brings the leaves to the mouth. There the food is sheared, chewed and then swallowed. In very rare cases red panda is found eating while in lying position.

Existence At Stake

Grazing and wood collection from the forests have reduced the habitats of red panda, thus threatening its survival. Even bamboo, its major diet, is also now used by local people for constructing houses as well as for fuel. Impact of tourism is also posing a threat to this animal.

It was recorded in Langtang National Park in Nepal that due to these disturbances the cub mortality rate rose to as high as 80 per cent, which is very alarming and a threat to the very survival to this animal. It was also recorded in 1988 that only a single cub was born in an area of 35 sq km.

Besides that other human activities like hunting for trophy and also for its skin, trapping the animal to smuggle it abroad alive etc. are also responsible for the extinction of red panda. Of course the Wildlife Act was implemented in 1972 to offer it security but that was not enough. The existence of red panda can be secured only by raising public awareness about this animal.

Presently red panda is found only in protected areas in India and Nepal. In India it is found in four reserves in West Bengal and Sikkim. It availability in Arunachal Pradesh is not known. In Bhutan its strength is perhaps less than 100. Its number in Darjeeling and Gangtok 300s are around only 13. Paducaja Naidu Himalayan Zoological Park in Darjeeling, since its inception, is the only zoo which has maintained continuously the red panda in captivity.

The Calcutta Zoological Gander was the first zoo to house red panda. It was followed by Prince Wales Zoo at Lucknow, the Hyderabad Zoo, the Kamala Nehru Zoological Garden at Ahmedabad and the Nandan Kanan Sactuary.

As per the available record the protected red panda reserves in India and Nepal are as follows:

TABLE 15.2

Protected Red Panda Reserves in India and Nepal

Location and Name	*Area sq km*
INDIA:	
West Bengal	
Singalila National Park	78.6
Neora Valley National Park	88.0
Sikkim	
Khangchendgonga National Park	850.0
Rhododendron Sanctuary	31.0

NEPAL:	
Langtang	1710.0
Lake Rara	106.0
Sagarmatha	1148.0
Shey-phoksumdo	3555.0
Dhorpatan	1325.0
Annapurna	2660.0
Total Protected Area	11551.6 sq km

References

M.C.K. Blaijenberg, (1984), 'When is a Carnivore not a Carnivore? When it is a Panda', *The Red Panda or Lesser Panda Studbook 3,* Royal Rotterdam Zoological and Botanical Gardens, Rotterdam.

C. Catton (1990), *Pandas,* Christopher Helm, London.

A.R. Glaston, (1988), *The Red Panda or Lesser Panda Studbook 5,* Royal Rotterdam Zoological and Botanical Gardens. Rotterdam.

Pocock, (1946), 'The Panda and the Giant Panda', *Zoo Life,* London.

16

Eastern Himalaya—Habitation of Salamanders

Introduction

A vertebrate of the amphibia family, Himalayan salamander, is the another name in the list of undangered species salamander is very sensitive to human disturbances. Its breeding cycle is greatly affected by the biotie disturbances in its habitat. Due to its shyness and comparative rarity few standy could be carried out on its life pattern. Urgent attention from researchers and emvironmentalists is needed. If salamander is to be saved from imminent extinction. Also some effective measures are to be taken to check polltion, preserve the habitat and finally to check the wildlife trade.

Representative of Caudata

In the cold climate of Eastern Himalaya, at a height of 1500-2250 m, Tylotoriton verrucosus or the Himalayan salamander is found. It is the unique representative of the Caudata among the amphibian family of the Indian subcontinent. Nepali name of salamender is 'pani kukur'. It has been given this name because of low noise it makes while snapping its jaw. In thai it is called 'mah nam.' Meaning of both these names is waterdog. Nepalis in North Bengal refer to this animal as 'gorho'. Because of its shyness andcomparative rarity, few studies have been able to focus on the life patterns of the species. Besides that other reasons are—it is darm in colour, slow and lazy. Moreover, like frogs and loads it does not produce prominent sound. Another reason is it remains active in its breeding pools only between late April and early September. Due to these reasons its presuance in the habitat is not felt by the environmentalists, researchers and fieldworkers. Its principal diet includes water buttes abd bugs, frog tadpoles, insect larvae earthworms, snails and slugs.

Salmander dwells in cool mountain lakes, everlasting and temporary pools and streams during summer months—from April to September. It spends the winter months—from October to March—under leaf litter, decaying logs and in mud burrows below the forest. Whereas in dry season it takes shelter among the bamboo stumps. There is feeds on the termites and woodlice that infest the rotting vegetation. It swims slowly with undulatory motions of its tail. Only occasionally it comes to the surface to breathe.

Structural Features

The body of this animal is cylindrical. The head is ringed by a hard porous ridge, with short longitudinal ridges along the top. Mostly dark in colour, the salamander has light brown limbs and tail. Its toes and fingers and the under surface of its tail are also pale in colour. The nuzzle is short but larger in diameter than the eye. The eyes are moderate in size with large upper eyelids. Large glandular tubercles occur in line along the body, terminating at the tail. The eardrum is distinct. The nostrils are located close to the tip of the blunt nuzzle having a space between them. They are semicircular and are closed by a small flap of skin. Palatine teeth are present in line with the internal nostrils along two ridges. Though they meet in front but widely diverge behind. Maxillary teeth are small and the tongue moderate in size. The size of the body is around three times the length of its head. It has four short stumpy limbs with 4 fingers and 5 toes. The long tail resembles that of the crocodile. The tail is flat and longer than its body. It has sharp uppes and lower margins that form median ridges both above and below it. Because of this overall resemblance this animal is also called the Crocodile Salamander.

The soft skin of the Himalayan salamander has fine tubercles and is covered with the slime secreted by the skin glands. The mid-dorsal ridge along the head is well developed and quite prominent. Two prominent rows of 15-16 knoblike porous glands lie to either side of the vertebral ridge, with the last three of these behind the legs. The anal opening is a longitudinal slit. Throughout of the year the males and females of this species are difficult to distinguish. However during the breeding season, the identification becomes easier. While the males expel milt when their abdomen is pressed hard, a highly distended belly distinguishes the females.

Comparatively a Safe Habitat

During the summer months of 1998 a survey of the salamander population was conducted in Jorepokhri. According to the survey a large number of salamander arrive at this site and breed in the pond. There this animal felt less disturbed and threatened. Generally monsoon is the best period to observe salamandars.

Normally, the salamanders come out of hibernation in order to breed, after the spring shawers. The males acquire bright colouration by early April. Courtship displays are very elaborate and last upto several hours. Thereafter, the male holds the female under her body and transfers a sperm sac or spermatophore into her cloaca. Egg laying commences in May and continues throughout the rainy season. Eggs are round in shape, greenish-yellow in colour and about 1 cm in diameter. Around 16 to 20 eggs are deposited into the water among the pond vegetation. The larvae which emerge about 20 days later measure between 9-11 mm in length and have large eyes and four external gills to either side of the head. Their bodies are grey with black netted patterns. Only their bellies are white. At first the larvae do not have to feed since their bellies are full of yolk. After two weeks, they start feeding on microcrustaceans and larval insects which inhabit the bottom sediments. Before growing into adults by October they pass through five growth stages over a period of three months.

Species in Great Distress

Tylotoriton verrucorus or the Himalayan salamander is listed in the endangered category of Schedule I of the Indian Wildlife Protection Act, 1972. Very little study has been conducted on this species in comparison to the work done worldwide on other salamander species. Since June 1998, studies on the habitat, ecology and breeding biology of the species in Darjeeling has been conducted by Shri Aurobindo Institute of Indian Culture (SAIIC), Shillong. This programme was funded by WWF, India.

But, unfortunately after 1998 this species is no more safe even in Jorepokhri, as it is gradually being converted into a tourist spot. The pond where the salamanders used to gather in large number for breeding is now converted into a swimming pool. Use of materials to build the swimming pool has adversely affected the entire habitat of the salamanders. Being unaware of the changes the animal arrives at the pool site to breed. Even it lays eggs too. Unfortunately those eggs never develop into normal embroys in that condition. The consruction Labourers catch the salamanders and shift them in small temporary refuge. Being in great distress there they pile up on top of each other. As a reuslt 50 to 60 are found dead daily.

Stringent Legislation

Thus it is now observed that human disturbances hve threatened the very existence of the crocodile salamanders. Due to alternation of the salamander habitat through cultivation, habitat modification and pollution their survival is at stake. Their distribution is now restricted mostly to the limited number of water bodies. Although no population estimates have yet been made, observations indicate that the probable reason for the species becoming endangered is the loss of water reservoirs. Disposal of domestic sewage into the water bodies and their pollution due to discharges of pesticides, fertilisers and detergents also take a heavy toll of the species population. Deforestation resulting in landslides and livestock grazing which loosens the soil causes siltation of the ponds and pools and is thus also responsible for destruction of salamander habitats. Introduction of carp and other freshwater fish into the breeding ponds also poses a serious threat to salamander eggs and larvae. Wild specimen collection is also conducted for research, and for food and medicine.

With the decline in the amphibian population and the threat of extinction that looms over the species, the pressing needs of the day are a strong legislative support for pollution regulation, habitat preservation, more stringent control over the wildlife trade and increasing awareness about their conservation. Lentic habitats in the Himalaya are gradually becoming dry. More specific steps must therefore be taken to halt human habitation, rockblasting, quarrying activity and other anthropogenic disturbances in the vicinity of the salamander habitat. Deforestation of the watershed areas should be actively prevented to preserve the water bodies so that the salamanders can breed. Water bodies recognised as salamander habitats should also be protected by strong legislation against drainage of water for agricultural activities, and for water supply to villages, urban areas and tourist facilities.

17

Trapping of Beautiful Butterflies in Darjeeling

Introduction

So far over discussion on conservation of Himalayan species has been focused mainly on the important and rare wildlife species of the Himalayan region. The insect species have gone unnoticed in our discussions.

This chapter is specifically on butterflies. Trade on rare insect species has grown worldwide. How thousands of beautiful butterflies are captured for commercial purpose thus threatening the existence of these species is explicitly described in this chapter. In the deep river cuttings of the Himalaya plenty of butterflies can be seen in summer. As we move towards east we will find more varieties of butterflies in the damper climate there.

Commercialisation of Butterflies

For several years trade on butterflies had been carried out by international traders with Darjeeling as the centre. They used to engage the local young ones to hunt the insects. Simplicity of local residents made the trade easier for the traders.

Early nineties was the park period of this trade till the arrest of a foreign national at Darjeeling in 1996 in that connection. As a result the local hunters who had so far been engaged in collecting the butterflies, fled to neighbouring states, countries or districts. They ever destroyed their instruments used for trapping butterflies. But this could not stop this illegal trade in Darjeeling. After a short lull period it started again. In due course of time, these youths became habitual offenders. They started demanding higher shares of profit. Finding it different to deal with these young ones, the traders used to engage the fresh youths. Besides that, local language was the another major problem for the traders. This problem stopped them from approaching the local trappers directly. So they had to depend on the dealers, who were mostly from Darjeeling. Anyway, gradually the trade spread to mountain areas of Ladakh, Sikkim and the Nilgiris and also in Andaman Nicobar Islands.

International traders had vast knowledge in this sphere. They often consulted relevant books to update their knowledge from time to time. On the other hand Indian dealers were also not far behind. The dealers used to visit the trapping sites to acquire practical knowledge. They even sometimes used to help in trapping the butterflies whenever there was shortage of local youths to do the job. Thus helped to learn the trapping methods practically. Soon they became middlemen and started purchasing the butterflies from the trappers. In course of time their knowledge on taxonomy, especially about butterflies, became super, even in some cases better than the knowledge of some scientists of this field. As per the available information Princeps janaka (Tailed Redbreast) if fairly common in Phalut region of Darjeeling.

Factors Determine the Price of Butterflies

Gradually the local residents became so expert in trapping that they could easily trap the butterflies alive without causing any damage. They had to be careful about causing the damage because the move damaged butterflies fetch them less cost. Of course cost depended upon some other factor too. Even finger marks on butterflies reduced the price of species like Papilio rhodifer or P. mayo.

Above everything, rare species, females and the one which is relatively difficult to trap are costlier. Rare species like Polyura scabieri fetched a very high price even for the damaged ones also. Though Hispirids generally do not fetch so much price. Rather in order to reduce weight of the baggage the trades used to avoid them. But the same variety of Andamans are preferred. These Andamanese butterflies fetch more price than the mainland species. Butterflies of Nicobar are the costliest ones.

Ways to Trap Butterflies

It is not so easy to trap butterflies. Bookish methods are applicable for certain species only. Local trappers, through their practical experience, have developed some innovative methods. They are more efficient than the Indian dealers. Other than using net to capture butterflies there are some other popular methods too. They are:

(a) Colour trap: To trap high flying butterflies brighter colours are appropriate. Bright-coloured papers can also be used. In order to entrap the butterflies a snare is made of colourful dupattas and attached to the end of a long culm of bamboo. Butterflies being attracted by the colour enter the net and get trapped. This process is more suitable to trap the butterflies of Papillionidae family. Colour varies according to the species for example, Pazala eurous (Six-bar Swordtail) is attracted by white net. Whereas blue-coloured net allows Achillides Krishna (Krishna Peacock), Papillio polyactor (Common Peacock), P. paris (Paris Peacock) and P. arcturus (Blue Peacock). The species like Teinopalpus imperialis and Achillides polyactor are tempted by green coloured net.

(b) Mixture trap: Ferment the mixture of raxi (a country liquor), gur (jaggery), rotten pineapple and banana for two days. Then scatter it into moist places near the streams. This mixture generally attracts butterflies from Nymphalidae family.

(c) Illusion trap: This trap is appropriate to capture the high-flying butterflies. For this process a damaged or dead butterfly is hung at a highly visible place. Butterflies of same species are attracted and are drawn to the trap. This process is suitable to capture papilio paris and P. buddha and is common in Nilgiris.

(d) Urine and Stool trap: For urine trap a wet and sandy area on the borders of a stream is to be selected. Pass urine there. It draws a number of species within an hour. On the other hand, a rocky site beside a stream is chosen for stool trap. These two methods are used to trap polyura dolo (Stately Nawab) P. athamas (Common Nawab) and P. endamippus (Great Nawab). Undoubtedly, the offensive smell of these trapping processes disturbs the senses.

(e) Vinegar trap: To trap Nymphalidae family, Dophla sp. and Euthalia sp. this method is used. Pour vinegar onto rotten wood. Leave it for a day. The trap is ready. This method is common in Thailand.

Beside all these methods, there is another common method. Butterflies generally follow a fixed route. Originally the route is chosen by a leading butterfly. Soon the same root is followed by a number of butterflies of that family. This process goes on. The trappers wait in that particular route and capture them easily. This method is used for trapping papilio sp.

Butterflies belonging to Lycanidae family are commercially not so important. But the specimens that behave in a particular manner are still preferred to be purchased. Specimen with right quality moves up to the top of a tree from the ground. Circles the free top. Then rests upon a leaf for a couple of minutes. At that position it observes using its antennae. It is very alert at this position. Again it moves around the tree top before it rests again and all of a sudden it disappears. This process is repeated. This butterfly is trapped while resting on the leaf. Being disturbed it tries to run away but unfortunately gets trapped in the net.

Trade Network

To begin with, the foreign traders started operating this illegal trade from Darjeeling. But soon they covered the whole country. The foreign traders acquired knowledge about the particular species before capturing them. Based on their knowledge they struck a bargain with the Indian dealers. The latter also followed the same policy and started bargaining with the village trappers. In order to earn more the village trappers started capturing rare and costlier species that too in good quantity. It had reverse effect. As the number of collection went up, the price went down considerably. So much so, the price of female teinopalpus imperialis (Kaiser-i-Hind) which was Rs. 2,000 each in the early nineties came down to as low as Rs. 300 to Rs. 400 each in recent years.

This trade soon spread over the habitats in some other countries. A trader was arrested for this offence in china and was fined heavily. Quietly paying the amount he fled to Latin America to carry on his trade there. There are reports about some odd incidents from Thailand and Philippine over the payment. This proves the existence of this illegal trade there too.

Due to legal complications carrying the butterflies out side India is difficult. Still, it is reported, that the international traders carry the butterflies personally saying that the specimens they were carrying were gifted to them by their friends in India, or has been purchased at an insignificant price. Considering that high price of the butterflies would alert the authorities, these traders never carried with them any document of transactions. All necessary papers of transaction were carried in the packings of Darjeeling tea, for which no question was asked so easy to take out of the country. Except that those specimen were sold through action by the international traders for a huge amount. any other end use of butterflies are not yet clear.

Conclusion

Millions of butterflies across the world have been either or trapped for illegal trade. As a result some species have already exhausted while many other species are on the verge of extinction. Disturbances in their habitat are also major factors behind this situation. Hebomoia glaucippe species are at the brink of extinction due to growing number of teastalls and felling of trees for cremation the dead in the Pankhabari area of Darjeeling.

18

The Lloyd Botanic Garden—Its Aesthetic and Recreational Value

Introduction

Darjeeling was discovered in 1828 by Captain G.A. Lloyd with J.W. Grant. The prestigious Lloyd Botanic Garden of Darjeeling was the vision of late Mr. William Lloyd, a well-known banker of Darjeeling. Rich in heritages status and conservation value, this Garden has been laid out in three levels, comprising three broad sections. Unfortunately due to an act of sabotage the scientific work at the Botanic Garden got a jolt in 1987. It reduced its valuable documentary records to ashes. Initially under the control of the Royal Botanic Gender, Howrah, the Lloyd Botanic Garden was transferred to the Department of Environment under the Government of West Bengal. We will know more about this Garden in details in this chapter.

Features of the Garden

The Lloyd Botanic Garden is the only conservation garden of its kind across the whole of Asia. The 567 ha of prime landscape was donated to the Government in 1878 by Mr. William Lloyd. Under the guidance of Dr. J. Anderson, the work to set up the prestigious garden as a branch of Royal Botanic Garden of Howrah started the same year. It was ready within a year. If lies within the heart of Himalaya.

The agerage elevation at the Garden is over 1800 m. Its collection represents flora of Eastern Himalaya, important plant specimens from the temperate and subtemperate Himalaya, and plants from other regions of the world with similar topography and climate. Its excellent collection of Himalayan flora range from stately conifers and broad-leaved trees, coloured rhododendrons and azaleas, to the smaller flowering plants like geraniums, lilies, etc.

Being maintained and developed by the Silviculture [Hills] Division of the Research Wing of the Forest Directorate, this Garden houses nearly 1500 species of plants in the categories of trees, shrubs, herbs, ferns and orchids which present the endless variety of forms and colour of deciduous and evergreen trees and shrubs. Besides the green sloping ground, rock garden and animal flowers beds the rich collection of this Garden is spread over the Garden's conservatory, herbariums, cactus house and orchid house, and laboratory. These are the priceless heritage collection of the Garden.

The conservation work of the Lloyd Botanic Garden have been guided by certain distinct scientific objectives. These are:

(a) to promote the scientific study of plant diversity, providing a mechanism for introduction and assessment of plants for forestry, medicine and horticulture purposes.

(b) to preserve, conserve and propagate plant genetic resources of indigeneous and rare exotic collection which have economic and taxonomic value.

(c) to generate awareness about the ecological value of trees and educate visitors about plants and about their own relation with nature.

(d) to establish and maintain a herbarium centre for the study of modern taxonomy and for easy identification of plants.

(e) to provide an environment of peace and tranquility to all visitors and nature lovers.

Lay-out and Administration of the Garden

At the heart of Darjeeling town below Eden Sanatorium is located the Lloyd Botanic Garden. Surrounded by Lebong Cart Road to its north, the Jail Road to its south, the Eden Sanatorium to the east and Victoria Road to its west, the Garden is spread over an area of just over 16 ha. Initially the Garden was under the jurisdiction of the Royal Botanic Garden at Shibpur in Howrah, with Mr. A. Jeffery as its first curator. In 1963 the Royal Botanic Garden was taken over by the Botanical Survey of India [BSI]. Botanic gardens at Darjeeling and other locations in West Bengal were left under the jurisdiction of the State Agriculture Department. In 1967, Lloyd Botanic Garden was transferred to the newly created Department of Environment under the

Government of West Bengal. The Divisional Forest Officer, Social Forestry [Hills] Division retained charge of the Garden. Finally, on July 28, 1994, the charge was transferred to the Divisional Forest Officer, Silviculture [Hills] Division.

Lloyd Botanic Garden has been laid out on three levels. The uppermost level represents the dominant broad-leafed species indigenous to the Eastern Himalaya. There are some rare specimens from the Western Himalaya also. The middle section represents the Himalayan coniferous species. In the lower section mainly exotic plants are found. A compilation preserved in the Herbarium of the Garden reveals the details of the species found there. They are: 195 shrub species, 60 species of herbs, 47 species of orchids, and 40 species of cactii. Side-by-side there are about 117 tree species and varieties of other flowering plants also. They represent around 20 countries of the world.

Remarkable Characteristics of the Garden

(a) Sir John Anderson Rock Garden: A rare combination of the rhododendrons, lilies, azaleas and prairies grows in Sir John Anderson Rock Garden. There are some medicinal plants also. Those are: *Cotonestor macrophylla, Meconopsis wallichii, Primula floribunda, Fragaria vasca, Saxifraga sermentosa, Ranumculus diffusum* and *Puya alpestris* with its tall, copper-green coloured spikes. Adjacent to the Rock Garden ascent the rhododendrons in thick bunch along the right hand side slopes. In spring season, they bloom in a succession of hues ranging from white through pink to crimson. Some old specimens of the Himalayan cherry *Prunus cerasoides* are found a little below.

(b) The Herbarium: Adjacent to the office complex is located the Herbarium. It is an Elizabethan-style building. Beside dried specimens of 175 species of Eastern Himalayan flora, the Herbariumpresently preserves some seed specimens of the hill species. Bentham and Hookes classification was followed to arrange the valuable collections. But the outbreak of five during the 1987 agitation records and to build a new specimen collection was initiated in 1997 as per the guidelines of BSI.

(c) The Orchid House: Towards the east of the Garden, there is a spacious glasshouse constructed along modern lines. This is called Orchiid House. It is located in the office complex. The Orchid House contains more than 47 orchid species that are found in the Himalaya.

(d) The Conservatory: To the north of Anderson Avenue there is an arched glass structure supported by iron frames. It is the famous conservatory of the Garden. It is, possibly, the most pleasant spot in the Garden. There are some beautiful flowers. Amidst some adorable ferns like Asplenium and Adiantum, those flowers are well arranged on stands at four levels. A semi-hardy long losting plant with very brihgt orange coloured flower is the added attraction of this conservatory. This flower is rightly calld Bird of Paradise flower (Strelitzia reginae). It is so called because the flower resembles the head of the exotic bird.

(e) The Cactus House: To the right of the Conservatory is the Cactus House. It contains around 40 species of cactii. Besides that there are several other succulents and a bonsai collection of pine, acer, rhododendron and other trees. Stag's horn ferns, *Boeya* and the hanging Zygo cactus are the unique specimens found here.

(f) Living Fossil Plants: The Dawn Redwood *[Metasequoia glyptostroboides]* is a species of immense botanical and geological interest. We know about it from the fossile record. It was found growing in the wild in the Sichuan province of China in 1994. There is a mother tree of this species to the left of the Conservatory in the Lloyd Botanic Garden. Another smaller specimen growing there was propagated from the mother tree. As the lower portion of the trunk of the mother specimen resembles the 'Phantom's Cave' of the comic-books, this is also known as the Phantom tree. Recent research has shown that during the cretaceous period, the species was widely distributed across North America, Asia, and Greenland and several Arctic and sub-Arctic islands. Each autumn, the tree sheds its branches along with its leaves.

It is also known from fossil records that the Maiden Hair tree (Ginkgo biloba) is the only surviving member of the family *Ginkgoales.* A.C. Seward [1938] remarks about the tree'as an emblem of changelessness, a heritage to the world of an age too remote for our human intelligence to grasp, a tree which has in its keeping the secrets of the immeasurable past'. For its edible seeds, it is widely cultivated in China and Japan. Its fan-shaped leaves change colour during autumn before they are shed.

(g) Plants of Medicinal Value: A large number of plants medicinal importance are found in Himalayan forests. On a small plot below the conervtory in the Lloyd Botanic Garden about 52 species of such plants have bene planted. Effort is on for their further extension and development. Local people are well aware about their usage.

(h) Other Attraction for Visitors: The large, outspread trunk of an old climber that attracts the attention of every visitor while walking into the conservatory is a specimen of Wisteria sinensis (Chinese wisteria). It is so large that presently it is supported on pillars. Belonging to the family of Leguminosae, it produces aromatic, deep lilac flowers in bunches. In 1878 this specimen was brought from a garden in candon in Chiina. This most impressive and the oldest climbers at the Garden derived its name from a distinguished professor of anatomy, Caspar Wister.

Another source of public attraction is the northern sector of the Garden. Lying to the left of Anderson Avenue this sector has the Tulip Tree *[Liriodendron elatus]*, mature specimens of the exotic oaks *[Quercus* sp.] and an eye-catching collection of *Camellia* sp. in a sunken garden. At the Garden's western end, there is an exuberant grove of Japanese bamboo. The green beautiful lawns as the northern extremity are home to the fascinating tree ferns and to *Cyathea spinulosa,* the prettiest of all ferns.

Aesthetic and Refreshing Value

The Lloyd Botanic Garden has a well-stocked library too. It has a collection of more than 100 books and journals covering various life sciences including botany, horticulture and taxonomy. It holds specialised interest for researchers and scholars. The old plant nursery located below the Curator's bungalow. Presently it is being renovated and upgraded with modern technological systems, which includes automated root-pruning stands, ultraviolet-resistant agro nets, root trainer hycopots and a compost unit.

Besides providing scientific facilities, the Garden has a beautiful picnic spot also. The spot located in a clearing on the avenue towards the Victoria Gate. There a pond with a fountain and dancing water lilies entertain the visitors. To view the beauty and tranquility of the Garden there is a pavilion above the fountain. It is a resting place too.

With its unique combination of aesthetic and recreational value with scientific and conservation values the Garden has attracted the visitors for more than 120 years. It is also a heritage spot and study centre for botany students and researchers from all over the world. The Garden is often visited by more than 100 species of birds and about 50 species of butterflies, damsel flies and dragon flies. This odds beauty to the resonant colours of the Garden.

Degradation of Loktak Catchment and Remedies

The only natural habitat of the rare and endangered Sangai or Brow-Antlered Deer is the Keibul Lamjao National Park. Under the Ramasar Convention [1990], Loktak, the largest freshwater lake of the North East, is listed among the most fragile wetland ecosystems of the world. It has a catchment of 980 sq km.

The lake is the source of water for electricity generation at the Loktak Hydro-Electric Project. Streams and rivers draining the Loktak catchment annually deposit huge quantities of silt into the lake. It threatens the delicate wetland system, the Keibul Lalmjao National Park and also the life span of the Hydro-Electric Project. As in some other parts of Himalayan region, the Loktak catchment is also adversely affected by inappropriate agricultural practices, overgrazing, ineffective control over felling and burning of trees. Lack of awareness and economic development exist among the people are the result of this devastation.

Implementation of the integrated Afforestation and Eco-Development Project through the Ministry of Environment and Forests, Govt. of India, offers some remedy. Identifying the priorities the project has taken some effective steps for—

- Developing forests to meet the demands for fuelwood and fodder demands of the local people
- Regenerating degraded forests
- Restoring vegetative cover on denuded hills
- Meeting the fodder needs of livestock
- Restoring the ecosystem in catchment areas
- Reducing soil loss and controlling torrents
- Raising awareness among the people

Most Unique Park in the Country—A Pride of Darjeeling

Under the Department of Education, Govt. of West Bengal, for the study and conservation of the Himalayan Fauna, a zoological park was established on 14 August 1958. On 21 November 1975, the park was dedicated by Shrimati Indira Gandhi, then Prime Minister of India, to the memory of Shrimati Padmaja Naidu, former Governor of West Bengal. Since then, it has been known as the Padmaja Naidu Himalayan Zoological Park. It ranks among the most unique in the country.

Against the glorious backdrop of the Kanchenjunga, this zoological parks is situated at an elevation of 2133.5 metres in Darjeeling. It is the only specialised zoo in the country. It has also received international recognition for its conservation effort because of its successes at captive breeding of the Snow Leopard and Red Panda. Padmaja Naidu Himalayan Zoological Park has been striving for the maintenance of ecological balance in the Eastern Himalayas with the following objectives.

- Conservation of natural resources and Himalayan animals, with research into animal behaviour and the ecology, biology and physiology of different high altitude fauna and flora
- Raising public awareness towards a role in conservation along with relevant training and education.
- Reintroduction of rare and endangered animal species into the wild
- Coordination of a conservation breeding programme

Padmaja Naidu Himalayan Zoological Park neither sells nor trades animals. Rather, when a need arises for conservation purposes, it supplies the animals free of cost. Padmaja Naidu Himalayan Zoological Park also procures those animals which fit its objectives. So, while all other zoos in the country still depend on the breeding of important species, Padmaja Naidu Himalayan Zoological Park has achieved notable success with all endangered animals in its collection. Substantial increase in the number of important animals has occurred over recent years. Red Panda and Snow Leopard populations have tripled. The number of enclosures at the zoo have approximately doubled during the last 5-6 years. This has enhanced the zoo's stature at the international level.

People who want to enjoy the sight of some rare Himalayan animals such as the Siberian Tiger, Tibetan Wolf, Orang, Yak, Blue Sheep, Himalayan Black Bear, Red Panda, Snow Leopard and Pheasant must visit the zoo.

19

Changing Process of Conservation Around the Great Himalayan National Park

Introduction

Wildlife conservation is a major universal problem. So many scientists, environmentalists, biologists, Forest Department, group of villagers and some other social groups are busy to meet the challenge. Each claiming to uphold the cause of conservation better than the rest. These particular groups or individuals pin-point the extraordinary cause of conservation and use it as the touchstone to explain themselves. Their practices are so intricate that the concern for conservation is difficult to recognize. Various social groups came out with different agenda for the conservation programme. As a result conservation emanate as a contested terrain where, not just nature as wildlife, but nature as the innate charracter of social being is staked and defended.

Generally, controversy around conservation is not clearly marked. Complication of conservation is clear when the career of ecodevelopment is examined. For wildlife conservation in India it is a new strategy. When the group of villagers, biologists, World Bank or Forest Department stake claim to conservation, they claim their rights over natural and social spaces.

This chapter reveals the changing perceptions conservation and degradation around the great Himalayan National Park, Mutual conditioning of action, knowledge and representation around conservation is also focused in this chapter. In making of a conservation project how state agencies and various groups disagree as well as collaborate is highlighted here. Moreover, how the conservation science and law led to the formation of the National Park and the role of World Bank in shaping state and community management structures with special emphasis on its role in making the Forest Department reinvent itself is also discussed in this chapter.

The Grant Himalayan National Park

Located in the Kulu district of Himachal Pradesh, the area of the Great Himalayan National Park has been described as a prime example of the North-West Himalayan biogeographic zone (Pandey 1998:4) Extending over an area of 765 sq km the area consists of high mountain ridges, rearing up to 5800 m. It is divided by the gorges cleft by three streams—the Tirthan, Sainj and Jiwa Nala—which ultimately flow into the Beas, and the separate watershed of the Parbati. These upper reaches of the Beas catchment are covered by moist temperate forests of oak and conifers, and high-altitude forests of birch and fir interspersed with sub-alpine pastures, crowned at the highest reaches by alpine meadows, scrub and permanently snow-capped peaks (Kothari *et al.* 1989:146). Rich flora and fauna are found in this Park. The list of fauna includes: several endangered species such as the musk deer, serow, brown bear, blue sheep, Himalayan tahr, western tragopan, cheer and monal pheasants. The range of herbs grasses and sedges that grow in the high-altitude pastures of the Park in the spring season serve as fodder for wild herbivores.

Other than being fodder, these seasonal herbs, grasses and sedges are commercially very important because of its medicinal value. From May to October the pastures inside the Park also provide fodder for some domestic animal like sheep and goats. Villagers also collect morel mushrooms from the shaded understorey of the temperate forest. These mushrooms (Morchella esculenta) are exported to Europe. Herbs such as dhoop (Jurinea macrocephala), kadu (Picrorbiza kurrooa), and nhaini (Valeriana hardwickii) are supplied by this region to the cosmetic and pharmaceutical industries.

Conservation Attempts Since Colonial Period

Practice of hunting during the colonial times intensified by British sport shooting for animal trophy and game birds. Besides that feathers of bird were used to crownthe ladies hats, and gentlemen use to collect them for natural history collection. This resulted in depletion of game species by 1870s. This alarming situation forced the British Government to regulate hunting by introducing the system of permit for hunting and pointing out the areas. So we find that conservation attempts in the area started since colonial period. After 1947, it was realised that the system of licenses did not work well. Easy accessibility of guns to protect crops and rising prices of musk in the international market endangered the existence of musk deer.

Moreover, the villagers used to trap the pheasants. When snowfall forced these pheasants to come down to lower altitudes in search of food, it becomes easy to trap them in that position. Besides meat, pheasants were fair game because of their feathers. The colourful crest of monal is still a valuable decoration for caps inKulu. The vilalgers also shot the musk deer for its pod, which is used in perfumery. Other than these, a little bit of poaching by shephards and their dogs is also reported.

For centuries the environment of the Park and its inhabitation have been subjects of state action. Hunting was mostly banned in Himachal Pradesh in 1982. In 1984, it was prohibited particularly in Great Himalayan National Park (Tucker, 1997: 23-24, 42).

In the conservation programme, after hunting the next focus was on forest use. Forest areas were divided in two classifications—timbers and tax. While foresters were interested in conserving timber stands, the Revenue Department was interested in expanding cultivation into forested areas, thus increasing its returns from agricultural taxes. While accommodating local demands for agricultural land, timber, grazing and other products the 1878, Forest Law further categorised forests such that timber availability for the colonial railways was maximised. Colonial forstry practices, as well as those which prevailed for forty years after Independence, placed commercial timber production above all else. For the colonial forester (or for that matter the Indian forester To have pure stands of deodar *(Cedrus deodara)* was the main object of forest management for the colonial foresters as well as Indian forests. When later on kail *(Pinus wallichiana* or blue pine) and chil *(Pinus roxburghii)* became valuable these were also added in their list. Though villagers' access to timber and grazing, was restricted, there was no check on the collection of medicinal plants, bamboo and fruit for villagers' personal use and sale. Even there was no restriction on collection by outsiders also. It was because demographic pressures were low enough to be treated as insignificant.

That grazing had been the reason behind the degradation of forests and limits had to be set on the number of animals to be allowed in different forests was the opinion of foresters in 1870s. So, the most important issue before the colonial state was the regulation of grazing. Permanently settling the grazing rights and the maintaining its proper records was not an easy task. It task more than five years to complete the process. The foresters wanted to settle the rights once for all. The Settlement Officer, A. Anderson predicted great difficulties in settling the rights. He tried to reconcile the views of local villagers with those of various sections of colonial administration. His argument regarding the issue was that the process must an ongoing one. He stated:

> The finality of the record of reserved forests, which has so many charms for the Forest Officer, is to me an objection. If the claims were few, mistakes would also be rare, but here the forests are close to homesteads and are the resort of the people for the supply of all their wants. Under the circumstances, a record, however carefully prepared, must also be incomplete, and to make it final would be dangerous (Anderson, 1894:26).

Marking the boundary of separate zones for different land use by different groups was necessary to minimise the area dispute. It was an important step to be taken for the settlement. But it fomented a new set of social divisions and disputes. Village leaders *(negis)* of the region told Anderson that they did not want open grazing in Class II forests. They wanted the rules enforced to keep away outsiders. Anderson in the event, prepared a detailed list of individual rights in different forests for each village and hamlet. It was adopted as a law in 1896. But pressures from villagers as well as from the Forest Department resulted in freezing the grazing rights at the turn of the century.

For securing access to timber as well as for increasing revenue codifying and controlling relationship between forest resources and the people depending on them was most necessary. It was also a part of the colonial mission of Improvement and Civilisation. Expansion of commercial agriculture,[3] and the introduction of new fruit and vegetable crops such as apples, plums and apricots could be noticed during this period. This period was also marked by the extension of roads. These initiatives were greatly accelerated after Independence. Road-building, dam construction and timber work attracted a number of traders and construction labourers towaqrds the Great Himalaya National Park. Moreover these brought more income to the area.

Measures to Evaluate Wildlife and Protected Areas

Responding to environmental movements of the North began during the period of Prime Minister Indira Gandhi. Deforestation, road-building and expanding populations increased environmental stress in the Beas basin by the late 1960s.

The Union government passed the Wildlife (Protection) Act in 1972. Under this Act certain ecologically important areas could be designated as National Parks and Wildlife Sanctuaries. This pattern of conservation emulated the 'national parks ideal' propagated worldwide from the USA. It was based on the idea that nature can be preserved from the effects of human activity by legislatively creating a bounded space for nature controlled by a government authority (Neumann, 1998:9).

The government of the newly-formed state of Himachal Pradesh[4] was keen to establish its own protected areas. It declared 8396 ha as the Tirthan Wildlife Sanctuary in 1976. International NGO such as the World Wildlife Fund and the Himachal Pradesh government launched the Himachal Wildlife Project in 1980 to evaluate the status of wildlife in the state. As part of this project, two biologists, A.J. Gaston and P.J. Garson, conducted a brief survey of the birds and large mammals in the reserved forests of Kulu. Based on their recommendation, the initial notification of the intent to create the Great Himalayan National Park was issued in 1984. Then in 1987, a ten-year management plan was prepared. In retrospect, it is clear that.

The Himachal Wildlife Project suffered from a glaring omission in its survey. It failed to examine human dependence upon the Park. The biologists did not consult villagers during their survey. They depended on the casual conversations they had with the porters accompanying the villagers and with shepherds met along the way. So they failed to collect the actual report about human and livestock pressures. As a result, the ten-year management plan assumed that local people were mainly responsible for degrading prime wildlife habitats. Based on that assumption it was suggested that they be provided alternative employment in exchange for giving up their traditional resource use. The opinions of biologists, as represented in the Himachal Wildlife Project, lent the weight of scientific legitimacy to the conservationist urges of the Himachal government.

Early researchers were highly impressed that the 'upper Sainj and

Tirthan valleys containe] both exceptional forests and a remarkable and complete array of the larger, more ecologically sensitive wildlife species typical of the Western Himalayas' (Garson, 1983:3). Despite its richness, scientists defined the region as ecologically delicate and under threat, especially due to grazing. As one study concluded:

> grazing has . . . modified the understorey vegetation considerably, reducing the amount and diversity of shrubs and ground vegetation. Along grazing routes in particular, large areas of meadow consist almost entirely of dock. . . and other nitrophilous herbs. The understorey is best developed in areas subject to little or no grazing. . . Comparisons of heavily and lightly grazed areas lead us to the conclusion that grazing by domestic animals is causing severe alteration of the natural forest flora. In most areas this reduces the suitability of the habitat for wildlife' (Gaston, Garson and Hunter, 1983:311).

Biologist P.J. Garson concluded that the upward spring migration of flocks to the high pastures probably disturbed the nesting habitats of pheasants and other species (Garson, 1983:7).

The survey report initiated the administration invoke the provisions of the Wildlife (Protection) Act for the area. Biological beliefs and the law concur on the issue of grazing. Under Section 35 (7) the administration prohibited grazing in the National Park. Other human use of the biodiversity inside a Park is permissible only if the Chief Wildlife Warden of the state authorises it on the grounds that 'such destruction, exploitation and removal of wildlife from the National Park is necessary for the improvement and better management of wildlife therein.' Other than scientific research and educational tourism there is a ban on all human activities. Legally, rights to natural resources inside the Park would have to be acquired by the state and the right-holders compensated, either with a one-time cash payment or by making provisions elsewhere. In the case of the Great Himalayan National Park, this process came unstuck at the start.

Thanks to Anderson's Forest Settlement Report of 1886. Villagers' rights had been recorded in minute detail in this report. But, surprisingly, it had never been revised. Villagers' confidence about their rights to resources inside the Park is mostly derived from this document. The Report acquired the aura of a sacred text; when questioned about their rights, illiterate village elders, who could never have read the volume, will say that on such and such page of the book the names of their ancestors are recorded and that is how they, the descendants, come to have rights. The rights accorded in 1886 now operate in a vastly changed context. Human and livestock populations, land use practices, lifestyles and links with the monetised economy are no more the same as existed more than a century ago. As per the law villagers' access to the Park, was curtailed. This step the Park authorities have been reluctant to grasp.[5] Himachali villagers being confident about their rights they disobey the the law of the Park. They are confident about their political backing too. In case of any major trouble they do not hesitate to call-up the concerned MLA who rebuke the Park Director on phone for harassing the villagers. This forced the Forest Department of the Great Himalayan National Park to use defensive.

And so ecodevelopment. If the Park were to be fully notified, the prospect of considerable resource displacement seemed inevitable. More than 150 villages on the western and southern edge of the Park and quite a large number of pastoralists from further afield stood to lose a crucial component of their economy. It was a threat at which brought forth murmurs of local protest. To avoid headaches like offering inadequate compensation and facing public wrath, trying to identify alternative resource sites in an already burdened landscape and so on—conservationists suggested that the Great Himalayan National Park be adopted for the ecodevelopment approach. Given the fraught relationship between Park authorities and people, ecodevelopment presented itself as a solution to all problems. To stop people from going into the forest create alternative sources of livelihood around the protected area. Give them an incentive. This will divert their attention away from their customary practices and they will surrender their rights. No longer will the Settlement of rights will no longer be such obstacle in the process of notifying a protected area. if these simple strategies are followed.

Conservation Process: Role of International NGOs and WWF

Ecodevelopment is like rural development. But it is mainly directed towards natural resource conservation with special emphasis on wildlife conservation within a protected area. Officially, ecodevelopment is described as 'a site-specific package of measures derived through people *(sic)* participation, which addresses all aspects of land use and other resources, in order to promote sustainable land use practices, as well as income-generating activities, which are not deleterious to Protected Area values' (Panwar, 1992, quoted in WII 1996:8). To divert biotic pressure from the Protected Area is the main objective of ecodevelopment (Kothari, 1996). This can be through planting trees to meet villagers' needs for fuel and fodder and starting cottage industries to provide other sources of income.

It is important to note that ecodevelopment operates within the parameters of the Wildlife (Protection) Act. It assumes that wildlife conservation is a priority which overrides people's rights to resources within the protected area. Unlike Joint Forest Management, another Forest Department initiative, which either maintains or augments 'local community' rights to non-timber forest produce and also promises local villagers a share of the harvested timber, ecodevelopment accompanies the state's alienation of villagers' rights.

After experimenting with the strategy of ecodevelopment in Africa see Neumann, 1997: 562), the WWF, an international NGO, first mooted the idea for India's protected areas around 1990. This idea got support at the 1992 United Nations Commission for Environment and Development (UNCED) Conference in Rio-de Janeiro. Following this World Bank created a special fund known as Global Environment Facility (GEF). The Indian Government, in consultation with the international NGOs and World Bank, identified possible protected areas where ecodevelopment could be tried out. In the first instance, two National Parks—the Great Himalayan National Park and Kalakad Mundanthurai in southern Nilgiris—were selected. International Development Assistance (IDA) funded this ecodevelopment programme. For the second phase, seven parks and sanctuaries were chosen. They included Ranthambhore in Rajasthan, Rajaji in U.P. Gin. Gujarat and Melghat in Maharashtra and funded by Global Environment Facility (GEF).

Since ecodevelopment affects only tiny populations around selected protected areas the World Bank's interest in wildlife conservation seems somewhat removed from its overall mandate of sustainable development. The Bank's interest in ecodevelopment possibly stems from its desire for social experimentation. Having burnt its fingers with projects requiring the forced displacement of large populations (such as the Sardar Sarovar Project), the World Bank now seeks to try out the softer option of voluntary' resettlement of livelihood rights. Also, initiatives such as Joint Forest Management and Participatory Irrigation

Management, have the backing of the Bank and bilateral donors. These enterprises are part of a larger framework in which the Indian economy is being liberalised as well as the Indian state is being redesigned to facilitate the expansion of private capital. Creating new institutions and procedures based on 'state and community partnerships' which provide 'good governance' is part of the Bank's agenda of 'reshaping civil society.

Coincidentally, such 'institutional divestment' has also been demanded by social activists and advocates of Joint Protected Area Management. Though they are otherwise highly critical of the environmental impact of World Bank and IMF liberalisation policies. They argue that instead of performing as a protector of public interest and as the guardian of socially distressed groups, has generally acted in the interest of elite groups and against exploited classes, so the shift towards community participation is a welcome one. The creation of new rural corporate bodies such as village forest protection committees, water users' associations, etc. will allow under-represented groups who are adversely affected by top-down state policies to shape conservation practice (see Kothari, 1996; Gadgil and Guha, 1995). However, the implications of 'institutional divestment' may be quite complex. The devolution of power may also imply the devolution of conflict (see Sundar, 2001). That is, the state may pass on the burden of enforcement to different social groups, transferring costs as well as creating new disputes in the process, pitting resident villagers against itinerant pastoralists or neighbouring villages against each other.

At the same time, the Indian state has been reluctant to surrender power. What we may be witnessing is not a democratisation of the state, but in fact a leaner, meaner state which still stays in control. As Tania Li argues:

> the effects of the decentralisation measures proposed by community advocates [and by institutions such as the World Bank] as a mechanism to roll back state power and strengthen communities need to be carefully considered. These measures have the capacity to further state territorialisation projects[6] and provide various other opportunities (e.g., through bureaucratic expansion, donor funding, international legitimation), for the intensification of rule (Li, 1997:75).

According to Li, the institutional arrangements entangle the countryside more deeply in state systems of information and control as every aspect of society including social stratification, local histories, conflicts and patterns of resource use and access is studied and made to serve in externally-driven planning operations. The reverse flow of information and resources does not take place; the freedom to set the agenda is not accorded to village-level bodies. Of course, these misgivings perhaps do not apply in the case of the Great Himalayan National Park. The new 'territorialisation projects' may not necessarily yield more power to the state. On the ground, projects are negotiated, subverted or may simply collapse under the weight of their totalitarian ambitions. In the case of the Great Himalayan National Park, ecodevelopment was launched in November 1994. Around 130 villages (2000 households) in a 5 km strip along the western edge of the Park were designated as an 'ecodevelopment zone'. The Great Himalayan National Park project is funded by a $2.5 million loan from the World Bank. The amount was to be disbursed over five years, repayable at four per cent interest. This money is to be spent on: (i) improving Park management (training, equipment, infrastructure); (ii) research (being conducted by the Wildlife Institute of India); and (iii) actual ecodevelopment. The loan conditions stipulate that the Park authorities must prepare detailed 'micro-plans' for every panchayat with the participation of villagers.

However, the exercise was entirely new to the Forest Department and micro-plans could not be made before 1994, despite a crash course in Participatory Rural Appraisal (PRA) techniques. Research gradually started in July 1995. In effect, this meant that research (which should have been an input into the planning process), planning and implementation, all started concurrently.

Ecodevelopment Process—Negative Response of the Villagers

Declaration of the area as National Park invited stern opposition from the local villagers. Starting of the ecodevelopment process intensified their opposition. They were under the impression that ecodevelopment would innitiate their rights. To reduce the tension, the Park authorities started an interim programme called TRUCO—Trust and Confidence-building Measures. Under this programme the Forest Department built bridges, repaired bridle paths, and improved water supply to gain the confidence of the local villagers. However, TRUCO was identified by some villagers as a move on the part of the Park authorities to bring themselves into areas of public life traditionally managed by the village panchayat. Villagers also felt that the Park authorities had to have some ulterior motive for their actions. Society for the Advancement of Village Economy (SAVE). SAVE started an active campaign against ecodevelopment. But when opposition leaders decided to piggyback on the agitation to attack the ruling party in Himachal Pradesh, the NGO retreated from direct action and concentrated on awareness building. To discuss the implications of ecodevelopment it brought out a pamphlet in Hindi. SAVE activists initiated a study, travelling with shepherds and medicinal plant collectors to examine their pattern of resource use, and concluded that if there was degradation, it was the result of unchecked collection by Nepali migrant labourers employed by traders. They stated that local practices were sustainable and villagers were mindful of natural limits.

That their dependence on natural resources within the Park is their lawful right as reflected in the Settlement Report and also the impression that ecodevelopment would effect their economy which, to a large extent, depend on the natural resources prevailing within the Park were the two main points on which the villagers based their opposition against the ecodevelopment programme. The SAVE introduced a new set of arguments into the debate. It points out that ecodevelopment has been initiated on the assumption that local resource use is at odds with the objectives of wildlife management. The scientific validity of this assumption has been challenged by SAVE. An alternative view proposed that traditional resource use by local villagers is based on sound ecological principles which have resulted in promoting, or at least maintaining, the interests of wildlife. By posing local management practices as superior to the scientific claims of the Park authorities, SAVE has shifted the debate away from a 'Park versus people' confrontation to a scenario in which villagers and wildlife have convergent interests which stand threatened by a Forest Department bent upon disrupting the fine ecological balance between the two. SAVE offers alternative interpretations of the two main activities regarded as degrading by the state, grazing and the collection of medicinal plants. SAVE's research states that villagers send their livestock for grazing, in herds of five hundred to a thousand, along with three or four shepherds and their dogs. Different villages have fixed routes, with specified numbers of days to be spent at places en route. The 1886 Settlement Report contains some of these routes. The routes are designed to make best use of available grass and sedge species along the way. Local villagers are aware of the lifecycles of different plants as well as their nutritional qualities. For instance, the highest pastures at Khandadhar become snow-free last and have the richest grass, a fact which is known and exploited by Sharan village through its grazing schedule. Other practices, such as bedding livestock down for the night near the small stone shelters where shepherds can sleep in some warmth, are designed to limit the spread of *Rumex* and other 'weed' species which reach higher densities in the presence of nitrogen-rich livestock droppings opines SAVE. By collating instances of such practices, SAVE argues that the shepherds have a good understanding of the environment that they use this innate knowledge enables them to use the pastures sustainably.

Grazing is indeed, a remarkable operation in its smooth functioning and efficiency. Villagers followed the grazing rules religiously. So there was hardly any incident in which villagers had to settle disputes with Park Authorities about the violation of rules and norms governing the grazing runs. Beside slicking to allotted routes and periods of stay, these rules also pertain to the maintenance of bridges and paths, etiquette regarding right of way, and so on.

Collection of medicinal plants from the Park area in another common practice of the local villagers. Successful collecting requires a detailed knowledge of the habitat, its aspect, altitude, topography, vegetation types, and so on. Beside that an idea where a particular plant may be found, and an understanding of moisture, soil and other micro-conditions favourable to particular plants are also necessary. And, of course, it presumes an ability to identify different species of plants. Like grazing, these skills are learnt on the job, passed on by the older men to the boys. Medicinal plants in the Park may be over-exploited has emerged as a matter of concern. In January 1997, many of the medicinal plants collected in the north-west Himalaya were included by the International Union for the Conservation of Nature (IUCN) in its Red Data Book. Patees and hathpanja were declared to be critically endangered', kadu 'endangered' and dhoop 'nearly threatened' (S. Singh: pers. comm.). It is also reported that dhoop roots which were as thick as the arm of the collectors who harvested them three years ago, are now no thicker than their fingers. According to local villagers, they harvest certain annuals only after they have flowered and set seed. However, some local traders employ migrant Nepali labours to collect plants from the Park. Though, being outsiders as well as unmindful of the plants lifecycle., the Nepalis are blamed for the over-harvesting, the real reason lies in the steadily rising demand for medicinal plants from pharmaceutical companies. Even local villagers also sub-let their rights to non-residents. There are reports that the villagers have started collecting some extra quantities of herbs to earn more money.

Being traditional stewards of the forest and its creatures, they have a close understanding of natural processes. Villagers feel that they have certain privileges too. They claim that the resources within thee Park are theirs. They won these by virtue of their longstanding customary rights. The rights that are recorded in the Colonial Settlement Report. As one elderly man put it, 'They call this a Park. Park means *bagicha* (garden). And one can pluck flowers from a garden'. A shepherd said, 'Just as God put the bird in the sky and made seeds for it to eat, so he placed our sheep on the earth and gave them pastures upon which to graze.' To express their opposition to the Park and to eco-development, villagers often use metaphors and analogies naturalising their relationship with nature and attributing to it a timeless quality. The identification of certain spots in the upper reaches of the Park as sacred sites endowed with mystic power adds another dimension of cultural signification to the Park landscape. It is an aspect which reinforces the feeling of continuity and timelessness suffusing villagers' perception of the environment. To further legitimise claims to resources within the Park, such an environment and sanctity of the Park people relationship is employed.

Changing View of Villagers

To go ahead with their action against ecodevelopment, NGO efforts drew specific attention towards the scientific validity of indigenous knowledge as well as the ecological sustainability of management practices based upon it. However, while asserting indigenous knowledge and challenging the scientific hegemony of the Forest Department, people have been compelled to construct a new myth, glorifying traditional culture and representing it as timeless and unchanging.[7] Although the 'eternal village' with its ancient institutions is a potent symbol to rally around, enabling village leaders to draw clearcut lines of difference in relation with the Forest Department and its outside experts, all villagers are not equally enthused by this representation of knowledge and community. The extent to which different people have supported the political use of 'community' is closely related to how they have responded to the opportunities opened up by changing institutional structures such as the market.

Increased link with the market through the improved road network, new cropping patterns where fruits and vegetables have gained ascendancy over barley, wheat and beans, combined with increase in human and livestock populations in the region modified the temperament and extent of people's relationship with the Park. Mostly larger landowners and those close to the road, took the maximum advantage of the new opportunities. They no longer depend on the Park. While livestock herds are getting smaller, grazing continues to be the mainstay of most poor and middle-income families. Dependence on resources inside the Park continues to be crucial for the survival of the weaker sections of village society. The collection of medicinal plants intensified. Since herb collection is a highly laborious activity, it is done mainly by boys and young men from poorer families.

The increasing differentiation of the village community has resulted in different political responses to ecodevelopment. It is the poorer and middle-income sections who provide the mass-base for opposing ecodevelopment. Among them, the invocation of indigenous knowledge resonates most in the minds of the older men. Majority of the young men even from the poorer families that mainly depend on resources within the Park, find grazing and herb collection very tiring. They preferred to do something else if they could. Education and market opportunities have turned their eyes away from the snow-clad rugged rocks of the Park towards the road and the urban settlements it links them to. For their livelihood the young men has shifted their preference towards Government job or tourism rather than depending on land and livestock.

Unable to come to grips with the of rural communities, the ecodevelopment programme has fallen back on the shortcut of dealing with existing administrative units, e.g., the revenue village, token representation for women and Scheduled Castes in the Village Development Committees, and assumptions about self-sufficient, subsistence-oriented, unchanging communities. The programmes follow a paternalistic attitude towards villagers. It refuses to consider that villagers may have a sophisticated knowledge of natural resource management. Villagers are treated as passive recipients of state initiatives and not as actors in their own right who bring their own agenda to a project, as well as their previous experience of dealing with the state (Mosse, 1986). Villagers were not at all inspired by the purpose and aims of ecodevelopment. They consider the project as a vehicle for short-term employment and for gathering resources such as irrigation pipes and sewing machines etc.

The ecodevelopment project is used by some villagers to extend the political power of their own faction or village. In spite of all such results, the planners of the project misunderstand the dynamic rural communities as passive, homogeneous and subsistence-oriented and thus do not recognize the complexity and vigour of the villagers. Whenever a project fails, instead of rectifying the loopholes in the project the planners blame the insubordinate villagers. But, at the same time it is also true that the villagers living in the ecodevelopment zone being form various class, caste, community and occupation are very different to address within an ecodevelopment project.

Introduction of new technologies and methods of production are constantly changing people's relationship with their natural and biological environment. Accordingly there is some change in their expectation about government programmes regarding their progress and prosperity. The modern technology has connected the villagers to distant markets in different states as well as with different social groups through a vast network. For example, it has linked the remote villagers in Himachal with Switzerland and Italy. Now those villagers are the leading exporter of morel mushrooms to the international markets. Of course everyone did not gain equally. With the changing situation, the villagers have changed their view. They are no more totally dependent on forest products or cultivation. Some of them even prefer government jobs or a job in tourism sector.

These connections often make it different to confirm the distinction between inside and outside, subsistence versus profit. In the Great Himalayan National Park the flourish in the medicinal plants trade has brought villagers material prosperity; when a farmer buys a Jersey cow with the proceeds from the sale of medicinal plants, is it for subsistence or profit? While the trade in medicinal plants connects villagers with national and inter-national pharmaceutical companies, the first link in the plants trade consists of some local villagers who double up as shopkeepers. The market is thus not an outside force but an intimate presence which is hard to banish from the village. While 'stakeholder analysis' puts villagers' and 'traders' in neat, separate categories, the situation on the ground is marked by messy continuities and practices which defy easy explanation (also see Li, 1997).

Involvement of Forest Department in Social Evolution

The attraction of receiving large funds from the World Bank was probably a factor crucial in leading. The Forest Department involved itself to the task of social evolution through ecodevelopment after it got sufficient fund from World Bank. Despite the rhetoric of community participation, any supporting dialogue with villagers was still completely contrary to the mindset of the Forest Department as many Forest Department officers consider local people to be simple and ignorant. The former Park Director narrates a story about the absurd fears that villagers entertained about ecodevelopment:

> Villagers told the Park Director that they had heard that, under the ecodevelopment programme, a special kind of leopard had been released in their forests. The leopard had been imported from America and its diet chiefly consisted of sheep and goats. The villagers said that they knew someone who claimed to have seen one such leopard and this person had reported that the leopard had an identification tag in its ear which said 'Made in America'. The Park Director had laughed heartily at this story and at the villagers. 'Bring me this man,' he said, 'and I will touch his feet. He must be extraordinarily brave if, on seeing a leopard, instead of running for his life, he went close enough to actually read what was printed on his ear-tag!.'

The above incident was narrated by the Park Director to impress that the gullible villagers will accept any action which would be in their interest. He demonstrates the power of the Forest Department's superior rationality to crush villagers' baseless fears. Yet the villagers' story is 'real' at another level. The leopard with the 'made in America' ear tag is a telling metaphor for an imported concept, ecodevelopment, consuming and alienating local resources.

That villagers may have knowledge and opinions about the objectives and strategies of natural resource management is not considered, let alone incorporated, into ecodevelopment plans. The only way in which local knowledge figures in the official discourse is in the outline of research to be conducted (WII, 1996) where, under 'Task No. 2: Assessment of the social context and socio-economic conditions of people using the Great Himalayan National Park' is listed a sub-task:

'Study ethnobiology of the area including knowledge of local species, their medicinal use, and involvement in local rituals, ceremonies, food, fibre, etc.' 'Local knowledge' is reduced to ethnobotany and studied in a manner which divorces it from its changing political and economic context. Such an approach has led to the diligent preparation of lists. A couple of villagers are employed as native informants to identify and collect plants and bring them to the outside researchers. The researchers refer to their taxonomic texts and figure out botanical names before adding the plants to their herbarium and noting the uses mentioned by the local informants. The list so generated is the only official acknowledgement of the existence of local knowledge about the environment. Yet the very manner of incorporating such knowledge into the ongoing 'scientific' research devalues it and makes it irrelevant to the process of formulating resource management plans.

Official attempt to bring in local knowledge has only resulted in the collection and filing away of a rapidly-drying mixture of various kinds of botanical odds and ends. But at the same time the critical contribution of indigenous knowledge in making the entire research programme possible has gone unacknowledged. The research component of ecodevelopment brings in several 'scientific experts' to study wildlife and suggest how the Park should be managed. To effectively do their research work on the range and breeding habits of the monal and western tragopan birds the scientific experts will need the help from local field assistants for their vast knowledge of bird habitat and behaviour is based on their previous experience of hunting and trapping. The study of wild ungulates, too, relies on the knowledge and skills acquired by villagers during the time they spent grazing livestock. Though some researchers casually admit that their porters and guides are more aware about the Park and its wildlife than them, their research has a preset agenda which cannot accommodate local world-views and management objectives. That is, the knowledge developed by villagers to manage a subsistence economy based on the integrated use of natural resources is now being acquired by 'scientific experts' as data to be incorporated into a new management regime aimed at optimising conditions for wildlife.[9]

Park authorities disregard for villagers and the perfunctory nod in the research agenda towards local knowledge (which is reduced to a travesty in actual practice), crop-up from the class-related prejudices of forest officers and scientists and, finally, an unwillingness within the Forest Department to surrender power. The desire to maintain old mechanisms of control is all the more strong because of the augmented funds that ecodevelopment places at the disposal of the Park authorities. The Park authorities did their best to assimilate ecodevelopment into their established modes of working despite the rhetoric of community participation.

However, change was propelled by two factors: firstly, the World Bank had stipulated that ecodevelopment involved a *quid pro quo.* In exchange for ecodevelopment investment in alternative livelihoods and facilities for the Park staff villagers would ultimately have to surrender their rights to resources inside the Park. Secondly, local villagers realizing threat to their resource rights in the Park reacted strongly. They also mobilised others against the project. Squeezed by local opposition and by their paymaster, the World Bank, the Park authorities were compelled follow an action extremely constrained nature.

The internal dynamics of the Forest Department contributed to the dilemma. The Forest Department has been primarily treated in sociological literature as a part of an elitist Indian state, facilitating and legitimising the exploitation of nature and labour by capital (Gadgil and Guha, 1992). However, less attention has been paid to the Forest Department as a *bureaucracy,* with a rational-legal system of rules. Weber categorised different modes of authority into the rational-legal, traditional and charismatic. Using this categorisation the Indian forest bureaucracy can be seen as combining two contradictory approaches to authority—the ostensibly rational-legal where decisions and processes are supposed to follow formal rules which guarantee consistency and efficiency, and the charismatic where power flows from the personal aura of a leader. Like other parts of the Indian state, the Forest Department responds best to demands made by a powerful leader.

Just as Indira Gandhi in her heyday could bypass all democratic procedures and stop Silent Valley from being destroyed, thereby winning accolades from eminent environmentalists such as Salim Ali and Sundarlal Bahuguna, so also can a charismatic leader galvanise Park staff into action (see Rangarajan, 1996:2393). In the case of the Great Himalayan National Park, the transfer of a powerful leader at the start of ecodevelopment led to the demoralisation of the forest bureaucracy. It created an unsustainable vacuum in which the project dwindled. As

Weber points out, charismatic authority is so personalised that it militates against the devolution of power. The recent appointment of a dynamic Park Director may mend matters, but the importance of the charismatic leader within a bureaucracy points to the weakness of the Forest Department.

The field staff of the Forest Department are regarded as mere *implementors.* They include range officers, forest guards, and wildlife watchers. They spend their energy and most of the time in ecodevelopment. They are specially trained to carry out their duty. However, it is more insightful to see field-level workers as policy-makers in their own right. Their talents and limitations are defined by their training and their conditions of work like pay, facilities, chances of advancement etc. Of course their relationships with colleagues, local traders and farmers are also considered. The institutional success of ecodevelopment depends on this layer of the Forest bureaucracy. Success or failure of a scheme, however ambitious it may be, depend on the field staff who give it a shape. But, unfortunately, the experiences are neither fed into future planning nor they are properly defined.

The shift to an ecodevelopment approach involves a complete turn-around for forest guards who have spent their careers guarding villages for checking of illicit felling. How do they adjust to a new set of relationships which involve giving up their punitive authority? According to the present Park Director, the biggest challenge in ecodevelopment was 'changing mental maps from enforcement to participatory management' (Pandey, 1998:8). However, the funds earmarked for training were used for a 'four months' training course ... attended abroad by the Director in the first year of the project, and one and a half months training course . . . attended abroad by the Deputy Director and a forest ranger' (Pandey, 1998:8).

By remaining within the boundary of bureaucracy, the field staff will introduce something new and improved idea is an unrealistic expectation. Ecodevelopment has added considerably to the responsibilities of the field staff. Besides continuing with his former duties, the forest guard is also the secretary of the newly-formed Village Development Committees. He has to record minutes, keep accounts and supervise the work of the committees. The committees are set up to prepare microplans and implement ecodevelopment programmes. There is severe understaffing in the Great Himalayan National Park with 30 per cent of the posts remaining vacant. Ironically, this is a consequence of the process of structural adjustment imposed by the IMF whereby governments are supposed to downsize. As a result the field staff are expected to exert themselves to compensate for their absent colleagues, and also they are expected to innovatively implement new programmes of ecodevelopment.

Role of Village Development Committee

Being squeezed between the paymaster, World Bank, demanding results and widespread opposition by the villagers, the field staff succumbed to the nearer pressure. The Park authorities found it easiest to take the line of least resistance. Instead of identifying the social groups or villages whose resource use was affecting the interest of wildlife conservation, the Park staff chose to allocate ecodevelopment funds on the principle of 'equity'. The rupees three crores to be spent on village-level ecodevelopment activities were equally divided between 2000 households in the 130 villages so that each household notionally received Rs 15,000.[10] Within some villages, people exercised an additional notion of 'equity'. They argued that households which had received loans from other government programmes such as the Integrated Rural Development Programme, should not be entitled to ecodevelopment funds as their 'quota' for *sarkari* money was full.

The Forest Department's and villagers' notion of equity as a principle for fund allocation minimised the disputes were minimised. They also separated ecodevelopment investment from its rationale of changing resource use towards greater ecological nurturing. Both villagers and Park staff skirted around the difficult issue of using funds for socially and ecologically satisfactory ends. Distribution of fund was not proper. Though those households who would be most effected if their entry to the Park was stopped, the money has been distributed equally and so sparsely that it will not give even one household sufficient means of livelihood. This creates a paradoxical situation. Though everyone is satisfied with the way in which the funds were distributed, yet no one is assisted in making their resource use more ecologically sustainable. The hope to go along with local sentiments has also led the Forest Department to compromise on its agenda of introducing a more egalitarian local governance structure.

The characteristics of a 'good' Village Development Committee (VDC) include internal co-ordination, ability to liaise with the government and NGOs, regular meetings, ability to take decisions, representation of all classes, special involvement of poor and deprived sections, ability to carry along all sections of the village.[11] Some of these characteristic are clearly contradictory. The Park staff do not have the resources to be able to resolve them. A good Village Development Committee includes women and Dalits. But if the forest guards want anything to be done, they have to approach —the *devta* committee the traditional power centre of the village.[12] This committee of upper-caste men have supreme command power over resources and rituals. Inviting the anger of the local *devta* committee is a recipe for disaster. So the field staff have chosen to follow one of the many points in the checklist of a good micro-plan—it must respect local customs and religious beliefs. Repair of local temples is considered to be most common ecodevelopment activity. It surely develops a close bond between Village Development Committeeand the government. The political expedience of working with existing structures of domination is easily disguised by citing the importance of respecting 'traditional' culture. Due to tight time-schedule of ecodevelopment and the impatience of senior staff to achieve their definite targets, field staff find it simpler and faster to work with the superior group of the village.

Administrative is the other characteristic of a good Village Development Committee. All decision regarding financial matters depen on budgetary allocations, audit and other bureaucratic rules. Civil works, purchasing vehicles and other tangible goods got more importance than the more difficult-to-audit expenditure on training, orientation and capacity-building. To encourage fuel-saving habits among villagers hundreds of pressure cookers were purchased with ecodevelopment money in the Great Himalayan National Park. But actually none of the villagers of the 130 villages in the ecodevelopment zone to whom the cookers were distributed go to the Park to collect fuel need. Distributing cookers may be a good move for encouraging energy efficiency but it is of no use from the point of view of changing Park-related practices.

Allocation of 'nature awareness programmes' is the other example of planning with administrative advantageousness in mind. Under this programme the villagers of the remote villages are provided with funds to purchase dish antennae and television sets so that, besides some entertaining programmes, they can watch some awareness programmes like modern agricultural practices, horticultural practices, animal husbandry programmes, family plannig programmes and other agriculture related programmes as well as mythological programmes.

Lack of commitment to the goals of ecodevelopment and the political and administrative advantageousness have reduced the ecodevelopment into a set of cosmetic measures and a token representation. That urgent need of the day is to bring about changes in resource use practice is not realized either by the villagers or the park authority. Ecodevelopment is treated just as a burden dumped on the shoulder of an ill-equipped as well as already over burdened guard. Increased demand for medicinal plants by the pharmaceutical companies is the basis of the present prosperity of the villagers. They have no stake in ensuring that medicinal plants are not harvested. In case any conflict arises around ecodevelopment it is managed and negotiated by villagers and park authorities. In the process often the objective of the exercise is vitiated.

Collaboration between Villagers and Park Staff

Exclusion of other significant stakehoders from the negotiation is the other important collaboration between villagers and Park authorities. Both of them are more than willing to exclude them because this will make their claim to be the sole user of the resources inside the Park stronger.

By virtue of recognition in the Settlement Report the farmer-pastorialists of driver Outes Siraj have the right to bring their livestock to graze in the Park. Moreover, these are some Nepali hired labourers and others, who, being migrants and newcomers there, have no right at all to natural resources of the Park. They are the ecological refugees from their place of origin. They are the poorest and most vulnerable among the groups dependent on the Park. But their resource distribution does not figure in any ecodevelopment plan.

The Forest Department considers the local residents as the defining factor in deciding who are to be included address the social complexities around ecodevelopment as well as to recognize the existence of new migrants, traders and pastoratosts who have shifted from distant places. There are basically two reason behind this non-recognition. Firstly, bureaucracy's uneasiness in dealing with these transient, amorphous groups. They are considered to be difficult to control. Secondly, their claims to natural resources are neither legally not socially acclaimed.

Extreme depravity of outsiders is a common refrain that runs through discussions between the Park authority and villagers in the ecodevelopment zone. For instance, 'we are always careful to leave behind a few plants so that they set seed, but it is these outsiders who don't leave even a blade of grass behind' . 'Outsider' can be anyone. On 'occasion, it can be one's kinsman from the next village, or a Nepali labourer, or a shepherd from Ani tehsil. Whoever, the 'outsider' is, it is always easy to make him a scapegoat. The entire blame for ecological degradation is laid on him. But if ecodevelopment does not come to terms with the significant presence of the 'outsider', it will remain unsuccessful.

Role of Forest Bureaucracy

The practice of ecodevelopment brings into prominence an inconsistency which lies at the heart of any planned intervention. Planning must order, discipline, simplify, standardise, and even invent rural society (Mosse, 1986:136). Still, the Forest Department's efforts to manage ecodevelopment are resisted by the intricacies of rural society and by the political economy of the forest bureaucracy itself. The official discourse of ecodevelopment with its invocation of science, law and the economic rationality of incentives and alternative investments, appears as a seamless, shining example of enlightened state practice.

On the ground, however, villagers conform ecodevelopment into their experience of other government programmes which they have seen come and go, and do their best to profit from it even as they resist the threat to their resource rights. And, ironically, they are joined in this by the forest bureaucracy which creates 'sincere fictions' around VDCs, microplans and participation, even as they reproduce established relations of power within the Forest Department and between Park staff and dominant villagers. But, finally, it is the practical knowledge of the villagers and Park staff that suffers patiently, not the official discussion of ecodevelopment.

The Forest Department and villagers through ecodevelopment, collaborate in the construction of a social group. But it is always incomplete. They refuse to acknowledge the role of the World Bank, the trade links which span the world, the presence of 'outside' pastoralists and migrants. They believe that science and law are unbiased and objective thus create a sanitised, closed world where it is possible to hope for conservation as a perfect solution. The closure applied by ecodevelopment is constantly under strain as Social groups from within as well as outside the circle of ecodevelopment try hard to break through the structures imposed on them. As a result, the closure applied by ecodevelopment is under strain constantly.

Absence of Baseline Survey

Four years after ecodevelopment started, researchers from the Wildlife Institute of India found that grazing had little or no deleterious impact on the Park (Mehra and Mathur 1998).[14] However, this discovery came rather late in the day, since the notification of the area as a National Park and the prohibition of grazing and other resource use had already taken place. Thus what should have been a tentative hypothesis about grazing and degradation was earlier elevated to the level of a self-evident truth, and combined with the force of the Wildlife Protection Act, to create enormous mischief in people's lives. Clearly lacking is a sense of accountability towards villagers. Both scientists and states including the World Bank, believe that their own opinions, must prevail over the concerns of villagers, however thinly they are based on facts.

The dubiousness of the data base upon which the threat to the wildlife is to be judged, is the other fact that also emanated after four years. Other than the study of grazing, only preliminary survey of flora has been done. The wildlife Institute of India has prepared a monitoring protocol for flora but the Park staff are yet to be trained on its use. Without a proper survey about wildlife it is not possible authentically determine the reasons that are threatening the existence of these wild animals. Under this condition generally human resource use is blamed for this major threat. For example, the study on birds has data only for three years, still it is argued that mushroom collection seriously disturbs birds in the breeding season and, therefore, it should be banned.

Moreover, without acquiring proper knowledge about the bird population or without knowing how much this numbers are affected by biological factors like disease, predator-pray population and weather condition, it is not proper to blame human actions to have threatened the existence of birds. Of course some conservationists are still ready to believe that continued human use of the Park area and its resources can be a threat to the wildlife conservation.

These 'pragmatic conservationists' (Rangarajan, 1996:2394) offer solutions which combine a more nuanced view of ecological processes with an appreciation of the importance of the long-standing livelihood rights of human populations. New developments in ecological theory and ecological history have been pointed out by some of the conservationists. They say that these new developments have proved that so-called 'pristine landscapes' are actually the products of some expressive human interventions. As Rangarajan observes, 'the problem would not be one of discovering [or restoring] nature in its pristine state but of working out which kinds of human activity significantly reduced or enhanced biological diversity. But to do this it [becomes] necessary to see the forests, scrub lands or marshes as existing inside human history' (Rangarajan, 1996:2401). Such a reorientation leads to the implementation of a more flexible approach so that scientists, foresters, villagers and other stakeholders can jointly set the terms of reference. In other words the relations of power are to be re-built to creat new systems of accountability. Then, whether it be the construction of roads or the rotation of pastures from which medicinal plants are collected, sustainable solutions can be revealed. This has been demonstrated by conservation efforts in Sariska, Kailadevi and elsewhere (see Kothari, 1996). However, in the case of the Great Himalayan National Park, the practice of ecodevelopment has suppressed the possibility of more dialogical outcomes. In this Park this practice failed as a model within which a consensus around conservation could be reached.

The most serious threat the Great Himalayan National Park has faced is the Larji Hydro Power Project. This project will lead to the denotification of a small part of the Park, but even that minor change will bring in, besides deforestation, new settlements and serious inroads into the remaining protected area. This type of schemes make a derision of the actual purpose of the ecodevelopment. If conservation is to succeed, it must be based on a shared understanding of what is in danger and what must be protected. The components of such a agreement is not be hard to find; transforming them into strategies of action is the toughest challenge.

Conclusion

Ecodevelopment must be seen not merely as a programme for wildlife conservation but as part of a project of rule that seeks to manage landscapes and people in order to keep the state in business (see Li, 1999). In order to fulfil the imperative of reproducing itself and keeping officials gainfully employed, the Forest Department draws legitimacy from a rational-legal and scientific aspects which are again based on shaky premises. Still t its need to secure control over resources and assert its ascendancy over villagers is compromised by its relationship with a World Bank bent on pursuing its own neo-liberal agenda of Grafting civil institutions to replace the state. But the demand of bureaucratic norms and procedures corrupt these rational-legal and scientific aspects of rule. Projects serve as vehicles. They enable the state leaders to satisfy their desire for foreign travel, consultancy fees, and personal gain. State projects of rule must also contend with unruly villagers who resist the curbing of their resource use, even as they collaborate with the Forest Department to expand their share of the spoils from ecodevelopment funds. Since the imperative of reproducing the state informs the grounded dynamics of ecodevelopment practice, it is no surprise that the project has been a failure. Fuller acknowledgement of implicit agendas, piercing the veil of 'sincere fictions' that different leaders weave together, may follow if there is greater scope for asking questions and enforcing accountability. The rapid rise of interest and activity initiated by ecodevelopment may provoke critical examination of the social relations at work in the Great Himalayan National Park and for restructuring resource management new alliances and understandings are made.

ACKNOWLEDGEMENTS

The research for this essay was part of a project co-ordinated by the Wildlife Institute of India on the 'Ecological Study of the Conservation of Biodiversity and Biotic Pressures in the Great Himalayan National Park: An Ecodevelopment Approach' I am deeply grateful to the WII researchers in Kulu, especially Badrish Mehra, Sanjay Singh and Pradeep Chaudhary, for their unstinted support. WII provided the opportunity for valuable discussions with several scientists and foresters, especially B.M.S. Rathore and Sanjeeva Pandey. I am also grateful to Iqbal Singh, Hukam Ram and Lata Sharma of SAVE, Farhad Vania and Ashwini Chhatre for their help. Beli Ramji and his family provided a home in Sharan village. Mahesh Rangarajan, Vasant Saberwal, Ashis Kothari and Rahul Ram kindly commented on draft versions of the paper.

Some of the ethnographic data in this paper has previously appeared in R. Jeffery and N. Sundar (eds), 1999. *A New Moral Economy for India's Forests?: Discourses of Community and Participation.* Sage Publications: New Delhi; and Roy Ellen, Peter Parkes and Alan Bicker (eds), 2000. *Indigenous Environmental Knowledge and Its Transformations: Critical Anthropological Perspectives.* (Studies in Environmental Anthropology, Vol. 5), Harwood Academic Press: Amsterdam.

Notes

1. This section is largely based on Tucker (1997).
2. For an interesting discussion of the contrast between subsistence hunting and sport-hunting under colonialism, see Pandian (1998).
3. The imperative of paying revenue spurred the cultivation of the opium poppy as a commercial crop. Although no revenue is exacted today, the opium poppy and cannabis are still grown commercially, albeit clandestinely.
4. Himachal Pradesh was carved out of Punjab and became a full-fledged state in 1971.
5. At the time of writing, there were reports that, in response to the Supreme Court order to settle rights in protected areas within a year of the initial notification, the Kulu District Magistrate had summarily declared the park out of bounds. However, there does not seem to be any move at present to enforce the ban.
6. Territorialisation projects are ways by which state institutions attempt to assert control over rural citizens and natural resources. Measures include defining zones of exclusion (e.g., national parks) and regulating how and by whom these areas can be used (e.g., only by those whose rights are recorded in the Settlement Report, and not subsequent migrants). These ways of mapping and rearranging peoples and places are always ongoing and contested (Vandergeest and Peluso, 1995).
7. Such a myth is now familiar in environmentalist writings about tribal and peasant struggles over nature and has been variously described as 'the myth of the ecologically noble savage', and 'the new traditionalist environmentalism' (see Baviskar, 1995; Sinha *et al.*, 1997).
8. Pierre Bourdieu discusses 'institutionally organised and guaranteed misrecognition' as a strategy for maintaining economic institutions (Bourdieu, 1977:171). The concept of collective misrecognition helps to understand how 'sincere fictions' about rural communities are devised to reproduce established relations of power between states and other social groups.

9. On the issue of the incorporation of 'indigenous' knowledge into 'western' science, see Grove (1998).
10. Funds are received by ecodevelopment units which consist of 100-150 households. Some of the money is spent on collective goods (e.g., road repair, field-leveling, irrigation) and the rest on grants to individuals (e.g., for setting up tailoring shops, poultry rearing, carpentry, etc.). A small sum of money is also ear-marked for eco-awareness programmes and for appointing an 'eco-volunteer' for liaison work with the Forest Department.
11. Parameters described during the ecodevelopment workshop for Park staff and ecovolunteers at Larji, June 25th, 1998.
12. The *devta* (deity) committee is in charge of managing the village temple. Devtas, through their appointed mediators, are a powerful force governing village life and the devta committee wields a great deal of influence through its collective decisions on matters of religious ritual, disputes over property and general village affairs.
13. Lapah village ecodevelopment microplan 1998.
14. This is consonant with findings from other places in the Himalaya which show that livestock grazing creates a mixed vegetation pattern which actually increases biodiversity (Saberwal, 1999). In the case of the GHNP, it appears that grazing has been integrated into the Park, with the domesticated livestock's manure fertilizing the pastures and creating greater floral diversity. There does not seem to be any competition between domesticated livestock and wild herbivores either.

References

Anderson, A., 1894, *Revised Forest Settlement for Kulu.* Government Press: Lahore.

Baviskar, A., 1995, *In the Belly of the River: Tribal Conflicts over Development in the Narmada Valley.* Oxford University Press: Delhi.

Bourdieu, P., 1977, *Outline of a Theory of Practice.* Cambridge University Press: Cambridge.

Gadgil, M. and R. Guha, 1992, *This Fissured Land: An Ecological History of India,* Oxford University Press: Delhi.

———, 1995. *Ecology and Equity: The Use and Abuse of Nature in Contemporary India.* Penguin: Delhi.

Garson, P.J., 1983, 'Comments on a Prospective National Park in Kulu District'. Typescript.

Gaston, A.J., P.J. Garson and M.L. Hunter, 1983, 'The Status and Conservation of Forest Wildlife in Himachal Pradesh, Western Himalayas' in *Biological Conservation,* no. 27.

Grove, R.H., 1998, 'Indigenous Knowledge and the Significance of South-West India for Portuguese and Dutch Constructions of Tropical Nature', in R.H. Grove, V. Damodaran and S. Sangwan (eds), *Nature and the Orient,* OUP: Delhi.

Kothari, A., 1996, 'Is Joint Management of Protected Areas Desirable and Possible?' in A. Kothari, N. Singh and S. Suri (eds.) *People and Protected Areas: Towards Participatory Conservation in India*. Sage Publications: New Delhi.

Kothari, A., P. Pande, S. Singh and D. Variava, 1989, *Management of National Parks and Sanctuaries: A Status Report*. Indian Institute of Public Administration: New Delhi.

Li, T.M., 1997, 'Boundary Work: A Response to *Community in Conservation* by Arun Agrawal' in A. Agrawal, *Community in Conservation: Beyond Enchantment and Disenchantment*. CDF Discussion Paper, Conservation and Development Forum: Florida.

———, 1999, 'Compromising Power: Development, Culture and Rule in Indonesia', in *Cultural Anthropology*. 14(3).

Mehra, B.S. and P.K. Mathur, 1998, *Livestock Grazing and Conservation of Biodiversity in the High Altitude Ecosystem: An Integrated Landscape Management Approach*. Wildlife Institute of India: Dehradun.

Mosse, D., 1986, 'The Social Construction of 'People's Knowledge' in Participatory Rural Development' in S. Bastian and N. Bastian (eds), *Assessing Participation: A Debate from South Asia*. Konark Publishers: New Delhi.

Heumann, Roderick, 1998, *Imposing Wilderness: Struggles over Livelihood and Nature Preservation in Africa*. University of California Press: Berkeley.

———, 1997. 'Primitive Ideas: Protected Areas Buffer Zones and the Politics of Land in Africa', *Development and Change*. 28(3).

Pandey, Sanjeeva, 1998, 'Biodiversity Conservation at India's Great Himalayan National Park: Lessons for Implementors'. Paper presented at the workshop on Ecodocumentation at the Wildlife Institute of India, Dehradun.

Pandian, M.S.S., 1998, 'Hunting and Colonialism in the Nineteenth-century Nilgiri Hills of South India' in R. Grove, V. Damodaran and S. Sangwan (eds), *Nature and the Orient: The Environmental History of South and Southeast Asia*. Oxford University Press: Delhi.

Rangarajan, M., 1996, 'The Politics of Ecology: The Debate on Wildlife and People in India, 1970-1995' in *Economic and Political Weekly*, Special number. September.

Saberwal, V.K., 1999, *Pastoral Politics: Shepherds, Bureaucrats and Conservation in the Western Himalaya*, Oxford University Press: Delhi.

Singh, I., 1997, 'Study of Grazing Management System and Status of Herb Collection', Draft report: Sainj.

Sinha, S., S. Gururani and B. Greenberg, 1997, 'The 'New Traditionalist' Discourse of Indian Environmentalism' in *The Journal of Peasant Studies*. 24(3).

Sundar, N., 2001, 'Beyond the Bounds?: Violence at the Margins of New Legal Geographies', in Nancy Peluso and Michael Watts (eds) *Violent Environments*. Cornell University Press: Ithaca.

Tucker, Richard, 1997, 'The Historical Development of Human Impacts on the Great Himalayan National Park'. Report submitted to the Wildlife Institute of India and Winrock International.

Vandergeest, P. and N. Peluso, 1995, 'Territorialisation and State Power in Thailand' in *Theory and Society*, no. 24.

WII (Wildlife Institute of India), 1996, *Forestry Research Education and Extension Project: Great Himalayan National Park: Research Proposal.* WII: Dehradun.

20

"Valley of the Lost Horizon"

Introduction

In 1934 Eric Shipton and Bill Tilman—two explorers— entered the beautiful domain of Nanda Devi the Blassed Goddess. It is regarded as a sacred area shining with its original splendour. In short, it is the master piece of nature's wilderness areas. Eagerly standing at the entrance of the sanctum Sanctorum Shipton astonishing by watched "beyond a well-defined lateral moraines, an expanse of undulating grass land stretched for miles, in lonely contrast with the desolation of the moraine—covered glacier and.... the pasturage is a sanctuary where a thousand of wild animals live unmolested, long may it remain so."

But, unfortunately, it did not remain so. Its sanctity was spoilt by the beginning of 1974. Quite a number of mountaineers, trickkers and graziers crushed the distinctive flora. Human being, by a false sense of self-honour, disfigured rocks and trees. They cut the trees to use as fuel for cooking, warming and compo-fires. They did not stop here. They even indulged themselve in killing animals not only for food but for commercial purposes too. The poachers kill the rare species like musk-deer.

These wilful destructions deformed the beautiful landscape of Nanda Devi. In this chapter the degeneration of flora and funa of this area is discussed, as it was revealed in a survey conducted in 1983.

Attempts to Reach Nanda Devi

The 800 km Nanda Devi Sanctuary is a cup-shaped valley. Surrounded by high peack in the heart of the Himalayas, this sanctuary is the home of many endangered mammal species and birds. The available literature divulge that W.W. Graham made the first attempt to enter the 'Nanda Devi Sanctuary' in 1883. He could go only up to a point where the Trisul Nullah feeds the Rishi Ganga. Now this area is known as Outer Sanctuary or the lower Rishi Gorge. Since then, it has been a cherished dream of mountaineers all over the world to descend into the sacred Nanda Devi Sanctuary and reach its base. After the first attempt several other attempts were made consecutively for 50 years. After failures by a number of expeditions from various countries, finally it was Eric Shipton and H.W. Tilman who pioneered a way into it in 1934 by negotiating one of the world's formidable paths. These two extraordinary travellers explored the so far unexplored world. They called the Nanda Devi sanctuary as 'Valley of the Lost Horizon'. The area has attracted many mountaineers, naturalists, wildlifers, ornithologists, botanists, zoologists and trekkers from several countries during the last two decades or so. On 6th November, 1982, the area was declared as a National Park. (Named as Nanda Devi National Park), the area is one of the twelve 'Proposed Biosphere Reserves' of the country.

How to Reach There

During the winter season the park area, which is popularly called the Rishi Gorge, remains under a thick carpet of snow. So it is easy to approach the area only for a limited period of about five months from June to October. The grazing meadows emerge from underneath and attract the herbivora during this period. One has to go via Rishikesh to Joshimath, Tapovan and Lata to reach the park from Delhi. Lata village is the last motor link for the Park. It is also the last link with the outer world.

'The Nanda Devi Trek' is the only main trek in the Park area. It is considered to be one of the world's toughest. There are only two entry points. One at the point of the junction of the Rishi Ganga rivers and Dhauli Ganga in west and other at Lata village road side in north-west of the park. They meet at the fourth camp site along the trek known as Dibrugheta and form the main trek which leads to the upper Rishi Gorge. the world's most charming wilderness. Considering the height, it is said that the trek from the former and the older entry point in the west is easier. But it involves more risk specially shortly before joining the north-west entry point at Dibrugheta. So it is generally avoided. Beyond Dibrugheta the track leads via Deodi, Ramani, Bhujgarh, Patalkhan or Tilchauni to Sarsopatal or the Nanda Devi Base Camp. The various diversions which lead to the base camps of different peaks, viz., Dunagiri, Changabang, Kalanka and Trisul, of the main trek are also shown in the following map. To give an idea about the magnitude of physical works and skill required to reach the Nanda Devi Base Camp a chart has been prepared. As shown in the graph, at various places by along the trek one needs the basic knowledge of mountaineering and has to fix the rope for a sure and safe reach.

Tography of the Park

Geographically situated between 30°16'-30°32' N latitudes and 79°44'-80°02' E longitudes the Nanda Devi National Park, covers an area of ca 800 km^2. It covers an altitude varying from ca 2100 m in the west to ca 7817 m in the east. The park is a cup-shaped valley. It is surrounded by almost above seventy white celestial peaks appearing like crystalised clouds in the deep blue sky. Among them the most popular ones are: Dunagiri (7066 m), Changabang (6864 m), Kalanka (6931 m), Nanda Devi East (7434 m), Nanda Devi West (7817 m), the highest peak of the Garhwal Region. Nanda Khat (663l m), Devisthan (6678 m). Maiktoli (6805 m). Mrigthuni (6855 m), Trisul (7120 m), Bethartoli (6352 m). These peaks form rim of the cup of boundary of the park. These high peaks encircle the major glacier system on the upper-half inner surface of the cup-shaped Park. Itconsists of Ramani Glacier, Changabang Glacier and North Rishi in the North and Nanda Ghunti Glacier, Ronti Glacier, Bethartoli Glacier, Trisul Glacier, South Rishi Glacier, and South Nanda Devi Glacier in the south. This glacier system is the permanent and continuous source of Rishi Ganga water.

Topographically, the Park area or the Rishi Gorge can be divided into two sectors, viz., the Upper Rishi Gorge (The Inner Sanctuary) and the Lower Rishi Gorge (The Outer Sanctuary). While the former is mainly a glacial zone intermingled with numerous alpine meadows, the latter is largely a thickly forested zone. The north facing slopes are more forested than the south facing slopes and conform to the general characteristic of the northern hemisphere.

Animal Life

Rich in animal life the Nanda Devi National Park and provides habitat to a number of endangered and threatened species of mammals and birds. Since July 1981 the authors have been working in the park for a Man and Biosphere (MAB) research project on "Status survey of endangered and threatened species of mammals and birds at Nanda

Devi National Park" B.S. Lamba, the then Joint Director of the Zoological Survey of India, Dehradun, was the leader of the research project. Presently, J.D. Department of Environment, New Delhi). During the period of their more than 100 days stay in the park (1981-83), the authors have so far specified as many as 15 species of mammals and 80 species of birds in their survey report. Some of the threatened and endangered species of mammals and birds that are still existing in the Park are as follows:

Mammals

1. The snow leopard *(Panthera uncia)*. It loiters at the higher altitudes of ca 3700 m-4700 m. During winter season, like some other animals, it moves downwards. Because of its noctural habit, no direct observation has so far been made by the authors. The International Fur Trade Federation has imposed, a total ban on the trade of its fur.

2. The brown bear *(Ursus arectos)*. The bare open peaks above the tree line are the usual haunt of the animal. Emerging from their winter sleep in the spring, they follow the melting snows up to their perpetual level. Inordinate hunting has threatened the existence brown bear.

3. The Himalayan black bear *(Selenarctos thibitanus)*. It dwells in the steep forested hills of the park. During summer it may be found near the limits of tree line at ca 3050 m-3600 m altitude. In winter most of them come down to the lower valleys at 1525 and even lower. like snow leopard, it is also difficult to observe because of its nocturnal habit. Nevertheless plenty of footprints have been registered at and around Dibrugheta and Deodi in the park. Due to excessive hunting in the past it is presently under threat.

4. The musk deer *(Moschus moschiferus)*. It is generally found within Birch (*Betula utilis*) forest above the zone of Pines. At times it comes down to lower levels. Perpetually it hides in thick cover and hence difficult to observe. Individuals and females with young ones have been observed at and around Dibrugheta, Deodi and Ramani Camp sites in the Park. Its musk, which is secreted by the abdominal gland of the males, is of high commercial value. As a result it attracts many hunters to kill them. This is a major thrat the existence of the musk deer.

5. The Himalayan Tahr *(Hemitragus jemlachicus)*. Of all the wild goats, it perhaps selects the most inaccessible ground to live in at an

altitude ranging from ca 3000 m to 4000 m. Comparatively, it is easier to observe when it comes out to graze in open clearings during mornings and evenings. Small groups of males, females with young and individuals were seen around Dharansi. Excessive hunting for its flesh and skin has endangered it.

6. The bharal or blue sheep *(Pseudois nayaur).* Found between the tree line and the snow line, it is sure to be seen in the upper Rishi Gorge. There it comes out to graze on alpine meadows in the forenoons and afternoons. Individuals and herds of 30-60 individuals have been observed. It is endangered for its distributional range. It is threatened on account of excessive hunting for its flesh and skin

7. The serow *(Capricornis sumatraensis).* It is found at an altitude ranging from 1850 m to 3100 m of thickly wooded gorges. It can be seen in the mornings and evenings when they come out to graze on the rank herbage of more open slopes. Authors have stated in their survey report that they have not observed the serow. But enquiries made from local people confirm their presence in fairly good numbers in the lower Rishi Gorge. They are endangered because of excessive hunting for their flesh and skin.

Birds

1. Himalayan golden eagle *(Aquila chrysaetos daphanea).* It is a resident of Himalayas. Generally found above 1850 m to 5500 m, it is observed twice near Lata Kharak (3750 m) and Dharansi (4200 m). Its existence is endangered.

2. Eastern steppe eagle *(Aquila rapax nipalensis).* Winter visitor. A threatened species, it tays at a wider altitudinal range from 3050 m to 7930 m in the Himalayas. It was sighted only once near Lata Kharak on 17th August, 1981.

3. Black eagle *(Ictinaetus malavaensis perniger). It is also a threatened species.* Resides over a wide altitudinal range from 305 m to 3355 m in the Himalaya's. At time it is seen flying near Belta Kharak (2700 m).

4. Himalayan bearded culture of lammergeir *(Gypaetus barbatus aureus).* It resides in Himalayas at 385 m to 7500 m. It is frequently sighted, throughout the park. An endangered species.

5. Himalayan snowcock *(Tetragallus himalayensis).* Resident of

Western Himalaya at 400 m - 5000 m. During winter migrates down to 2100 m; observed and heard in small flocks of 3-20 at and around various base camps in the park. A threatened species.

6. Western tragopan or western horned pheasant *Tragopan melanocephalus).* Although a Himalayan resident, it has not been observed by the authors. The only possible habitat is the dense forested area below Dharansi. An endangered species on account of excessive hunting for its flesh.

7. Himalayan monal pheasant or impeyan *(Lophophorous impejanus).* Resident of Himalayas, found between 2600 m and 5000 m. Individuals, pairs and families have been flushed and heard on steep grassy slöpes and cliffs. It is endangered due to excessive hunting for flesh and the beautiful crest of male birds.

8. Koklas pheasant *(Pucrasia macrolopha macrolopha).* Resides between 1500 m and 4000 m in the Himalayas. Often flushed and heard in forested valleys of the Park at altitudes 2700-3450 m. Endangered because of excessive hunting for its flesh.

There is no published record above invertebrate fauna, and insects, of the Park. If the richness of flora is any indication of the insect life, the entomological component of the fauna should be tremendous. So it is necessary that the study of invertebrate fauna is to be given top priority in future studies in the Park.

The Flora

Like fauna, the rich and impotant flora species of the Nanda Devi National Park are also work mentioning here. More than 200 species of plants are known to exist in the park. Among them, about 15 species are of rare plants are known to occur in and around the park area. The prominant ones are: *Aconitum spp., Circeast agrestis, Epipogium aphyllum, Listera sp., Meconopsis aculeata, Nardostachys granadiflora, Orchis latifolia, Podophyllus hexandrum, Saussurea obyallata,* etc.

Considering as per altitude, the vegetation of the park can broadly be classified into three categories: 1. Sub-alpine forest, 2. Moist alpine scrub, and 3. Alpine meadows.

1. The sub-alpine forest comprises of three different canopy formed of Abies pindrow—A. spectabilis, Rhododendron campanulatum and

Betula utilis tree species. The middle canopy is of comparative small tree-like shrubs Rosa, Viburnum and Jasminum etc., Various herbs and grass species cover the ground. At an height of ca 3500 m, this forest type is generally met with in the lower Rishi Gorge. In east it is restricted only upto Ramani.

2. The moist alpine scrub forest consists of several pure stands of *Betula* and Rhododendron species.

3. The alpine meadows do not have tree species. They are composed of numerous shrubby plants, e.g., Rhododendron—*anthopogon, Junipers* and Salix species, herbaceous plants, e.g., *Aconitum, Cynanthus* and *Polygonum species* and grasses e.g., *Dantnonia, Festuca* and *Poa* species.

Observations on Status of Mammals at the Park

While working on a Man and Biosphere (MAB) project "Status survey of endangered and threatened mammals and birds at Nanda Devi National Park" the author had an opportunity to make some field observations on status and distribution of mammals of the Park. In all, records of 14 species of mammals belonging to 13 genera, eight families and five orders were obtained from the park.

In fact the present communication is one of the several outcomes of seven, 2-7 weeks, expeditions made to the area from 1981 to 1985. The altitudinal range explored varies from 2750 m to 4950 m. During this period more than 700 observation-hours were spent in all the three major vegetation-types—Sub-alpine forest, Moist alpine scrub and Alpine meadows—of the Park.

A detailed systematic account of their status and distribution is given below:

Status and Distribution

CLASS : MAMMALIA

ORDER : PRIMATES

FAMILY : CERCOPITHECIDAE

GENUS : *PRESBYTIS* Escholtz, 1821.

1. *Presbytis entallus schistaceus* (Hodgson). Langus (Entellus Monkey). Ellerman and Morrison-Scott. (1951). Checklist: 205.

Status	:	*Common*
Distribution	:	Restricted to the Lower Rishi Gorge (The outer sanctuary). Often seen in birch forests and other patches of forest in the park.
ORDER	:	CARNIVORA
FAMILY	:	URSIDAE
GENUS	:	*SELENARCTOS* Heude, 1901.

2. Selenarctos Thibetanus (G. Curier), Asiatic Black Bear. Ellerman and Morrison-Scott. (1951). Checklist: 239

Status	:	*Uncommon*
Distribution	:	Mostly found in the Lower Rishi Gorge. Leaves plenty of footprints in the soft ground.
GENUS	:	*URSUS* Linnaeus, 1758.

3. *Ursus arctos* Linnaeus, Brown Bear. Ellerman and Morrison-Scott. (1951).

Checklist	:	236
Status	:	Rare
Distribution	:	No authentic record of its existence in the park is available at present.
FAMILY	:	MUSTELIDAE
GENUS	:	*MUSTELLA* Linnaes, 1758.

4. *Mustela Kathiah* Hodgson, Yellow-bellied weasel. Ellerman and Morrison-Scott. (1951). Checklist: 259.

Status	:	*Common*
Distribution	:	Everywhere in the park. Often seen among boulders at 4000-4400 m
FAMILY	:	FELIDAE
GENUS	:	*PANTHERA* OKEN, 1816

5. *Panthers pardus* (Linnaeus), Leopard. Ellerman and Morrison-Scott. (1951). Checklist:. 316.

Status	:	*Uncommon*
Distribution	:	Found in the Lower Rishi Gorge. Difficult to observe on account of its nocturnal and

shy habits. Leaves plenty of footprints in the river bed and soft ground

6. *Panther uncia* (Schreber). Ounce or Snow Leopard. Ellerman and Morrison-Scott. (1951). Checklist: 320

Status	:	*Rare*
Distribution	:	Mostly found in the Upper Rishi Gorge (Inner Sanctuary). Very difficult to sight this highly endangered field on account of its nocturnal and shy habits, and its limited populations and distribution. Often leaves plenty of footprints in the soft ground and snow.
ORDER	:	ARTIODACTYLA
FAMILY	:	BOVIDAE
GENUS	:	*CAPRICORNIS* Ogilby, 1837

7. *Capricornis Sumatraensis* (Bechstcin). Serow. Ellerman and Morrison-Scott. (1951). Checklist: 399

Status	:	*Common*
Distribution	:	Confined to lower part of the lower Rishi Gorge. It prefers thickly wooded gorges and can often be seen in the mornings and evenings when it comes out to graze on the range herbage of the more open slopes.
GENUS	:	*Hemitragus Hodgson.* 1841

8. *Hemitragus Jemlahicus* (H. Smith). Himalayan Tahr. Ellerman and Morrison-Scott. (1951). Checklist: 403.

Status	:	*Uncommon*
Distribution	:	Mostly found above the tree line on the cliffs and ledges of Dharansi Col. and its adjoint ranges in the park.
GENUS	:	*NAEMOTHEDUS* H. Smith

9. *Naemothedus goral* (Hardwicke), Goral. Ellerman and Morrison-Scott. (1951). Checklist: 401.

Status	:	*Uncommon*
Distributor	:	Found only on the outer slopes of the ridges

		forming park boundary in the NW of the park.
GENUS	:	*Pseudois* Hodgson 1846

10. *Pseudois nayaur* (Hodgson), Bharal or Blue Sheep Ellerman and Morrison-Scott. (1951). Checklist: 410.

Status	:	*Very common*
Distribution	:	Throughout in the Upper Rishi Gorge (the Inner sanctuary) area of the Park Frequently seen on the alpine meadows in the mornings and evenings when they come out for grazing.
FAMILY	:	CERVIDAE
GENUS	:	*MOSCHUS* Linnaeus 1758

11. *Moschus moschiferus* Linnaeus. Musk Deer or Kastura. Ellerman and Morrison-Scott. (1951). Checklist: 353

Status	:	*Common*
Distribution	:	Throughout the park in the sub-alpine birch forest. Difficult to observe on account of its concealing habits. Leaves several "Latrine" sites (distinct groups of pellets).
ORDER	:	LAGOMORPHA
FAMILY	:	IEPORIDAE
GENUS	:	*OCHOTONA* Link, 1975

12. *Ochotona Roylei* (Ogilby), Royle's Pika. Ellerman and Morrison-Scott. (1951). Checklist: 450.

Status	:	*Very common*
Distribution	:	Throughout the park. Frequently observed among boulders between 3500 m and 4450 m.
ORDER	:	RODENTIA
FAMILY	:	MURIDAE
GENUS	:	*ALTICOLA* Blanford, 1881.

13. *Alticola roylei* (Gray). Royle's High Mountain Vole. Ellerman and Morrison-Scott. (1951). Checklist: 617.

Status : *Uncommon*

Distribution : Occasionally seen feeding on grasses, at Upper Deodi (3850 m) in the park.

GENUS : *MUS* Linnaeus, 1758

14. Mus Booduga (Gray), Little Indian Field Mouse. Ellerman and Morrison-Scott. (1951). Checklist: 609.

Status : *Uncommon*

Distribution : Occasionally seen at Latakharak (3700 m) in the park.

References

Tak, P.C. and Kumar, G., (1981a), Nanda Devi National Park: A suggestion for *Cheetal—J. Wildl. Presv. Soc. India,* Dehradun 25(1): 38-39.

Tak, P.C. and Kumar, G., (1983b), Nanda Devi National Park: The Home of several endangered mammals and birds. *Science Reporter—A CSIR's monthly publication,* New Delhi 20(10): 569-574.

Tak, P.C. and Kumar, G., (In press). Nanda Devi National Park: An update. *Indian J. Forestry.* Dehradun.

Tak, P.C. and Lamba, B.S., (1985), Nanda Devi National Park: A contribution to its Mammalogy *Ibid.,* 8(3): 219-230.

Tak, P.C. and Lamba, B.S., (1984), Field observations on abundance of some smaller mammals at Nanda Devi National Park. *Ibid.* 7(3):242-244.

21

Ecological Disturbances and Conservation of Wildlife Garhwal Himalaya

Introduction

Geographically situated between 29°27' and 30°33' N latitudes and 77°48' and 80°06' E longitudes, the Garhwal region is, administratively, comprised of five districts—Uttarkashi, Chamoli, Tehri, Pauri and Dehradun of Uttar Pradesh, Garhwal Himalaya in Uttar Pradesh played a vital role in the conservation of flora and fauna.

The five districts of Garhwal Himalaya have three national parks out of four, one tiger reserve out of two and the only biosphere reserve in Uttar Pradesh. Besides these, the largest glacier—Gangotri glacier—of the region is situated in the district of Uttarkashi.

Wildlife of Garhwal Himalayas with special reference to endangered and threatened species and its conservation has been discussed in details in this chapter.

Impact of Rapid Advancement

Due to Variety of topography and climate as well as its Geographical location, India is considered to be world's second largest country in having a large variety of wildlife. It inhabits nearly 350 species of mammals, 1200 species of birds and 450 species of reptiles in different type of inhabitants spreading from lofty Himalaya to the deserts of Rajasthan and also around the ocean surrounded boundary in west and south.

The impact of rapid advancement in the field of science and technology and rising population has resulted in ecological disturbances. Construction of roads, hydroelectric dams, and railway lines, conversion of forest lands into agricultural land, rising demand for wood and other forest material for forest based industries, fodder for domestic livestock, resulted in deforestation. Moreover, large scale hunting of wild animals for their flesh, skin and valuable fur used in leather industries, musk of musk deer and horn of rhinoceros for perfume and medicinal industries etc., has threatened the existence of species like the cheeta, lion, tiger, rhinoceros, swamp deer, snow leopard, gharial, monal, chir and tragopan etc.

Being aware of the ecological disturbances the Britishers started raising voices for conservation of wildlife. As the habitat of these threatened species with a small number of inhabitants was still there. the damage was thought to be not beyond repair. It resulted in the creation of the first national park of the country—Corbett National Park—in Garhwal Himalaya of Uttar Pradesh in 1936. In the last 54 years, conservation areas were identified all over the country on the basis of habitat available for endangered species. Legal status was also given to them. Today, approximately 300 sanctuaries, 60 national parks, 16 tiger reserves, 3 biosphere reserves and 10 proposed biosphere reserves exist in India.

Garhwal Himalaya's Role in Conservation

In Garhwal Himalaya, on the basis of researches done by wildlife scientists, and the rising voices of naturalists and wildlife lovers, legal protection to the wild animals has been given by the forest department in the form of conservation areas. It has different types of habitats between altitudes of 1000 and 27,000 feet. It inhabits nearly 60 species of mammals, 360 species of birds and 45 species of reptiles. Out of these species 33 species of mammals, 5 species of reptiles and 10 species of birds are found to be very important as they are game species. Altitudinal distribution of these species and their present status in the five districts of Garhwal have been shown in Tables 21.1 and 21.2 respectively.

The Garhwal Himalaya of Uttar Pradesh has played a leading role in the conservation of its flora and fauna. It has the distinction of having many internationally famous conservation areas. Corbett National Park, the first national park and tiger reserve of India; Nanda Devi Biosphere Reserve, among the first three biosphere reserves; Valley of Flowers National Park, the first national park of India to protect flora and among one of the ten proposed biosphere reserves are the prominent ones. Five Districts of Garhwal Himalaya have three national parks out of four, one tiger reserve out of two, and the only biosphere reserve in UP. For location of conservation areas and other information see 21.1 and Table 21.2. Wildlife Institute of India is also situated in Dehradun District of Garhwal. Special projects and breeding farms have been launched for the conservation of seriously threatened species. For the tiger, two reserves, Corbett National Park and Dudhwa National Park have been declared. For gharial and crocodile, a breeding farm at Kukrail, near Lucknow, has been developed. Here eggs collected from different habitat of gharial and crocodile are hatched. The baby gharial and crocodile are released in the rivers when they are able to survive in nature. For musk deer, a breeding farm has been established at Kedarnath Sanctuary in Chamoli District. Musk deer fawn are collected from different areas and then are brought here breeding. Elephants, as they consume large amounts of food, are threatened. Due to the construction of a hydroelectric dam at Kalagarh, large fresh water lakes have sprung up engulfing 80% of savannah grassland and sheeshum island which are the main habitats of elephants. For the conservation of the elephant population, the Rajaji National Park is established. It is said to be the largest in Uttar Pradesh. According to the Wildlife Institute of India it plays an important role as the corridor for the movement of elephants from Corbett National Park to Rajaji National Park.

In spite of these efforts, the lack of knowledge and public awareness, interaction between scientists and planners and the problem of poaching has proved to be the biggest hurdle in the path of conservation. It is recommended that regular interaction between scientists of different fields and planners of the area should take place regularly for the identification and creation of the best conservation areas as well as their better management.

For the success of the conservation programme, public awareness plays a vital role. Through the media like radio, television, newspaper and magazine students in schools and colleges as well as common people should be taught about the importance of wildlife resources and their conservation.

TABLE 21.1

Table Showing Information about Wildlife Conservation Areas of Garhwal Himalaya

Name and Legal Status	*Year of Establishment*	*Area in sq km*	*Location District*	*Altitude in Mt.*	*Latitude Longitude*	*Visiting Months*	*Nearest Town with Infrastructural Facility*
1. Nanda Devi National Park	1980	800	Chamoli	3400 to 7817	30°16' N to 30°32' N 79°44' E to 80°02' E	J une to September	Joshimath
(Now Biosphere) Reserve)	1988	2000		2000 to 7817	30°17' N to 30°41' N 79°40' E to 80°3' E		
2. Valley of Flowers National Park (Proposed Uttarkashi Biosphere Reserve)	1981	87.55	Chamoli	3600 to 6479		June to September	Joshimath
3. Corbett National Park	1936	521	Pauri	400 to 1200	29°13' N to 29°51' N 78°33' E to 78°46' E	November to June	Ramnagar
Corbett Tiger Reserve	1973	521	Nainital				
4. Rajaji National Park	1983	8205	Dehradun	302 to 1000	29°15' N to 30°51' N 77°52' E yo 78°22' E	November to June	Haridwar and Rishikesh
5. Govind Pashuvihar Wildlife Sanctuary	1855	953	Uttarkashi	1300 to 6315	30°55' N to 31°18' N 77°49' E to 78°22' E	May to October	Uttarkashi
6. Kedarnath Sanctuary (Musk Deer Breeding Farm)	1972	967	Chamoli	1500 to	30°25' N to 30°45' N 78°55' E to 79°26' E	April to October	Gopeshwar

TABLE 21.2

Present Status of Important Wildlife Species in Garhwal Himalaya

S. No.	Name of Species (Local Name)	Name of Species (Zoological Name)	Dehradun	Tehri	Uttar Pradesh	Pauri	Chamoli
Mammals							
1.	Rhesus Macaque	Macaca mulatta	1	2	3	2	4
2.	Common Langur	Presbytis entellus	3	2	2	1	2
3.	Tiger	Panthera tigris	4	4	4	3	4
4.	Leopard	Panthera pardus	3	3	2	2	2
5.	Snow Leopard	Panthera uncia	4	4	4	4	4
6.	Fishing Cat	Felis viverrina	3	3	3	2	3
7.	Leopard Cat	Felis benoalensis	3	3	3	2	3
8.	Jungle Cat	Felis chausi	2	2	2	2	2
9.	Himalayan Palm Civet	Paouma larvata	2	3	2	2	2
10.	Common Mongoose	Herpestes edwardsi	2	2	2	2	2
11.	Jackal	Canis aureus	2	2	2	2	2
12.	Red Fox	Vulpes vulpes	3	3	3	2	3
13.	Brown Bear	Ursus arctos	6	6	5	5	5
14.	Himalayan Black Bear	Selenarctos thibetanus	2	2	2	2	2
15.	Common Otter	Lutra lutra	3	3	3	2	3
16.	Himalayan Yellow-throated Marten	Martes flaviguala	2	3	2	2	2
17.	Yellow Bellied Weasel	Mustela Kathiah	3	6	3	6	2
18.	Indian Porcupine	Hystrix indica	3	3	3	2	3
19.	Himalayan Mouse-Hare	Hare ochotona roylei	2	3	2	6	1
20.	Blacknaped Hare	Lepus nigricollis nigricollis	2	2	3	2	3
21.	Bharal	Pseudois nayaur	6	6	4	6	2
22.	Himalayan Tahr	Hemitragus jemlachicus	4	6	4	6	2
23.	Goral	Nemorhaedus goral	2	3	2	3	2
24.	Serow	Capricomis sumatraensis	4	4	4	4	3
25.	Nilgai	Boselaphus tragocamelus	3	6	6	3	5
26.	Samber	Cervus unicolor	3	4	4	2	3
27.	Swamp Deer	Cervus branderi (duvauceli)	5	6	6	5	6
28.	Chital	Axis axis	3	4	4	1	6
29.	Hog - Deer	Axis porcinus	3	6	6	2	6
30.	Musk Deer	Moschus moschiferus	5	5	3	5	4
31.	Muntjack or Barking Deer	Muntiacus muntjak	3	3	3	2	3

Contd....

S. No.	Name of Species (Local Name)	Name of Species (Zoological Name)	Dehra-dun	Tehri	Uttar Pradesh	Pauri	Cha-moli
32.	Indian Wild Boar	Sus scrofa	2	3	3	2	3
33.	Indian Elephant	Elephas maximus	2	6	6	2	6
	Reptiles						
34.	Monitor Lizard	Varanus monitor	2	3	3	2	3
35.	Fresh Water or Swamp Crocodile	Crocodylus palustris	5	6	6	3	6
36.	Gharial	Gavialis gangeticus	5	6	6	3	6
37.	Python	Python molurus	3	4	4	3	4
38.	Snake Birds		2	2	2	2	3
39.	Snow Partridge	Lerwa lerwa	3	3	3	4	3
40.	Himalayan Snowcock	Tetreogellus himalayensis	4	6	4	6	3
41.	Indian Black Partridge	Francolinus francolinus asiae	2	2	3	1	3
42.	Western Tragopan	Tragopan sp	4	5	4	4	4
43.	Himalayan Monal Pheasant	Lophophorus impejanus	3	4	3	4	2
44.	Whitecrested Kaleej Pheasant	Lophura leucomelana hamiltonii	2	3	2	1	2
45.	Indian Red Jungle Fowl	Gallus gallus murghi	2	3	3	1	3
46.	Koklas Pheasant	Pucrasis macrolopha macrolopha	4	4	4	3	3
47.	Chirpheasant	Catrsus walliclni	4	4	4	4	4
48.	Indian Peafowl	Pavo cristatus	2	3	3	2	6

Abundant-1 Common-2 Rare-3 Endangered-4 Locality Extinct-5 Not Present-6

Threat to Wildlife

The large variety of wildlife depicts India's richness in biological diversity. It is said that there are approximately 500 species of mammals, 1200 species of birds, 300 species of reptiles and more than 20,000 species of insects in India.

But their existence is adversely threatened by the progress and prosperity of human race. Continuous erosion of their habitat, use of forest resources for personal gain and hunting of wildlife for commercial purpose resulted in abrupt decline in their number.

Now in India most of the wildlife is concentrated only in conservation areas. Comprising of 54 national parus, including 18 tiger reserves, and

247 sanctuaries, the conservation areas are spread all over India. In the Garhwal region alone there are four national parks. They are: Nanda Devi, Valley of Flowers, Corbett and Rajaji. While some of the conservation areas like Nilgiri, Namdapha and Nanda Devi have already been declared biosphere reserves, some other area are still under consideration for being declared as biosphere reserves.

Geography of Garhwal Region

Administratively Garhwal region can be divided in five districts. They are: Uttarkashi, Chamoli, Tehri, Pauri and Dehradun of Uttar Pradesh. India. The region is geographically situated between the coordinates of 29°27' to 30°33' N latitudes and 77-48' to 80°06' E longitudes, the altitude of Garhwal region rises from ca 303 m (1000') in the south to ca 7816 m (25,645') in the north. As the altitude increases the forest type changes from Tropical Moist Deciduous Forest at lower altitudes through Subtropical Pine Forest, Himalayan Moist Alpine Scrub to Dry Alpine Scrub at higher altitudes. A narrow belt of bhabar (porous detrial piedmont) skirts the hills to the south. The largest glacier, Gangotri glacier of the region of ca 29 km length. It is situated in the district of Uttarkashi. The Ganges is formed by the confluence of the Bhagirathi and the Alaknanda at Devprayag, and it controls the drainage.

Wild Species

There is different of opinions about the definition of the world wildlife. It varies from one authority to another. Dictionary defines the word as all the animals, birds, and insects that naturally inhabit a particular region. According to by Mitra (1980) "One should consider all animals and plants in their wild state as wildlife". For all pratical purposes this definition stands valid. stands valid. This conveniently leads us to the following two sections, viz., the flora (plants) and the fauna (animals). Further, according to the present author, as a subject wildlife is one of the most difficult, multi-disciplinary, applied field of science. It deals mainly with the manipulation of population (temporally). This can be done through studies, especially, studies such as population dynamics, limiting factors, climatic factors etc.

Going through the Indian zoological literatures we come to know that there are a number of faunal lists of the Garhwal hills. These literatures are mainly prepared by some renowned zoologists, naturalists and conservationists. Prominent among them are: Dang, 1964a, b; Lavkumar, 1978; Lamba, 1985; Lamba Tak and Kumar, 1984; Tak and Kumar, 1983, 1987; Tak and Lamba, 1984m 1985 and Read, 1979.

Dr. P.C. Tak, with his more than twelve years' research experience in the field of ecology and wildlife behaviour, travelled through the fields of Nanda Devi, Chamoli between 1981-85 and 1989; Nayyar Valley Catchment Area, Pauri in 1983; Dokriani Baniak/Glacier, Uttarkashi in 1983; Gangotri Glacier, Uttarkashi in 1980. He also stayed continuously for three years (1976-79) at Dhikala in Corbett National Park, Pauri. As we have known from his survey report as well as the information we could collect from the zoological literatures imphy that about 60 species of mammals, more than 400 species of birds and around 40 species of reptiles inhabit the Garhwal region.

Endangered and Threatened Species

Based mainly on four factors the International Union for Conservation of Nature and Natural Resources (INCN) recognised the various categories of rare animals under these two categories—endangered and threatened. Those four factors are: a) present and past distribution, b) decline in number of populations in course of time, c) abundance and quality of natural habitats and d) biology and potential value of the species.

Endangered: Some casual factors have endangered the existence of certain rare species. If these factors continue to operate then the survival of these species is doubtful. Continuous reduction in their population and shrinkage of their habitat will lead to a day when these species will totally extinct.

Threatened: The term 'threatened' is used in the conservation context for species which are in one of the following three categories: viz., Endangered, Vulnerable or Rare. If the casual factors continue operating the species are likely to move into the endangered category in the near future

There are about 81 threatened species of mammals, 47 birds and 15 reptiles in India. Put of that 14 mammals, 8 birds and 4 reptiles are found in the Garhwal region. Some of them are listed below alongwith some explanation on their present status and distribution in the region.

A. Mammals

Class: Mammalia

Order: Carnivora

Family: Canidae (Dog family)

1. **Canis aureus aureus** (Linnaeus)—Asiatic Jackal.

One of the most familiar animals of the country, it is closely related to the wolf. This animal can live in almost every environment. Nocturnal in habit, it is well established at hill stations 1220-2135 m (4000'-7000').

Status: Vulnerable; on account of its persecution for pelt, its habit of raiding sugarcane fields, and extensive use of pesticides in agricultural fields.

2. **Vulpes Vulpes montana** (Pearson)—Hill Fox or Red Fox.

It is an animal of dry, arid and semi-arid regions. It is found in birchwood and near cultivation in the Garhwal Himalaya.

Status: Vulnerable; due to indiscriminate hunting and trapping for its handsome winter pelt.

Family: Ursidae (Bear family)

3. **Ursus arctos** (Linnaeus) (Himalayan) Brown Bear.

Usual haunt of this animal are the bare open peaks above the tree line. It is now rarely seen in inaccessible parts of the Garhwal region.

Status: Endangered; as it has been hunted excessively in the past.

4. **Melursus ursinus** (Shaw)—Sloth Bear.

Lives in forested tracts of the region, at the base of the Himalaya, at comparatively much lower altitudes.

Status: Vulnerable; due to the destruction of its habitat for human use, and was killed in the past for the medicinal value of its bile.

Family: Felidae (Cat Family)

5. **Felis chaus affinis** (Gray)—Jungle Cat.

Prefers bushy jungles, scrubs, forests of the region. It preys upon small animals of all kinds. It raids poultry also.

Status: Vulnerable; due to shrinkage of its habitat and its excessive killing for pelt in the past.

6. **Felis bengalensis horsfieldi** (Gray)—Leopard Cat.

Inhabits forested areas far and near human habitations of the region, and raids poultry also.

Status: Vulnerable; on account of its habitat destruction, and its persecution for pelt and habit of raiding poultry.

7. **Panthera pardus** (Linnaeus)—Leopard or Panther.

Inhabits a variety of habitats like the hills, plains, thick forests and scrub jungles of the region. It approaches the human settlements also. Despite being a medium sized cat, it is very strong. It may kill even large sized prey animals.

Status: Vulnerable; due to excessive hunting for its graceful pelt, reduction of its preys and habitat shrinkage through its distributional range.

8. **Panthera tigris tigris** (Linnaeus)—Tiger.

This extremely powerful animal inhabits in dry, open jungles and humid, evergreen forests of the Garhwal region. It can kill all kinds of prey. It may kill and eat even fowl, fish, reptiles etc.

Status: Vulnerable; on account of excessive hunting for its magnificent skin, reduction of its preys, and habitat shrinkage. Thanks to 'Project Tiger'. Due to care and maintenance of this animal, under this project, its number has now risen considerably.

9. **Panthera uncia** (Schreber)—Snow Leopard or Ounce.

It loiters in the higher altitudes ca 3500-4700 m. In winter it visits lower elevations upto 2000 m or so. Because of its inaccessible haunts and its nocturnal habits it is very difficult to observe.

Status: Rare: It has been over-exploited for its valuable fur used for making luxury items such as costly dresses, gloves, cushions etc. However, in 1971, the International Fur Trade Federation imposed a total ban on the trade of its fur.

Order: Artiodactyla

Family: Cervidae (Deer Family)

10. **Moschus moschiferus** (Linnaeus)—Musk Deer.

Though mostly it is found within birch forests above the zone of pines, at times it comes down to lower levels. It always conceals itself in thick cover and hence difficult to observe.

Status: Endangered; due to large scale commercial exploitation of 'musk' which is secreted by the abdominal gland of the males. But the chance of further depletion is reduced. As the method of collecting musk without killing the animal has come into practice.

Family: Bovidae (Sheep and Goat Family)

11. **Nemorhaedus goral goral** (Hardwicke)—Goral.

Lives throughout the region between ca 900 m and 2500 m. Often seen near human habitation.

Status: Vulnerable; as it is being hunted by man for flesh and is preyed upon by various carnivores.

12. **Hemitragus jemalhicus jemalhicus** (H. Smith)—Himalayan Tahr.

Of all the wild goats, it perhaps selects the most inaccessible ground to live in at an altitude between ca 3000 m and 4000 m. Comparatively, it is easier to observe when it comes out to graze in open clearings during mornings and evenings.

Status: Endangered because of excessive hunting for its flesh and skin, and shrinkage of its habitat.

13. **Pseudois nayaar nayaar** (Hodgson)—Bharal or Blue Sheep.

It is found between the tree line and snow line throughout the region.

Status: Endangered on account of excessive hunting for its flesh and skin, and also for shrinkage of its habitat.

14. **Capricornis sumatraensis sumatraensis** (Bechstein)—Serow.

It prefers an altitude ranging from 1850 m to 3100 m of thickly wooded gorges, e.g., Rishi Gorge in Nanda Devi Area.

Status: Endangered because of excessive hunting for its flesh and skin, and for also for shrinkage of its habitat.

B. Birds

All the eight species of birds listed below are either endangered or threatened. They are found in Garhwal region wherever their habitat exists. Four out of these eight birds belong to the birds of prey group. The remaining four belong to the pheasant group. The alarming status of Western Tragopan in the Garhwal region, in conservation context, invites the species-specific survey or rather exploration.

Class: Aves

Order: Falconiformes

Family: Accipitridae (Hawks, Vultures)

1. **Aquila chrysaetos daphanea** (Severtzov)—Himalayan Golden Eagle.

2. **Aquila rapax nipalensis** (Hodgson)—Eastern Steppe Eagle.

The Most Jajestic Mountain Chain

Himalaya, 'the abode of snows,' is the world's loftiest mountain chain. This majestic mountain chain is the global heritage of all mankind. From time immemorial it has attracted sages and saints. Mention about the devine nature of its snow-covered peaks, icy glaciers, life-supporting rivers and its fertile valleys can be found in ancient texts and literatures. Devotees from all over the country visit the places of pilgrimage in Himalaya.

But the process of some developments in the name of modernization have irresistibly disturbed the nature's balance. Besides that increase in population in the region is another major cause of ecological disturbances in the area.

Furthermore, Himalaya has proved to be a powerful magnet for those who have quest for adventure. This game has attracted a large numbers of them from all over the world.

Young and Still Growing

The Himalaya was formed sixty million years ago when a plate of the earth's crust carrying the Indian land mass travelled 5000 kms from near the South Pole and collided with Laurasia. While the most substantial rise took place in the past 38 million years and the final uplift has occurred in the past one million years. Geologically Himalaya is considered to be still young. Even now it is said to be growing. Exercising a very powerful influence upon the entire subcontinent, the tectonic forces, which brought the Himalaya into being, created a remarkable complex environment. They are the major factors in determining the climate of India, the beneficient monsoon rains and affording protection in the winter from the cold Siberian winds. They also produce their own climate, ranging from near polar in the higher reaches to tropical humid in the foothills of its eastern segment. The winter snows are stored by the Himalaya for feeding the ever lasting northern rivers.

Climate and Geologic Variety

The climate and geologic variety has contributed towards the creation of diverse ecosystems ranging from the cold desert of Ladakh to the tropical rain forests of Arunachal Pradesh, and encompassing an enormous wealth of natural living and non-living resources. Through their altitudinal zones, the latitudinal gradation from the equator to the poles are recaptured over short vertical distances. The sharp zonation in vegetation and consequently animal life results in extreme diversity of plants and animals. Fertile tropical rain forests to timber-line and alpine meadows can be surveyed in vertical heights of less than three thousand metres. Their complex assay of ridges and valleys multiplies the surface area contained within narrow geographic reaches and provides conditions equivalent to an archipel ago of islands scattered over the sea.

Tropical swampy forests, deciduous forests, coniferous forests, rhododendron forests, alpine meadows and often, cold as well as hot deserts, occur from the *terai* to the snowline within the Himalaya. The rain shadow effect is pronounced, and even the northern or southern aspects of slopes have a decisive influence over vegetation. The indigenous human communities have evolved rich cultures adapting themselves to often hostile environment with remarkable vigour and resilience. Such an immense variety of resources is the greatest and also the most vulnerable asset of the Himalaya, for this young mountain chain is extremely fragile and highly susceptible to any thoughtless interference in the natural balance of its sensitive ecosystems.

Highway for Living Beings

The biological diversity in Himalayan ecosystems co-exists with human, cultural diversity. Besides primitive forms of life the difficult terrain provides sanctuaries for human civilizations and cultures also. They have provided refuge for ancient vegetation types such as the tropical rain forests of north-eastern India and even for the mythical

Yeti and his counterparts in other snowy mountains. The Himalaya was considered to be a highway for living being to cross its barriers and travel along and spread across its length and breadth. Of course all living being could not cross the barriers. The Eurasian group of animals like the wild sheep and goat moved eastwards along its border. They reached almost upto South China. Whereas the Malayan squirrel family and the Macaques moved westwards. Though the camel and horse groups failed to cross the barrier southwards, the tiger succeeded. The same effects on distribution area also noticed for a number of prominent plant groups too.

The Aryans moved slowly through the lower passes in the west. They pushed the Dravidians southwards, who in turn pushed the Negrito and Australoid human races into renote forest strongholds. The Mongolians, however, crossed the Himalaya. Mingling with the North Indian groups, they formed the Indo-Mongolians. So, Himalaya is said to have the richest wealth of living form within its territory. The wealth which is considered to bedelicate.

Life Not So Easy

In the environment of these high mountains the life is not to easy and amiable. During the last thousand years man has learnt to cultivate and raise domesticated animals on the harsh hill slopes. Terrace cultivations, natural gravity irrigation, well adapted cultivated crop varieties and distinctive life-styles have gradually developed without undermining the delicate web of the Himalayan mountain ecosystems. Populations were never so high. So human demands were limited within self-sustaining levels of such ecosystems. Life was slow. The meager productivity of the high valleys and slopes was somehow enough to meet their demand.

Hill Communities in Peril

Extension of communications and increasing urbanisation, resulted in an inflow of a large number of people from outside into the Himalayan region. This put the delicate ecosystem of the region into adanger. Unadapted to outside processes the communities and cultures within such ecosystems often lose their identity and quality.

Extensive exploitation of resources, hydro-power, minerals, timber, plants and animal produce has not only affected the mountains, but also the vast plains below. This is the condition of Himalaya today. This has put the very foundation of human societies in the Gangetic and Brahmaputra Valleys in peril.

Unsustainable 'Development'

In order to meet their basic requirements of food, fuel, fodder, fertiliser, fibre, shelter and water the hill communities utilised the natural resources of the Himalayan region for centuries in such a way that the balance of the nature of the region is not disturbed. Lifestyles of such communities were adjusted with the surroundings though mountains are not the best habitats for man, what with physical restraint reduced their capacity to support large populations.

The basic life-support systems of soil, water and biota are the limiting factors in mountain ecosystems obliging hill communities to use their environment and resources economically and sensibly without depleting the resource capital. The steep slopes and rapid change of weather make the soil flinsy. At the sametime the stern environment of the region makes the forests of the area very delicate. Here the animals are vulnerable because they can survive only within a narrow range of conditions. Simple lifestyles and isolation has not moulded the people of this region to resist the outside pressure. This makes the people here very delicate. There were always, thus, few people in the hills who subsisted on minimum resources from large areas. However, with 'development', the situation changed. Large number of people with a new philosophy of life based on the market economy entered the mountains. They introduced an exploitative approach which disturbed the prevailing balance. Forests were taken away from community ownership for meeting commercial requirements of timber in the plains. Besides constructing roads through the maintains to carry timber, plant and mineral resources. This, naturally put huge hydropower schemes to support industries were also launched. At the same time, the human and domestic animal population of hill communities also increased. This, naturally put, heavy pressures on the limited natural resources of the mountains. With the disappearance of vegetation extensively, water and fuel resources dwindled. As a result soils rapidly lost their productivity. This has led to a large scale migration of the hill people to the plains for economic survival. Those who were left behind had to travela long distances for fuel and water. Realising that survival depended upon the protection of the life-supporting water and soil resources provided by the forests the remaining people in the hill area, comprising largely of women, started the CHIPKO movement in the early seventies in Central Himalaya.

CHIPKO—An Action for Survival

The roots of this movement extent deeply in India's forest culture which flourished in the woods. The sages who lived in *ashrams* with their students realised the problems of mankind. They developed a philosophy which believed in the coexistence and interdependent of all forms of life: They preached and that all creations such as birds, beasts, human beings, rivers and mountains have life, and all life is sacred and merits worship. They also advocated that austerity and wisdom are worthy of respect. This cultural legacy was gradually supplanted, under the impact of colonial rule, by materialistic values. Earlier in the century, the people of Uttarakhand opposed the policy of reservation and commercial exploitation of Himalayan forests and as patch-work settlement, panchayat or community forests were created. After Independence, the hill people pledged to activate the friendly relationship between the forests and forest-dwellers. They wanted to stop exploitation for easy availability of raw materials for local forest-based small industries and revision of forest settlement. The change in land use pattern brought about by substituting mixed natural forests by monoculture pine plantations leading to accelerated soil erosion, as well as rapid increase of population, forced a large member of men to seek employment in cities in the plains. As a result, the women of the community had to take the burden of managing the family and caring for the aged, the children and cattle. Even they had to take the responsibility of carrying on agricultural operations. To collect such basic necessities of life they had to walk a long distances.

Effort to educate the locla peopple about their right over the Himalayan forest resources was initiated under the leadership of various followers of Mahatma Gandhi and Sarvodaya workers. Through foot-marches and folk song they tried to motivate the people. The popular slogan was—"What do the forests bear? Soil, water, the pure air. Soil, water and pure air are the basic of life." Women played the leading role in this movement for survival. Soon it spread rapidly in villages. By 1974 efforts were made to stop tree felling by hanging them. Not only that the women of Advani village in Tehri Garhwal tied sacred threats round the trees which were marked for felling. They decided to stop the tree felling even at the cost of their lives. In 1977, demonstration against auctioning the forests were organized.

Effect of CHIPKO Movement

An alternative plan of development in the Himalaya was provided by the CHIPKO movement. It emphasised that the main products of the forests—soil, water and oxygen—are all vital for human survival. So there was, urgent need for exercising some strictness in the use of forest products. Therefore, the state government was compelled to impose some check on felling of trees in the areas exceeding 1000 mts atltitude and 30 degrees slope. The hope for the future lay in reafforestation through the villagers' own efforts rather than through state ownership and action, as the people were in the best position to protect and plant trees that fulfil their basic needs of oxygen, water, food, fodder, fuel, fertiliser, fibre and small timber. The aim of CHIPKO movement was not only to preserve the forest resources but also to protect the entire life support system—the option for human survival. It was felt that for the success of the movement small groups of humanitarians, scientists, social activists and compassionate literary men were needed. CHIPKO's search for a strategy for survival has indeed global implications. Gandhi's mobilization for a new society, where neither man nor nature is exploited and destroyed, was the beginning of this civilization's response to a threat to human survival.

Effect of Expedition on Biological Wealth

Tourism and mountaineering aregaining popularity in the mountains during the summer months. This puts pressure on the limited biological wealth of the Himalaya. Foreigners were allowed to enter many prohibited areas. Continuous expeditions to a number of popular peaks seriously affected the biological balance of the region during the crucial growth period in the few summer months. For example, the large numbers of expeditions to the Nanda Devi Sanctuary has led to rapid exhaustion of unique natural resources of the area. It has also affected quiet a member of indigenuous species that have developed for centuries under the natural protection there. That at high altitudes the birch tree attains a diameter of two centimetres in six years and a juniper bush takes twenty years for the same growth reveals the fact that this human activity has adversely affected the growth of plants too. In order to save the dilapidating situation, some protective measures to restore the biological and natural assets of the region are the need of the day. As a measure towards that, except for purely scientific works such as National Parks or Biosphere Reserves, all types of human activities in the Nanda Devi Sanctuary and valley of flowers have been prohibited.

Effort to Restore Himalayan Ecosystem

Besides being our shelter for observation and recreation, many societies depend on mountains for their survival. We have noticed in previous para tourism and mountaineering may cause havoc unless they are planned carefully keeping in mind the delicacy of ecosystem. We must judge to what extent the system can be exploited. Roads and huge dams are being constructed without realizing their impact on the environment. But before irreparable damage is done, a resource survey along with planned, integrated development with river catchment areas as units, are most urgently needed to protect the Himalayan ecosystem.

Youngest and Highest Chain of Mountains

Considered to be the world's youngest and highest, the Himalayan mountain range links up three different biological regions of the world. They are: a) very rich Malayan tropical biogeographic region, b) cold and temperate Eurasia, and c) Indian Peninsula. It has some connection even with the African-Arabian region. Plants and animal have found an asylum in very valuable niche of the Himalaya. Today these should be identified and declared as protected biosphere reserves, so that evolution of life there can continue safely in natural surroundings. As a pecautionary measure entry of large numbers of visitors in the popular areas should be restricted before it is too late.

Plans of Direct Relevance

For the people in the hill region very carefully drawn up plans of direct relevance to them need to be implemented. To implement a programme for developing the mountain areas, the delicate natural environment and the cultural balance the hill people have developed for their survival in that rough surroundings should be taken care of.

22

Jigme Dorji National Park Largest Protected Area of Bhutan

Introduction

Jigme Dorji National Park is largest protected are of Bhutan. It covers an area of 4280 km. Situated between 27°40'-28°15'N latitudes and 89°5' - 90°20'E longitudes, it encircles the upper catchment of Bhutan's three major rivers—Pho Chu, Thim Chu and Paro Chu. This Park is said to be one of the least spoilt ecosystems in eastern Himalaya. The only means of transport of this region is yak. It is also the sole livelihood for the inhabitants of the Jigme Dorji National Park.

This chapter deals with the wildlife, floral aspects and vegetation of this region as revealed in the survey report which was conducted by Nature Conservation Section, Forestry Services Division in Collaboration with WWF Bhutan Programme in 1994-95.

Area and Features of Jigme Dorji National Park

Jigme Dorji National Park (JDNP; was officially declared as Wildlife Sanctuary in 1974. The area consisted of three separate but adjoining sanctuaries—Laya, Gasa, and Jigme Dorji. The Park is characterized by varied topography viz. snow clad mountains, glaciers, glacial lakes, flower laden alpine meadows, upland and low lying river valleys and vast wilderness with diverse flora and fauna. Height of the area ranges from 1300 m, near Rim Chu bridge in Mo Chu valley to over 7000 m. The highest point is Chomolhari (7314 m). It is one of the most sacred mountains of Bhutan. Conservating from biological, geo-hydrological, ethnic and aesthetic points, conservation of this park is very important.

Some of the forests of the Jigme Dorji National Park give shelter to rare species which are endangered or extinct in other parts of the region. This park represents one of the least spoilt ecosystems in the eastern Himalaya. Of course it is not free from anthropogenic pressures. If not planned carefully, the construction of motorable roads, and bigger buildings in remote areas such as Lighshi and recently created Gasa Dzongkhag are likely to destroy the natural resources. Moreover, the prevailing landuse practices that have been followed by the ethnic groups in these mountains are against the conservation interests.

For the people of this mountain region the only mens of livelihood and transport is the yak. Habitat degradation in some areas of the park due to continuous grazing by large yak herds has adversely affected the delicate mountain ecosystem. This has resulted in problems of crop damage by wild herbivores because not much fod is left for them in the forests. To save the deteriorating situation as well as to protect the economic and social interests of the people residing within the park, the Nature conservation Section (NCS) within the Dorestry Service Division, Royal Government of Bhutan, has been given the charge to work-out an effective conservation plan for the Jigme Dorji National Park.

A biological inventory has been prepared by the NCS in collaboration with WWF Bhutan Programme as a first step towards conservation planning for JDNP. Basic inventory plans were drawn by Mackinnon (1991), RGOB (Anonymous, 1993). It was followed by brief faunal surveys (Yonzon, 1992 and Wikramanayake, 1993). Socio- economic surveys were conducted by WWF (Johari, 1993). Other assessments and reports highlighting the conservation issues of the area include Thinley *et al.* (1993), WWF (1989), Bhadra and Alirol (1986), Blower (1986), DANIDA (1988), Dunbar (1981), Miller (1987, 1988), Sherpa *et al.* (1992), World Bank (1988), etc. Recently, a boundary evaluation team has thoroughly surveyed the area. It demarcated the southern boundary including some of the crucial wintering ranges of important animals such as takin *(Budorcas taxicolor);* Bhutan's national animal (Mongar, 1993). With special emphasis on floral structure and composition of vegetation, animals and their habitats, the Nature Conservation Section carried-out a biodiversity survey during June-July 1994 and again during May-June, 1995.

The methodology adopted for the survey findings on the range of floral and faunal diversity in upper parts of JDNP is described below.

(a) The travel routes were selected in such a way that most of the habitat at types were covered and least overlapped with the previous survey area routes.

(b) Throughout the travel route transects and study plots of known altitude the vascular plants were recorded systematically. Plant specimens were field numbers and other relevant information such as rarity and also use, local name, habit and habitat as followed in the previous survey Grierson and Long's (1983-1991) *Flora of Bhutan* Vols. I & II (i), and Polunin and Stainton's (1987) *Concise Flowers of the Himalaya* were used in the field to identify some of the species. But due to lack of enough document, all the species of large genera such as *Carex, Poa, Saussurea, Juncus, Gentiana,* etc., could not be identified in the field. As a result, such species as well as some others were collected in duplicate or triplicate. They werepressed in the blotters and dried for further processing, proper identification and incorporation in the NCS herbarium. Specimens of common, well-known or weedy species were not collected due to logistic reasons.

(c) The elderly people in the villages and yak herders camps were requested to show the plants of their day to day use along with their local names etc., for ethnobotanical information.

(d) With the help of landuse maps prepared by Land Use Planning Project (LUPP), MOA (1993) and some broad categories identified by Negi (1983) the vegetation and habitat types were recognized and confirmed. The use of various vegetation types by wild animals specially by larger mammals and birds) were verified using direct or indirect evidences as well as local information.

(e) The vegetation transects were laid in the subalpine forests, alpine scrub and alpine meadows. Each transect was 2.5 km long with sampling points 250 m apart. As far as possible, narrow altitudinal belts (within 200 m) were followed to sample one vegetation type. For subalpine forests circular plots of 12.5 m radius were laid at each sampling pcint. All the tree species, their number, height, and total canopy cover was recorded. A nested plot of 5.65

m radius was laid within each plot used for trees to record the shrub species and their number. Herbs, grasses and other ground flora were recorded within four subplots of 2 m x 2 m area systematically placed at North, East, West and South points on the periphery of 5.65 m radius plots. Alpine scrub (krummholz, riverine scrub and scattered shrubland) and meadow vegetation was sampled within the plots of 5.65 m radius. Calculating the member of shrub species in the alpine zone was very difficult because their growth in the zone was very slow at the same time spread all over the zone. So percentage cover of each shrub species within the plot was recorded. For alpine meadows all the dwarf shrubs, herbs, grasses and sedges were recorded within 2 m x 2 m square plots which were systematically laid as in alpine scrub and subalpine forests. The site characteristics (soil type, depth, moisture, etc.), wildlife evidences (feeding signs, dung, burrow, etc.) and geographical location (using Global Positioning System GPS) were noted for each plot. All the data are being entered in the computer at NCS headquarters for further analysis.

(f) Mammals and birds were recorded systematically along the travel routes as well as transects. Prater (1984), Ali and Ripley (1983), and Flemming *et al.* (1984) were used for identification of mammals and birds in the field.

Extensive Range of Plant Species

In total 1,434 species belonging to 144 families of higher plants (Gymnosperms and Angiosperms) were recorded within the survey areas of Jigme Dorji National Park (Appendix). The floral diversity is summarized as follows:

Group	*Genera*	*Species*
Gymnosperms	9	13
Angiosperms	554	1421
Total	563	1434

The total number of tree, shrub, climber and herbaceous species (including grasses and orchids) were 138, 240, 47 and 1009 respectively. Five dominant families (in terms of species richness) are as follows:

	Family	*Genera*	*Species*
(i)	Asteraceae	32	105
(ii)	Rosaceae	20	78
(iii)	Poaceae	38	69
(iv)	Ranunculaceae	17	64
(v)	Cyperaceae	7	50

The following families were represented by single genus and species: Ephedraceae, Taxaceae, Anthericaceae, Musaceae, Pontederiaceae, Annonaceae, Cactaceae, Cannabaceae, Chloranthaceae, Circaeasteraceae, Diapensiaceae, Droseraceae, Hippuridaceae, Illiciaceae, Meliaceae, Myricaceae, Nyssaceae, Orobanchaceae, Santalaceae, Saururaceae, Simaroubaceae, Staphyleaceae, Sterculiaceae, and Tetracentraceae. The following genera had more than 10 species: *Carex* (31), *Saxifraga* (29), *Rhododendron* (27), *Pedicularis* (24), *Gentiana* (23), *Corydalis* (20), *Juncus* (19), *Potentilla* (18), *Saussurea* (17), *Primula* (16), *Rubus* (14), *Rhodiola* (13), *Kobresia* (12), *Arisaema* (11), *Salix* (11), *Aster* (10), *Astragalus* (10), *Meconopsis* (10), *Bistorta* (10), *Aconitum* (10), *Anemone* (10), *Thalictrum* (10), *Lonicera* (10).

Common ferns recorded within the sample plots were: *Araiostegia delavayi, Arthromeris wallichiana, Athyrium pectinatum, Cyathea spinulosa, Cryptogramma crispa, Dennstaedtia scabra, Dennstaedtia* sp., *Dicranopteris* sp., *Diplazium polypodioides, Drynaria mollis, Drynaria propinqua, Dryopteris* sp., *Gleichenia linearis, Lepisorus loriformis, Leucostegia immersia, Leucostegia* sp., *Oleandra wallichii, Onychium japonicum, Onychium* sp., *Polystichum aculeatum, Pteridium aquilinum, Pteris cretica, Pteris quadriaurita, Sphenomeris chinensis, Thelypteris* sp., etc. Systematic collection and study of fern flora is yet to be done in Jigme Dorji National Park.

Plants of Medicinal Value

Large varieties of medicinal plants are found in the high altitude areas of Bhutan. The people of the region collect various wild medicinal herbs to cure their common diseases. Moreover, frequent collection of medicinal plants from Laya and Lingshi areas for Indigenous Hospital in Thimphu, is also an evidence that high altitude areas have more variety of medicinal plants.

Plants of Day-to-Day Use

Besides medicinal plants, there are a number of species of day-to-day use. For example, Cypripedium himalaicum (a rare lady's slipper orchid) is used in Laya area for making vegetable curry. In Lingshi area, leaves of Salvia nubicola are chewed as paan, while its roots are used as substitute for areca nut.

Cordyceps sinensis is an important alpine fungus. Local people collect this species in large quantity as they fetch high price by illegally selling them in the international market. An evergreen Himalayan tree is Taxus baccata. It needs special conservation efforts throughout its range.

Cupressus corneyana is the national tree of Bhutan. Besides being a valuable timber, its leaves are used in incense. Furthermore, its smoke is used to purify the air during auspicious occasions.

Rare Indigenous Plants

Various plants of high conservation value in Jigme Dorji National Park are: e.g., *Podophyllum hexandrum,* species of *Aconitum, Delphinium, Herminium, Pleurospermum, Gentiana, Corydalis, Parnassia, Polygonatum,* etc., for their valuable alkaloides and various medicinal properties; species of *Allium, Fritillaria, Lilium,* etc. as wild gene pool for future crop research; *Rheum nobile, Pterocephalus hookeri,* species of *Aster, Senecio, Saussurea, Rhododendron, Geranium, Meconopsis, Epilobium, Anemone, Potentilla, Pedicularis,* etc., as potential horticultural crops of ornamental value; *Circaeaster agrestis, Triosteum himalayanum, Helwingia himalaica, Diapensia himalaica, Corylopsis himalayana,* etc. as rare endemic species for Bhutan.

A detailed study of plants listed under schedule I of Bhutan's Forest *and* Conservation Act (1995) states that almost all the herbaceous species are for alpine and subalpine regions. They have very specific microhabitat requirement of threatened species e.g., *Podophyllum hexandrum* (Himalayan May Apple), grow among rocky, *Berberis-Juniperus* shrubberies. Frequent fires and regular collection of shrubs from such areas may cause the local extinction of some species. Hence, there is need to identify the populations of rare plants and their microhabitat in order to prove monitor them.

Effect of Attitudinal and Topographic Variations

Quite a number of habitat types exist in Jigme Dorji National Park. It is the result of extensive altitudinal and topographic variation. For a proper conservation planning the Nature Conservation Section and WWF, Bhutan, have identified eight different habitat types. The different habitat types have given birth to different vegetation communities. They are discussed below:

(a) ***Alpine scree slopes:*** Areas close to snowline, such as high passes, based slopes and loose rocks (4800-5500 m) are characterized by very specific vegetation cover. They resemble with cold deserts in appearance. Only herbaceous plants such as species of *Draba, Corydalis, Saxifraga, Arenaria, Lagotis* etc., grow in such habitats. The rocky outcrops abound in lichens such as *Thamnolia vermicularis, etc.,* and mosses.

(b) ***Alpine meadows:*** Large number of valuable medicinal plants are found in alpine meadous, especially in gentle slopes scale grazing by domestic livestock as well as wild herbivores their number is presently declining. Besides over grazing, burning, camping and usage of fuelwood are also responsible for their degenaration. These plants communities vary according to stability of slope also. The most important a areas are: Kobresia redge, Potentilla-Geranium, Trachydium Euphorbia stracheyi and Pediculari Juncus. They mostly grow in moist areas.

While both the domestic animals there—yaks and horses—graze most of the saith facing alpine scrubs and adjacent meadows, the meadows inter spersed by streams are specially preferred by yaks for gazing.

Species typical of overgrazed alpine meadows are: *Prunella vulgaris, Trollius acaulis, Primula sikkimensis, P. rosea, Carex nivalis, Rhododendron lepidotum, Senecio chrysanthemoides, Polygonum viviparum, Plantago erosa, Taraxacum officinale, Euphrasia himalaica, Rumex nepalensis, Capsella bursa-pastoris, Scirpus* spp., etc. Preferred palatable plants include *Cortia lindleyi, Trachydium roylei, Selinum vaginatum,* species of

Astragalus, Oxytropis, Polygonum, Sibbaldia, Festuca, Poa, Agropyron, Carex, Kobresia, etc.

(c) ***Alpine scrub:*** Areas between timberline (3700 ± 200 m) and alpine meadow (4200 ± 200 m) are characterized by scattered bushes of various species. Depending upon the exposition (sunny vs shady slopes) two distinct categories of alpine scrub vegetation can be identified in JDNP:

(i) ***Alpine dry scrub:*** Nearly all the south and south-east facing slopes between the alpine meadows and timberline are dominated by various shrubby species such as *Ephedra gerardiana, Juniperus pseudosabina, Lonicera myrtillus, Rosa sericea, Rhododendron lepidotum, Cotoneaster microphylla, Salix lindleyana, Chesneya nubigena,* etc. These slopes are also preferred by domestic livestock for grazing. *Thermopsis barbata* and *Rumex nepalensis* are the typical plants of overgrazed slopes.

(ii) ***Alpine moist scrub:*** Besides all the shady moist slopes, the valleys between 3700-4200 m support dense thickets of stunted trees and shrubs. Riverine *Salix-Myricaria* scrub, dense krummholz (stunted forest) of *Rhododendron-Lonicera- Sorbus,* and Ericaceous dwarf shrub communities of boundary area above krummholz zone are the three different shrub communities that are identified in this area.

(d) ***Subalpine forests:*** In the Himalayan mountains subed pine forest are said to be one of the important ecological zone. Situated between ca 3300 and 3700 m, this zone exhibit characteristics of both temperate and alpine regions. Terminating at timberline towards upper limit forming the natural edge, these forests merge with cool temperate forests towards lower elevation. The following communities are the characteristics of subalpine forests:

(i) ***Birch-rhododendron forest:*** This community have species like *Rhododendron* (*R. hodgsoni, R. campanulatum, R. barbatum,* etc.,) along with birch *(Betula utilis)* and occasional fir *(Abies densa)*. The steeper areas with prolonged snow deposition causes typical gnarled appearance of birch and rhododendrons.

(ii) ***Juniper woodland:*** South facing subalpine slopes are mostly dominated by juniper (*Juniperus pseudosabina)* woodland. These forests have relatively low tree species. However, the cover of mosses, shrubs and herbs are considerable. Because of its valuable timber and fuelwood Juniperus species gets cut and burnt frequently reducing. As a result, the forest cover is drastically reduced every year. *Rhododendron lepidotum, Gaultheria trichophylla, Phlomis* spp., *Thalictrum chelidonii,* etc. are the common species of the area.

(iii) ***Fir forest:*** Though fir *(Abies densa)* is not a typical subalpine species as it is found in association with other broad-leaved and coniferous tree even at lower elevation (2500 m), it forms a considerable forest cover on northern aspect as high as treeline interspersed by patches of birch-rhododendron, bamboo, etc. depending upon the canopy opening and site characteristics.

Rheum australe, Aconitum heterophylloides, Pleurospermum angelicoides, Angelica glauca, Thalictrum spp., etc. are the number of rare and endangered herbs including medicinal plants that are found in the subalpine forests.

(e) ***Mixed conifer forest:*** Particularly on north facing rocky slopes between ca 2000-3300 m there is a mosaic of forest. Though mainly dominated by oure or mixed conifer species, fir (Abies densa) triumph at higher elevation (more than 2600 m). The small batches of broad-leaved species, bamboo brakes and rhododendron thickest within the mixed confer forests vary upto a large extent in richness and physiognamy therefore influencing the wildlife harbitat. Larch (Larix griffithiana)—another confer species—is found mostly along the ravines towards interiors of Pho Chu and Mo Chu.

Pure to mixed patches of Tsuga dumosa (homlock), Juniperus recurva and pica spinulosa (spruce) as well as some broad-leaved species like maples (Acer caudatum, A. pectinatum, A. campbelli) Prunus rufa, Oaks (Quercus spp.), Sorbus, Betula etc. can be found at lower elevations. To describe vividly the smaller varieties mentioned above, large scale satellite imageries along with extensive ground survey are necessary.

(f) ***Pine Forests:*** In middle elevations, at a height of 1500-3000 m, blue-pine (Pinus wallichiana) and chir-pine (Pinus roxburghii) for large patches.

Blue-pine—In the lower western part of Jigme Dorji National Park, at a height of 2200-3000 m, the wide spread blue-pine forests are situated. Mainly thorny species are found there. Some of them are: Prinsepia utilis, Rosa sericca, Berberis spp., Zanthoxylum armatum etc. Being dry these forests are prone to fire. The ground vegetation is dominated by grasses.

Chir-pine—Very limited area with chir-pine falls under Jigme Dorji National Park. It grows only at lower elevation at a height of 1500-200 m. Distinguished shrubs are: Indigofera dosua, Buddleja asiatica, Rhus paniculata, Wood fordia fruticosa. They cover only some parts of lower Pho Chu and Mo Chu valleys.

(g) ***Temperate broad***—Broad-leaved forests are found along the major river valleys and gentle slopes. Spread over areas between ca 1400-3300 m, this belt can be divided into two zones—

(i) ***Cool temperate broad-leaved forests***—Among the large variety of broad-leaved species in these forests many are deciduous nature. Generally spread over areas above 2200 + 200 to ca 3200 these forests consist of deciduous trees like—Acer, Prunus, Sorbus, Betula alnoides, Carpinus viminea, Quereus semecarpifolia etc. Besides occasional bamboo thicket— Yashania maling and Y. hirsuta, the cool temperate broad-leaved forests are rich in understorey vegetation also. Prominent species are: Rosa, Rubus, Viburmum, Symplocos, Lindera, Euomymus Rhododendron etc.

(ii) ***Warm temperate broad-leaved forests***—Situated along lower Mo Chu at a height of 1200 to 1500 m, these forests are limited in extent. They spread over the warm and moist valleys. Distinguished vegetation communities that are found in these forests are evergreen elements of subtropical areas. They are members of Lavraceae (Persea, litsea spp.) Moraceae (Ficus spp.), Euphorbiaceae (Macaranga, Daphniphyllum himalense, Bischofia Javavica), Quercus glauca, Toricellia tiliaefolia, Mussaenda roxburghii, Schima wallichii, Alangium chinensa, Engelhardtia specicata etc.

(h) ***Riverine vegetation:*** This vegetation is found throughout the belts along the perpetual rivers and streams at a height of 1400-4200 m. The distinguished communities are—

(a) Hippophae-Myricaria at a height of 4000-4200 m

(b) Willow scrub at a height 200-3600m

(c) Debregeasia-Boehmeria-Ficus at a height of less than 2000 m.

Considerable Variation in Richness

Both within and between habitat types one can notice a considerable amount of variation in the species richness. A detailed analysis of plant species diversity and distribution is yet to be done. General trend of tree, shrub, and herbaceous species richness in various habitat types are showing in Table 22.1.

The Table 22.1 shows that in altitude the tree and shrub species richness decreases. The mixed conifer and cool temperate broad-leaved forest. equal species richness. The striking feature of this data is a significant number of herbaceous species in alpine scrub. The higher meadows, on the other hand, represent mainly herbs and few grasses. Their number also declines with increase in altitude.

Wildlife Habitats

The vegetation categories also form the habitation for animal communities during various seasons. The characteristic plant and nutrition found in various habitats are given in Table 22.2. It is noteworthy that species use very narrow range of habitats *viz.*, Himalayan musk deer, panda, tragopans, etc., while others use a wider range e.g., sambar, wild boar, wild dog etc.

TABLE 22.1

Average Number of Species Per Transect in Various Habitat Types of JDNP of 10 Plots Each)

WBL = Warm temperate broad-leaved, CBL = Cool temperate broad-leaved, MCF = Mixed, SAP = Subalpine forests, AS = Alpine scrub, AM = Alpine meadow.

	Habitat type	*Elevation range (m)*	*Tree species*	*Shrub species*	*Herb*	*Others*
1.	WBL	<2000	22	18	16	14
2	OBL	2100-3300	13	12	10	17
3.	MCF	2000-3500	14	16	16	13
4.	SAF	3500-3800	4	5	30	6
5.	AS	3800-4200	1	16	46	18
6.	AM	>4000	0	6	38	9

* Other include ferns, climbers, epiphytes, etc.

We recorded about 160 species of birds (Table 22.3) and 28 species (Table 22.2) along the survey route. Golden Bush Robin *(Erithacus ch* Crowned Black Finch *(Pyrrhoplectes epauletta),* Firetailed Myzo *pyrrhoura)* were among the rare birds sighted in the subalpine birch forests. Frequent sightings of other species such as blood pheasants, direct and indirect evidences of mammals such as Himalayan musk and takin in subalpine forests indicated good habitat conditions in these areas. On the contrary, most of the alpine pastures showed signs of degradation (eroded slopes, preponderance of weedy species, and human pressure). Though, blue sheep, the most common ungulate was sighted 5 times but all the groups were restricted only to the higher, rocky slopes. This means the better pastures are occupied by the domestic livestock (mainly yaks) and blue sheep are pushed in submarginal habitats. Continuous presence of yak in all the lower pastures also negatively affects wintering ranges of blue sheep. Evidences of other ungulates such as barking deer, sambar and goral indicated that their density may be very low in JDNP. A status survey of their population and habitats will be desired for effective management of the park.

TABLE 22.2

Vegetation and Habitat Characteristics of JDNP

Vegetation/ Habitat type	*Altitude/ Aspect (m)*	*Characteristics*	
		Floral	*Faunal*
Perpetual snow/glacier	>5500	—	—
Alpine scree slopes (exposed and rocky areas)	>4800	*Draba, Androsace, Arenaria, Lagotis, Saxifraga*	Snow leopard Blue sheep Snow cock
Alpine pastures	4200-4800	Grasses, sedges, *Potentilla, Geranium, Primula, Trachydium*	Blue sheep Marmots, Pica, Red fox, Tibetan wolf
Alpine scrub	3800-4200		
Dry scrub	S, SE	*Ephedra, Rosa, Cotoneaster*	Blue sheep
Moist scrub	N, NW	*Rhododendron, Sorbus, Lonicera*	Musk deer Takin
Subalpine forests	3300-3800		
Fir	N,W	*Abies densa*	Musk deer
Birch-rhododendron	N, NW	*Betula utilis* *Rhododendron*	Takin Serow,
Juniper	S	*Juniperus*	Wild dog
Mixed conifer forest	2500-3300		
Hemlock-Spruce	N, NW	*Tsuga dumosa*	Sambar
Larch	E, SE	*Larix griffithii*	Wild pig
Oak-conifer	SE, S	*Q. semecarpifolia* *Abies densa*	Barking deer
Pine forests	1500-2800		
Chîr-pine	S, SE	*Pinus roxburghii*	Goral, Y-thr. martin,
Blue pine	S, SW, NE	*P. wallichiana*	Sambar, Leopard
Temperate broad-leaved forest		1400-3300	
Cool temperate (>2000 m)	S, SE N	*Q. semecarpifolia* *Acer* spp., *Sorbus, Prunus* Bamboo brakes	Barking deer, Red panda Leopard, Black bear
Warm temperate (<1600+100 m)	Valleys	Lauraceae *Q. griffithii* Araliaceae	Barking deer Sambar, Pig, Tiger, Langur
Riverine vegetation	1500-4500	*Hippophae-Myricaria* scrub Willow scrub	Lynx Takin Otter Weasel

TABLE 22.3

Checklist of Birds Recorded During Biodiversity Survey of Jigme Dorji National Park

(Habitat types: RC = River courses, FE = Forest edges, CR = Cliffs and rocks, FC = Forest canopy, SS = Scree slopes, SAP = Subalpine forests, TF = Temperate forests, TDS = Temperate degraded scrub and canopy openings, HB = Human habitation, AP = Alpine pasture, WBF = Warm broad-leaved forest, ADS = Alpine dry scrub with scattered juniper or dwarf rhododendron Alpine moist scrub-riverine willow or dense rhododendron, BR = Birch-rhododendron forest Frequency: C = common, F = frequent, S = sparse, R = rare)

	Common Name	*Scientific Name*	*Habitat*	*Frequency*
1.	Goshawk	*Accipiter gentilis*	FE	R
2.	Sparrow Hawk	*Accipiter nisus*	FE	S
3.	Lammergeier	*Gypaetus baratus*	CR	F
4.	Himalayan Griffon	*Gyps himalayensis*	FE	F
5.	Common Buzzard	*Buteo buteo*	FE	F
6.	Black Eagle	*Ictinaetus Malayensis*	FC	S
7.	Tibetan Snowcock	*Tetraogallus tibetanus*	SS	F
8.	Blood Pheasant	*Ithaginis cruentus*	SAF	S
9.	Satyr Tragopan	*Tragopan satyra*	SAF	S
10.	Impeyan Monal	*Lophophorus impejanus*	SAF	S
11.	Kaleej Pheasant	*Lophura leucomelana*	TF	F
12.	Woodcock	*Scolopax rusticola*	TF	S
13.	Ibisbill	*Ibidorhynca struthersii*	RC	C
14.	Wedge T Gr pigeon	*Treron sphenura*	TF	F
15.	Hill Pigeon	*Columba rupestris*	TDS	F
16.	Rock Pigeon	*Columba livia*	HB	C
17.	Snow Pigeon	*Columba leuconota*	AP	C
18.	Oriental Turtle Dove	*Streptopilia orientalis*	TDS, AP	C
19.	Spotted Dove	*Streptopilia chinensis*	HB	C
20.	Slaty-H parakeet	*Psittacula himalayana*	WBF	F
21.	Large Hawk cuckoo	*Cuculus sparverioides*	TF	F
22.	Eurasian Cuckoo	*Cuculus canons*	TF	C
23.	Common Hawk cuckoo	*Cuculas varius*	WBF	C
24.	Oriental Cuckoo	*Cuculus saturatus*	TF	F
25.	Small Cuckoo	*Cuculus poliocephalus*	SAF	F

Contd....

	Common Name	*Scientific Name*	*Habitat*	*Frequency*
26.	Mountain Scops owl	*Otus spilocephalus*	TF	F
27.	Jungle Owlet	*Glaucidium radiatum*	TF	S
28.	Large Pied kingfisher	*Ceryle lugubris*	RC	F
29.	Hoopoe	*Upupa epops*	HB	F
30.	Great Hill barbet	*Megalaima virens*	TF	C
31.	L Green barbet	*Megalaima zeylanica*	WBF	F
32.	L Scalybd. W. pecker	*Picus squamatus*	TF	F
33.	L Goldenbk. W. pecker	*Chrysocolaptes lucidus*	WBF	C
34.	Sm. Y-naped W. pecker	*Picus chlorolophus*	WBF	F
35.	Black-Naped W. pecker	*Picus canus*	WBF	F
36.	Homed Lark	*Eremophila alpestris*	AP	F
37.	Nepal House martin	*Delichon nipalensis*	RC	C
38.	House Swift	*Apus nipalensis*	HB	C
39.	Alpine Swift	*Apus melba*	RC	C
40.	L Whiterumped swift	*Apus pacificus*	RC	C
41.	Gray-Blacked shrike	*Lanius tephronotus*	TDS	C
42.	Maroon Oriole	*Oriolus traillii*	WBF	C
43.	Ashy Drongo	*Dicrurus leucophaeus*	WBF	F
44.	Black Drongo	*Dicrurus adsimilis*	FE	C
45.	Hari-Crested drongo	*Dicrurus hottentottus*	FE	F
46.	Wh- Bellied drongo	*Dicrurus caerulescens*	FE	F
47.	Him. Treepie	*Dendrocitta formosa*	HB	C
48.	Y-Billed chough	*Pyrrhocorax graculus*	AP	F
49.	R-Billed chough	*Pyrrocorax pyrrhocorax*	ADS	C
50.	Jungle Crow	*Corvus splendens*	HB	C
51.	Common Myna	*Acridotheres tristis*	HB	F
52.	Jungle Myna	*Acridotheres fuscus*	FE	F
53.	Red -billed Blue magpie	*Cissa erythrorhyncus*	TF	F
54.	Yellow billed Blue Magpie	*Cissa flavirostris*	TF	F
55.	Nutcracker	*Nucifraga caryocatactes*	CF	C
56.	Scarlet Minivet	*Periplocotus flammeus*	FE	
57.	Orange-Bid. leafbird	*Chloropsis hardwickii*	WBF	
58.	Upland Pipit	*Anthus sylvanus*	AP	

Contd....

	Common Name	Scientific Name	Habitat	Frequency
59.	White Wagtail	*Motacilla alba*	RC	
60.	Grey Wagtail	*Motacilla caspica*	RC	
61.	Fire-Tailed sunbird	*Aethopyga ignicauda*	AMS	
62.	Nepal Sunbird	*Aethopyga nipalensis*	SAF	
63.	Mrs Gould's sunbird	*Aethopyga gouldiae*	TF	
64.	Long-Tailed minivet	*Periploclotus ethologus*	FE	
65.	Wh-Cheekked bulbul	*Pycnonotus leucogenys*	HB	
66.	Black Bulbul	*Hypsipetes madagascariensis*	FE	
67.	Red Vented bulbul	*Pycnonotus cafer*	HB	
68.	Striated Bulbul	*Pycnonotus striatus*	TF	
69.	Scaly-br wren Babbler	*Pnoepyga albiventer*	US	
70.	Red-headed Babbler	*Stachyris ruficeps*	US	
71.	Robin Accentor	*Prunella rubiculoides*	AP	
72.	Rufous-Br Accentor	*Prunella strophiata*	AP	
73.	White-capped Redstart	*Phoenicurus erythrogaster*	RC	
74.	Black Redstart	*Phoenicurus ochruros*	ADS	
75.	Wh-capped Redstart	*Chaimarromis leucocephalus*	RC	
76.	Plumbeus Redstart	*Rhycornis fuliginosus*	RC	
77.	Grandala	*Grandala coelicolor*	SS, AP	
78.	Gray Bushchat	*Saxicola ferrea*	HB	
79.	Pied Bushchat	*Saxicola caprata*	TDS	
80.	Speckled Mt. thrush	*Zoothera dauma*	TF	
81.	L-tailed Mt. thrush	*Zoothera dixoni*	TF	
82.	Blue Whistl. thrush	*Myiophoneus caeruleus*	RC	
83.	Wh-collared Bl. bird	*Turdus albocinctus*	T/SAF	
84.	Gray-winged Bl. bird	*Turdus boulboul*	TF	
85.	Eurasian Bl. bird	*Turdus merula*	TF	
86.	Wren	*Troglodytes troglodytes*	SS	
87.	Brown Dipper	*Cinclus pallasii*	RC	
88.	Wh. Breasted dipper	*Cinclus cinclus*	RC	
89.	Spotted Forktail	*Enicurus maculatus*	RC	
90.	Chestnut Cr. warbler	*Ceicercus castaniceps*	US	
91.	Striated Prinia	*Prinia criniger*	US	
92.	Large Niltava	*Niltava grandis*	T/SAF	

Contd....

	Common Name	*Scientific Name*	*Habitat*	*Frequency*
93.	Wh-throater L thrush	*Garrulax albogularis*	TFE	F
94.	Striated L thrush	*Garrulax striatus*	TF	C
95.	Streaked L thrush	*Garrulax lineatus*	TDS	C
96.	Red-headed L thrush	*Garrulax affinis*	TDS	C
97.	Black-faced L thrush	*Garrulax erythrocephalus*	T/SAF	F
98.	Red Tailed milna	*Minla ignotinca*	TF	F
99.	Fire-Tailed myzornis	*Myzornis pyrrhoura*	SAF	R
100.	Y-naped Yuhina	*Yuhina flavicollis*	TF	F
101.	Rufous Vented yuhina	*Yuhina occipitalis*	BR	F
102.	Black-Capped sibia	*Heterophasia capistrata*	TF	F
103.	Green-Backked tit	*Parus monticolus*	TF	C
104.	Coal Tit	*Parus ater*	TF	F
105.	C nut-beld Nuthatch	*Sitta castanea*	TF	F
107.	Verditer Flycatcher	*Muscicapa thalassina*	WBF	C
108.	Slaty Bl flycatcher	*Muscicapa leucomelanura*	TF	F
109.	Grey-headed-flycatcher	*Culicicapa ceylonensis*	TF	C
110.	Y-bellied Fantail FC	*Rhipidura hypoxantha*	T/SAF	C
111.	Ferrugi. Flycatcher	*Muscicappa ferruginea*	SAF	R
112.	Wh-brd Bl flycatcher	*Muscicapa superciliaris*	TF	F
113.	Ornge-Gorg. flycatcher	*Muscicapa strophiata*	TF	F
114.	Brown Leaf warbler	*Phylloscopus collibita*	BR	F
115.	Str-footed Bush warb	*Cettia fortipes*	TDS	F
116.	Tickell's L warbler	*Phylloscopus affinis*	TF	F
117.	Olivaceous L warbler	*Phylloscopus griseolus*	SAF	F
118.	Dull green L warbler	*Phylloscopus magnirostris*	TF	C
119.	Crowned L warbler	*Phylloscopus reguloides*	TF	C
120.	Orange-fl Bush robin	*Erithacus cyanurus*	SAF	C
121.	Golden Bush robin	*Erithacus chrysaeus*	SAF	R
122.	Him. Tree sparrow	*Passer montanus*	HB	C
123.	House Sparrow	*Passer domesticus*	HB	C
124.	Crimson Rosa finch	*Carpodacus rubescens*	AMS	F
125.	Common Rose finch	*Carpodacus erythrinus*	ADS	S
126.	Himalayan Gold finch	*Carduelis spinoides*	CF	F
127.	Greyheaded Bull finch	*Pyrrhula erythaca*	TF	S

Contd....

	Common Name	Scientific Name	Habitat	Frequency
128.	Red-headed Bull finch	*Pyrrhula erythrocephala*	ADS	S
129.	Gold-cr Black finch	*Pyrrhoplectes epauletta*	SAF	R
130.	Tibetan Snow finch	*Montifringilla adamsi*	AP	S
131.	Wh-Winged grosbeak	*Mycerobas carnipes*	ADS	C
132.	PK-browed Rose finch	*Carpodacus rhodochrous*	ADS	F
133.	Scarlet Finch	*Haematospiza sipahi*	WBF	F
134.	Juniper Finch	*Propyrrhula subhimachala*	SAF	S
135.	Speckled Piculet	*Picumnus innominatus*	WBF	S
136.	Darjeeling W. pecker	*Dandrocops darjellensis*	CF	C
137.	Golden thr. barbet	*Megalaima asiatica*	WBF	F
138.	Himalayan Swiflet	*Collcalia brevirostris*	SAF	F
139.	Collared Owlet	*Glaucidium brodiei*	TF	O
140.	W-tailed Gr. pigeon	*Treron sphenura*	TF	F
141.	Eurasian Jay	*Garrulus glandarius*	TF	O
142.	Nut Cracker	*Nucifraga caryocatactes*	CF	F
143.	Ferruginous Flycatcher	*Muscicapa ferruginea*	TF	F
144.	Small Niltava	*Niltava macgrigoriae*	TF	O
145.	Little Forktail	*Enicurus scouleri*	RC	O
146.	White Wagtail	*Motacila alba*	FE	O
147.	L Pied wagtail	*M. maderaspatensis*	RC	O
148.	Grey Bushchat	*Saxicola ferrea*	FE	O
149.	Hill Prinia	*Prinia atrogularis*	FE	O
150.	Whiskered Yuhina	*Yuhina favicollis*	TF	F
151.	Fire br. Fl. pecker	*Dicaeum ignipectus*	TF	O
152.	Strkd Spider hunter	*Arachnothera magna*	WBF	S
153.	Plain L. thrush	*Zoothera mollissima*	FE	F
154.	Him. Red br. falconer	*Microhierax caenrulescens*	FE	R
155.	Nepal Rosefinch	*Carpodacus nipalensis*	AP	F
156.	Rufous Hd parrotbill	*Paradoxornis ruficeps*	SAF	S
157.	Daurian Redstart	*Phoenicurus auroreus*	AS	S
158.	Red Billed leiothrix	*Leiothrix lutea*	TF	F
159.	Blyth's Leaf warbler	*Phylloscopus reguloides*	FE	F
160.	Changeable Hawk eagle	*Spizaetus nipalensis*	TF	S

Contd....

Conclusion and Recommendations

The wide altitudinal and topographic variation of JDNP in reflected in the form of rich flora and vegetation types. Relatively less disturbed areas were richest in species but areas of comparable altitude and heavy disturbance showed less vegetation cover and less species. However, the species number alone should not be taken into consideration as several opportunistic weeds occupy the disturbed habitats adding to species number.

APPENDIX

Flora of Jigme Dorji National Park: Summary

Group/Family	*No. of genera*	*No. of species*
A: Gymnosperms		
1. Cupressaceae	2	4
2. Ephedraceae	1	1
3. Pinaceae	5	7
4. Taxaceae	1	1
B: Angiosperms-Monocots		
5. Alliaceae	1	7
6. Anthericaceae	1	1
7. Araceae	7	19
8. Asparagaceae	1	2
9. Commelinaceae	2	4
10. Cyperaceae	7	50
11. Dioscoreaceae	1	5
12. Haemodoraceae	1	3
13. Hypoxidaceae	2	2
14. Iridaceae	1	5
15. Juncaceae	2	23
16. Juncaginaceae	1	2
17. Liliaceae *(s. l.)*	16	35
18. Melanthaceae	2	4
19. Musaceae	1	1
20. Orchidaceae	21	29
21. Poaceae	38	69
22. Pontederiaceae	1	1

Contd....

Group/Family	*No. of genera*	*No. of species*
23. Potamogetonaceae	1	3
24. Smilaceae	1	5
25. Zingiberaceae	5	11
C: Angiosperms-Dicots		
26. Acanthaceae	7	9
27. Aceraceae	1	7
28. Actinidiaceae	2	3
29. Alangiaceae	1	2
30. Amaranthaceae	3	6
31. Anacardiaceae	4	7
32. Annonaceae	1	1
33. Apiaceae	13	25
34. Apocyanaceae	2	2
35. Aquifoliaceae	1	5
36. Araliaceae	10	11
37. Aristolochiaceae	2	2
38. Asclepiadaceae	2	3
39. Asteraceae	32	105
40. Balsaminaceae	1	8
41. Begoniaceae	1	3
42. Berberidaceae	2	12
43. Betulaceae	4	6
44. Bischofiaceae	1	1
45. Boraginaceae	12	17
46. Brassicaceae	23	40
47. Buddlejaceae	1	3
48. Buxaceae	2	2
49. Cactaceae	1	1
50. Caesalpiniaceae	3	4
51. Campanulaceae	4	14
52. Cannabaceae	1	1
53. Caprifoliaceae	5	18
54. Caryophyllaceae	8	28
55. Celastraceae	2	4

Group/Family		*No. of genera*	*No. of species*
56.	Chenopodiaceae	3	5
57.	Chloranthaceae	1	1
58.	Circaeasteraceae	1	1
59.	Convolvulaceae	2	3
60.	Coriariaceae	1	2
61.	Cornaceae	4	5
62.	Crassulaceae	3	20
63.	Cucurbitaceae	3	3
64.	Daphiphyllaceae	1	2
65.	Diapensiaceae	1	1
66.	Dipsacaceae	4	6
67.	Droseraceae	1	1
68.	Elaeagnaceae	2	4
69.	Ericaceae	8	43
70.	Euphorbiaceae	6	14
71.	Fabaceae	17	32
72.	Fagaceae	2	10
73.	Fumariaceae	3	22
74.	Gentianaceae	7	43
75.	Geraniaceae	1	7
76.	Gesneraceae	3	4
77.	Grossulariaceae	1	6
78.	Haloragaceae	1	6
79.	Hamamelidaceae	3	3
80.	Hippuridaceae	1	1
81.	Hydrangeaceae	2	5
82.	Hypericaceae	1	4
83.	Illiciaceae	1	1
84.	Juglandaceae	2	2
85.	Lamiaceae	18	30
86.	Lardizabalaceae	2	2
87.	Lauraceae	7	15
88.	Lentibulariaceae	2	2
89.	Linaceae	2	2

Contd....

Group/Family	*No. of genera*	*No. of species*
90. Loranthaceae	3	3
91. Magnoliaceae	3	5
92. Malvaceae	2	2
93. Melastomaceae	2	3
94. Meliaceae	1	1
95. Menispermaceae	3	3
96. Mimosaceae	1	3
97. Moraceae	2	9
98. Myricaceae	1	1
99. Myrsinaceae	3	6
100. Nyssaceae	1	1
101. Oleaceae	2	4
102. Onagraceae	2	9
103. Orobanchaceae	1	1
104. Oxalidaceae	1	3
105. Papaveraceae	2	11
106. Parnassiaceae	1	4
107. Philadelphaceae	2	3
108. Phytolaccaceae	1	1
109. Piperaceae	2	4
110. Pittosporaceae	1	1
111. Plantaginaceae	1	2
112. Podophyllaceae	1	1
113. Polygalaceae	1	1
114. Polygonaceae	9	40
115. Primulaceae	3	22
116. Pyrolaceae	1	1
117. Ranunculaceae	17	64
118. Rhamnaceae	4	6
119. Rosaceae	20	78
120. Rubiaceae	6	9
121. Rutaceae	7	9
122. Sabiaceae	2	2
123. Salicaceae	2	14

Contd....

Group/Family	*No. of genera*	*No. of species*
124. Santalaceae	1	1
125. Saururaceae	1	1
126. Saxifragaceae	4	38
127. Schizandraceae	1	2
128. Scrophulariaceae	10	38
129. Simaroubaceae	1	1
130. Solanaceae	5	6
131. Stachyuraceae	1	1
132. Staphyleaceae	1	1
133. Sterculiaceae	1	1
134. Symplocaceae	1	6
135. Tamariscaceae	1	2
136. Tetracentraceae	1	1
137. Theaceae	2	4
138. Thymelaeaceae	2	4
139. Ulmaceae	3	4
140. Urticaceae	12	31
141. Valerianaceae	2	4
142. Verbenaceae	4	4
143. Violaceae	1	4
144. Vitaceae	5	5
Total	563	1434

References

Blower, J., (1986), *Nature Conservation in Bhutan. Project Findings and Recommendations.* Project BHU/83/022. UNDP/FAO, Rome.

DANIDA, (1988), *Proposal for a National Conservation Strategy for Bhutan.* DANIDA. Copenhagen, Denmark.

Dunbar, G.A., (1981), Upland vegetation in Bhutan: extracts from a field diary. *Annual Journal of Royal New Zealand Institute of Horticulture* 9: 57-66.

Flemming, R.L. Sr., Flemming, R.L. Jr. & Bangdel, L.S. (1984). *Birds of Nepal.* Nature Himalayas, Kathmandu, Nepal.

Grierson, A.J.C. and Long, D.G. (1983, 1984, 1987, and 1991), *Flora of Bhutan.* Vols. I (i, ii, iii), and II(i). Royal Botanic Garden, Edinburgh.

Johari, R., (1993 *ed.*), *Review of Findings and Recommendations for Socio-Economic*

Surveys conducted in and around Jigme Dorji Wildlife Sanctuary, July 1993. Project report BHU/WWF 9665.

Mackinnon, J., (1991), *National Conservation Plan for Bhutan. Master Plan for Forestry Development.* Dept. of Forestry, RGOB, Thimphu.

Miller, D.J., (1987), Rangelands of the Himalayan Kingdom of Bhutan. *Rangelands* 9 (6): 257-59.

Miller, D.J., (1988), *Wildlife Management and Problems with Livestock Grazing in the Himalaya.* Paper presented at the Symposium 6 Asian-Pacific Mammalogy, Beijing, China, July 1988.

Mongar. T.B., (1993), *Reconnaisance Survey Report of Jigme Dorji National Park*, August 1993. Unpubl. Project Report BHU/WWF/6421.

Nakao, S. and Nishioka, K., (1984), *Flowers of Bhutan.* Asahi Shimbun Publication Co., Tokyo.

Negi, G.S., (1983), *Forestry Development in Bhutan. Report on Remote Sensing, Land use and Vegetation Mapping.* UNDP/FAO, Rome.

Prater, S.H. (1984), *A Book of Indian Animals.* Bombay Natural History Society, Bombay.

Rawat, G.S., (1994), *A Preliminary Floral and Ethnobotanical Survey of Royal Manas National Park. Bhutan.* Interim Report, WWF Bhutan programme.

Sherpa, M., Wangchuk, S. and Mongar, T.B., (1992), *Developing a System of Protected Area in Bhutan.* Paper presented at World Parks Congress, Caracas. 1992.

Wikramanayake, E., (1993), *Biodiversity Surveys and Management Plan Preparation for Jigme Dorji and Manas National Parks.* Interim Report, Project BHU/WWF/6421. Thimphu, Bhutan.

World Bank, (1988), *Bhutan: Development Planning in a Unique Environment.* World Bank, Washington, D.C.

World Wildlife Fund, Inc., (1989), *Strategy for Environmental Conservation in Bhutan.* Unpubl. Report. WWF. Washington, D.D.

Yonzon, P., 1992, *Strategies of Wildlife Inventory in Conserving Bhutan's Biodiversity.* Project BHU/WWF/6689, Thimphu.

23

Global Trade in Wildlife

Introduction

The ecological and environmental degradation and fast decline of habitats due to human activities in the name of development and progress throughout the Himalayan range and their effect on wildlife and nature as a whole, have so far been discussed in previous chapters of this book. In this chapter readers' attention has been drawn towards wildlife trade realities from both the global and the Indian scenario as expressed in an article by Ms. Fahmeeda Hanfee who works for TRAFFIC-India, the trade monitoring division of WWF.

It is estimated that the global trade in wildlife worth approximately US $ 20 million is done every year, mainly with USA. This trade in wildlife and illegal poaching are the matters of great concern not only for TRAFFIC by for the conservationists all over the world too. Existing laws and treaties to safe guard the protected areas will not be so effective unless we can protect the biodiversity which is presently prevailing in our protected areas and national parks. For the successful achievement of the concept of a protected are, emphasis on anti-poaching as well as raising public awareness about the concept are the need of the day.

Establishment and Mission of TRAFFIC-India

This division of WWF, India, was established in January 1982. The wildlife trade monitoring programme is looked after by a part of this organisation. TRAFFIC-India works in confromity with the principles of the World Conservation Stragegies.

The specific mission of TRAFFIC-India includes the conservation of biological diversity in India, monitoring trade and other activites that are causing harm to animals and plants etc. Identification of areas of use which may be injurious to any species and control of trade of such species are also the aim of this organisation. Others major targets of TRAFFIC-India are: investigate, monitor and report the illegal wildlife trade activities, provide technical basis for the establishemnt of programmes and policies of wildlife trade and at the sametime educate and train the agencies regarding these programmes and policies for their successful implementation.

Transboundary Issues

The global trade in wildlife includes about 40,000 primates, USA being the main consumer. To ban the hunting poaching and trade in wildlife there are some laws and treaties like—Wildlife Protection Act etc. In spite of that tigers and some other cat species, elephants, musk deer, bears and shinos are poached for several purposes of human satisfaction—commercial or personal. The major commercial activities being—wild life derivatives, pets memento and medicine. The remains of three lions were found on March 3, 2007. Again on March 31, 2003, skinned remains of three Aziatic lions and a hyend were found in the Gir Wildlife Sanctuary. Skins and claws of the lions were missing.

The above mentioned laws and treaties are enforced by some agencies like State Forest and Wildlife Department, Indian Customs, Indian Police, Coast Guards, BSF, CBI and Revenue Intelligence. These agencies are attached with the Directorate of Wildlife Conservation, Ministry of Environment.

But the forest policies of adjoining countries may differ from one another. So may the law and order. Under this situation, often effective implementation of anti-poaching and conservation activities become different when transboundary issue come forth. In Nepal, licences were issued for hunting the sheep (prendois nayaur whereas in India we cannot even think about it. This animal is a very good indicator species for snow leopard *(Panthera uncia)* existence, as it is a major prey species for the animal. Very often, involvement of foreign nationals makes it difficult for a country to take action against them. This makes the border areas more prone to illegal wildlife trade activities. For example, the Royal Manas National Park of Bhutan has a village inside whose economy depends mainly on orange orchards. These oranges are transported for sale through the Indian Manas. The newly constructed road is used for transportation purposes. This road goes from the Indian Manas to Panbang village in the Royal Manas-Bhutan. This has already raised concerns among conservation. In a statement in her article, Ms. Fahmeeda hanfee reveals a case of Rhino *(Rhinoceros unicornis)* horn seizure two weeks ago in Thimphu. The person who was caught disclosed during interrogation that the rhino horn was obtained from India. This confirms the possibility of poaching and transportation in the Manas area.

To check illegal wildlife trade, the management of National Parks which share a common boundarys should be carried out jointly by the countries involved. Unless joint effort is initiated at the high official level of the respective countries this trade will continue to flourish. A Government data about wildlife seizure in and around Dudhwa in Indo-Nepal border and one in Siliguri discloses that between 1994-1996 there were about 20 seizures of various items. They included tiger and leopard skin and bones and also ivory. A transboundary Joint Meeting between Nepal and India was said to have taken place in January 1997.

She also stated that the Director of the Kanchanjunga National Park—India expressed the possibility of shepherds being involved in the wildlife trade. In her musk trade study she found this to be true in the case of the Uttar Pradesh Himalayas also.

Our Permeable Borders

Our borders are very permeable. As a result, overgrazing, construction, and similar activities on one side of the border result in a loss of forests on the other side. In Valmiki and Dudhwa Tiger *(Panthera tigris tigris)*. Reserves this problem is very common. There where a whole belt of dying Sal *(Shores robusta)* trees was observed. Some cases of encroachment along the Dudhwa-Nepal strip have been noticed. Thanks to the Nepalese authorities, some of these have been removed recently. In Valmiki also, it has been reported that about 5,400 acres of forest land have been encroached upon because the boundary is not well defined. This shows how permeable our borders are.

Endangered Animals

Among the animals involved in the wildlife trade the following are the most endangered ones:

Rhinocerous: Rhino is hunted for its horn. The horn mainly goes to Yemen and Oman for making carved handles for traditional daggers. The African rhino is mainly used for this purpose. One-horned Asian rhinoceros is mainly used in China, Taiwan, South Korea, and Tibet for traditional medicines. In Taiwan and China the Asian rhino horn, also called 'fire horn' costs five-ten times more than African rhino horn or 'water horn'.

Tiger: For some purpose or the other every part of a tiger is used While skins and heads are used as trophies, and claws as talismans, bones and skulls of a tiger are used for medicine. For potions and balms its fat is used. In 1993 alone, 475 kg of bones and 13 tiger skins were seized. It was estimated that 47 tigers must have been killed to account for these alone.

Musk and Bear Bile: Within Indian the most endangered animals are: Musk deer *(Moschus chrysogaster)*, Himalayan Black Bear *(Selenarctos thibetanus)*, and Brown Bear *(Ursus arctos isabellinus)*. Their distribution is limited to the Himalayas. Because of its valuable must pods, the musk deer is poached throughout the Himalayan belt. The musk pods are used as a perfume base and in medicines.

Fur Items: Another important item of trade is fur items. The wild cat fur trade in India deals with at least 20 species. some of them are: snow leopard *(Panthera uncia)*, lynx cat *(Felix lynx isabellina)*, fox *(Vulpes spp)*, and otter *(Lutra spp)*. Kathmandu was a major trade centre for fur items. They are openly displayed in the shops. TRAFFIC India carried out three surveys in Kathmandu to quantify the fur trade. Thanks to the Nepalese authorities, now there are no more displays in the Kathmandu market of these items.

Reptile Skins: REptile skins are used for manufacturing wellets, belts, shoes and many other items. When the trade was legal, it is estimated that India used to export skins of shakes, monitor lizards and crocodiles for worth US $ 60 million. Today, though it is not permitted but still illgaly it is going on. Even quiet a number of turtles are killed for their flesh and for curios.

Besides animals, plants are also exported from India, especially medicinal plants. India and Brazil are the largest exporters of medicinal plants. In about 2,500 medicinal plants that are found in India. 2,000 are used in traditional medicines. Around 150 species are used commercially on a fairly large scale. It is believed that a plant collected from the forest is more effective than a cultivated one. This creates problems for plant conservation. Besides total ban on 46 species of plants for export, an export ban exists on all plants obtained from the forest, except the ones with no objection certificates. These certificates are issued only from Mumbai, Delhi, kolkata, Chennai, and Cochin. Orchids are also traded for their ornamental value. Orchids are suggled with cht flowers. Generally, they are wrongly decraled as lilies or some other exoitic flower. Kalimpong, Shillong, and Trivandum are the Orchid trade centres in India.

Discussion

Dr. Kattel gave the participants some information about the Transboundary Joint Meeting between Nepal and India especially dealing with these issues in January 1997. There was a bird city in Patna which was visited by some journalists from Kathmandu, and they brought some of the trade in birds to light.

Ms. Fahmeeda Hanfee, TRAFFIC-India, informed the participants that there was a study of live birds being carried out, covering this area, and forest officials had been informed about the trade. One place in Patna was not only famous for live birds but also for live animals. The report would be out in a month or two, and the recommendations would be implemented. Forest officials would then be pursued to get them to take action.

Mr. Brian Penniston of the Mountain Institute wanted to know who were buying the orchids and where the ultimate market was. Was it a domestic or an international market and were they buying these as cut flowers or plants?

Ms. Hanfee replied that the orchid trade had both markets, domestic and international. They were basically traded for ornamental value. They were traded both as flowers and plants, but there was a lot of fraudulence in this trade.

Mr. Javed Hussain, WWF-Thailand, asked the Chinese delegates whether the demand for animal products was increasing or decreasing in China since many of the animal (tiger) products were consumed in China. Was there any government policy to ban these products? What was the current position of China?

The Representative from China, Mr. S. Ban Zong, replied that this was a part of culture in China and it could not be stopped immediately. Due to the endangered wildlife species, the government, including the

Tibetan Government, had made regulations to restrict some use of wildlife. Some were absolutely prohibited such as the rhino horn. Officially it was not permitted, but illegally it was another matter. Musk deer products and bear bile were permitted legally. There was a deer breeding farm in Tibet designed to breed domestic deer to supply to the market. There were bear breeding farms in northeastern China and southwestern China.

Prof. Pei further added that there should be a good balance between conservation of nature and conservation of culture, and this takes time. In China, about 5,000 medicinal plants were being used, out of which 1,000 were regular and 500 were used on a large commercial scale. Five years previously, the annual export of medicinal plant products was worth US $1.6 billion. Regarding animal products, 400 different animal species were used in medicine, out of which 50 were for regular use. According to scientific research and statistics, there were no species used in Chinese medicine that were already extinct. The best conservation strategy for plants and animals was to use them. When the demand was there, people automatically would find a way to develop or multiply them. This was just general information. Illegal trading and black marketing should not be considered a good thing. Substitute materials had been produced in China for many years, but this had not been successful. The only one which had been successful was buffalo horn as a substitute for rhino horn.

It was also added that an effective and practical management system must be developed so that poor people did not get poorer and the rich did not get richer.

Remarks from Sikkim

Mr. P.K. Basnet

It is my privilege to present my remarks on the ecology efforts to this Regional Workshop on the Conservation of the Kanchanjunga Mountain Ecosystem. Kanchanjunga Mountain Ecosystem is a unique system. This system has an area covering three different political boundaries, namely, those of India, Nepal, and Tibet (China). On the Indian side, Kanchanjunga falls in the State of Sikkim and is designated as the Kanchanjunga National Park. The special feature of this park is that it has a large number of flora and fauna which makes it extremely rich in biodiversity. My friends from Sikkim have already described the

Kanchanjunga National Park in greater detail with regards to area, irrigation, vegetation, habitat, flora, and fauna and also on socio-economic activities of the people living in and around KNP. The State Government of Sikkim and the Central Government are working on raising KNP's status to a Biosphere Reserve. In Sikkim, the best efforts are being made to conserve this unique mountain ecosystem. The system, although separated by political boundaries, is one whole system, and conservation and management issues should be dealt with through common complementary action. The transboundary issues of high altitude grazing, wildlife movement, tourists, conservation of plants, and so on have to be looked into professionally. Measures for implementation should be adopted after common study in a site-specific manner. In order to do this, there should be site visits and study tours by experts involved in research management in Kanchanjunga in each country. This should be followed by a workshop in which specific transboundary issues common to this country are identified. Following this, a comprehensive research and management programme should be developed. I am extremely pleased that we have come to one table to discuss transboundary issues of the Kanchanjunga Mountain Ecosystem. The efforts made by ICIMOD are highly praiseworthy. We had a good beginning and I strongly believe that we will be able to conserve this unique mountain ecosystem for years to come. Let us dedicate ourselves to conservation in this area. Thank you very much.

Discussion

Replying to the question about how many people/population there would be inside the proposed biosphere reserve area, Mr. Sharma replied that, currently, the Kanchanjunga National Park had 1,760 sq km and 10 families living inside the area. There was very sparse settlement along the fringe area. Now another wildlife sanctuary had also been included which had no population. With this addition, they were making a compact single biosphere reserve. For the settlement on the fringe areas, they were making a buffer zone. The actual physical identification was being carried out currently by the Department of Forests.

Mr. Brian Penniston of the Mountain Institute wanted to know about the advantages of declaring a biosphere reserve. The strategic thinking behind this was not very clear, e.g., there were no specific funding sources to provide money only for the biosphere reserve. If the advantages were to increase local participation in the system of the Government of India, were there any other ways to do so without declaring it a biosphere reserve?

Mr. Sharma replied that in National Parks there was no designated buffer zone, but in the biosphere reserve there was a core zone and a buffer zone. The core zone was an area totally protected from any outer interference. On the fringe areas of National Parks, people had been in some ways, traditionally, drawing resources from the core zone. So the idea of various activities which were implemented in the buffer zone was to keep them away from the core zone. He also added that funds were provided to the buffer zone for protection and regeneration of the degraded area as well as for the total protection of the core zone.

Mr. Narayan Poudel, Chief Ecologist from the Department of National Parks and Wildlife Conservation (DNPWC), requested the Indian participants to share their experiences of Joint Forest Management, and how much of it was applicable to buffer zone management, with the participants. He also asked if community development activities and income-generating activities really motivated the local people to contribute towards conservation and whether the protected area system really increased the biodiversity of the area?

Mr. Sharma replied that Joint Forest Management had been introduced by the government in some areas, and it had been very successful in the plains of West Bengal but had not been successful in the hilly regions. Regarding conservation, if provided with economic incentives and close cooperation with the community and, with the proposal coming from the community, itself, then it could really work well. He also added that, since people from different areas had different demands and needs, and their requirements were dependent on the resources, conservation could not be generalised and needed to be site-specific.

Mr. Mingma Sherpa of the WWF-Nepal Programme, commented that the enlightening thing about the biosphere reserve was the total aspect of the inclusion of people, core areas, buffer zones, and of living action research. The idea of how to actually involve people was very interesting and this might be an interesting concept for the future which both the China and Nepal side could also propose. There was no specific legal action in setting up a biosphere reserve. It also drew the attention of the international community. Transboundary parks between the United States and Canada had also been set up as biosphere reserves; this could also be done here on the level of regional cooperation. There could be cooperation on research, tourism, training, management, technology transfer, and sharing of information on data. So, it was worth pursuing.

Remarks from India

Dr. R.K. Rai

I would like to mention that the Government of India is in the process of designating this area a biosphere reserve. Old institutions such as the Zoological Survey of India and the Botanical Survey of India, which have been working for 70-80 years, and the G.B. Pant Institute of Himalayan Environment and Development, which has been working for seven to eight years, are giving support to the process. The area harbours unique biodiversity. These institutions are already assessing the biodiversity status and conservation needs of the area. Once we designate the area a biosphere reserve, it will have a better conservation status. Presently, the proposal put by the Sikkimese Government mentions that there is one existing National Park and Maenam Wildlife Sanctuary. These areas will be a part of the core zone. The total area of the biosphere reserve is going to be 2,665 sq km.

Once we designate the area as a biosphere reserve, the Government of India expects a management action plan from the State Government for implementation, and 100 per cent funding will be given by the Government for carrying out approved activities. I am told by the representative of the Sikkimese Government that they will be submitting the reports very soon. Once the management action plans are finalised and approved by the Government of India, funds will be provided for research, training, education, protection, and many eco-development activities in the region. The basic philosophy of the biosphere reserve programme from our side is that all the human activities in the core zone are to be more or less banned. These areas are supposed to be preserved in their totally natural form. Activities encouraged in the buffer zone are intended to keep people away from the sources in the core area, because people who are living there ultimately depend upon local resources. Once people have alternative sources of livelihood, then the dependency on forest products will decrease. Basically floriculture, horticulture, and beekeeping can be encouraged. Our major focus is on people's participation. In which way can we convince the people to participate in conservation activities? For this purpose, whatever activities are initiated, some economic benefits must accrue to the people, only then will the locals be attracted towards the programme. Unless local people become involved, all the efforts of the government will fail. Nothing will materialise at the field level. For this purpose, the people need to be educated, and they should be provided with alternative sources of livelihood. These are the two focussed programmes which are implemented in all the biosphere reserves. Till now, we have been largely involving the Forest Department since the areas are basically managed by the Forest Department. In the buffer zone, there are some areas which are outside the control of the Forest Department, therefore, we are also considering the involvement of other departments like agriculture, fisheries, and so on. There is a lot of scope for exploiting resources for the benefit of the people without having any damaging effects.

There are major problems in the hills such as infrastructure, communications, roads, and so on. These problems are such that, if you develop infrastructure, they have their own repercussions and if you do not develop infrastructure, again there are repercussions. Our national policies on any major development activities of major natural resources, hydroelectric, and other projects used to be decided only by a group of officials from the Central Government, State Government, and some experts. Now the Government has decided to have a public hearing in which the officials will go to the field and discuss these matters with the local people. A lot of information about various problems and solutions comes from the local people. Such areas have great potential for horticulture and floriculture. These projects are such that they do not have any impact on overall conservation efforts.

In the central Himalayas, there is a problem of absentee farmers. Wherever there is free labour available, only then do people practice agriculture. Wherever money has to be spent, it is not considered to be a beneficial activity. We will be designating this area as a biosphere reserve very soon. As our friends from Sikkim already mentioned, it will cover a large area of Sikkim — almost more than 25 per cent. In fact, in buffer zones, the restrictions are not too many. Traditional activities will continue with little modification. There are some activities which are detrimental to conservation efforts. People will be educated to change their way of doing things. Once we come out with the final recommendations, we have to be very practical. We should come up with recommendations which can be accepted by both the government as well as the people.

Thank you.

Remarks from Nepal

Dr. Bijaya Kattel

A biosphere reserve has three major components, namely, protected areas, people's participation, and regular monitoring. So, once this Kanchanjunga area is established as a whole in three different countries, probably then we can call it one biosphere reserve. That is collective work. Today, we have already heard of a biosphere reserve and national park, and now we come to a conservation area. Why it has been proposed as a conservation area is that the experience of Nepal has so far been an experience in which people's participation is maximised. I will describe briefly the status of this area on the Nepal side. This area has already been proposed as a conservation area and any day it can be declared as such. Different research has taken place and a database has already been created courtesy of the WWF Nepal Programme on flora, fauna, and socioeconomic conditions (inclusive of traditional rites and rituals of the local people, especially for grazing and pasture).

Participation through the private sector is very critical. Can we rely only on private sectors for conservation? It is a big question and I do not have the answer. This forum could come up with some sort of answer, if not all answers. When we talk about conservation, law enforcement comes as one of the major components because many illegal activities have to be stopped. Whom do we trust? Do we give law enforcement responsibility directly to institutions? Therefore, a participatory approach must be there, and we have had several experiences in Nepal in which local committees can contribute to conservation. But, law enforcement should be with the government agencies which can be minimised into a small unit. We can proceed on two different levels for management: the macro-level and micro-level. On the macro-level, the principal question is the establishment of the area as a conservation area or protected area. The other thing we need is collaboration and cooperation with major NGOs. So far, the WWF-Nepal Programme has been deeply involved along with other government agencies and ICIMOD has created the database. A big database has been collected which would be very useful for cooperating countries. On a micro-level, we can go through different local-level institutions such as the 'Kanchanjunga Development Committee', and many others. We also heard that, in Sikkim, India, there are many committees. What legal base do we have? Because, unless and until we create a protected area with a legal base, we will be desperate and confused. There are too many laws in Nepal, e.g., National Park and Wildlife Conservation Act and Forest Act. Along with these laws there are three regulations which I would like to mention here which are directly related to the proposed protected area. Conservation Area Regulations 2053 (1996), which are very recent. Forest regulations 2051 (1994), which are two years old, and Buffer Zone Regulations 2052 (1995). These three sets of regulations have provisions for an area to have been created and a mechanism developed for its management. Ecoregional conservation is the focus of this morning's session, so I have called it Eco-conservation on a Regional Basis. Nobody clarified what eco meant when somebody put that word there, because unless and until we combine 'economy' with 'ecology', I don't think sustainability will be there.

We can have uniformity of management on two levels. On a regional level, we can gather and exchange ideas and come up with some policy guidelines. The implementation of those policies is carried out at the local level in the field. In Nepal, there are many traditional rites. Although, the public lands are government lands which were nationalised in the mid-50s, a lot of people in the Kanchanjunga area are still managing and prescribing who can take the cattle into which area and which sheds they can use. The lands are locally managed, despite the fact that the government has its own regulations and policies. Therefore, incorporating the cooperation of those people in conservation is essential.

A trans-frontier action committee could be formed from this meeting I appreciate the steps taken by ICIMOD to organise this meeting and WWF for supporting it. The government alone cannot do things like this, so initiatives should be taken by other institutions. This action committee should facilitate conservation action. Eco-tourism should be taken as an enterprise. We are all talking about eco-tourism, and I don't know what it really means I know what is there, has been documented. But those people who are really running the show are right. I may make some money from tourism. Is that all we are talking about? Control of illegal activities is very important because I personally was involved in many raids. Kanchanjunga may not be directly affected by this but, if we have a Joint Action Committee, it would provide us with an opportunity to exchange information on all the management and the implications of illegal activities in different areas. Income-generating activities should be promoted. Training and capacity building are very important. We should be able to go to the local level and train them in conservation, and that needs research. If we do not have any information, we cannot talk. If the G.B. Pant Institute establishes some sort of database on GIS applications, they can share it with ICIMOD, for example, which has enough information stored. This kind of exchange is very important. I want to share our experiences on core protected area and buffer zone management. In the Annapurna Conservation Area, which was established with a different concept from that of simply a protected area, we are still undergoing experiments and we have yet to see whether those concepts which we had adopted are evolving and changing every decade, for example. In the last two decades, we have carried out changes in three different phases, so we have to see where we will be in the next decade or two. Thank you.

Discussion

Mr. Brian Penniston, Representative of the Mountain Institute, remarked that the Annapurna area was delegated to the King Mahendra Trust. He further questioned what would be the new situation for the Kanchanjunga area, which department would have the mandate to control all the activities there, and what would be the relationship of the department with local bodies such as the VDCs (Village Development Committees)?

Mr. Kattel replied that the Annapurna Conservation Area was delegated to King Mahendra Trust and, with experience, it had been conceptually adapted. The government could not give it completely into the hands of the local people because of law enforcement which was a very important component. Law enforcement should always be the prerogative of the government, he stated, but local people should be mobilised to provide, through local institutions, the core unit of the law enforcement group with minimum staffing. This had been the concept.

Mr. Javed Hussain of the WWF-Regional Office, Thailand, remarked that they had been talking about biodiversity, cultural diversity, and social diversity, and, if that was the structure, there could not be one single model. There would be diversity of actions and models so there was not going to be a single prescription which could apply everywhere. In the workshop group there should be some commonality. They should focus on local diversity and localised approaches.

Mr. Kattel answered that this was very important because they needed to be site-specific. Things which were successful in the lowlands of the *Terai* might not be effective in the highlands. Unless and until one small unit was institutionalised, growth would not take place.

Outcome of Group I

The participants in Group I discussed 'A Common Framework for Biodiversity Assessment and Monitoring'. The six participants discussed different aspects of the issues: The outcome of this discussion is provided below.

Participants	Mr. Gut Lepcha	Dr. R.K. Avasthe	Mr. Pradip Regmi
	Mr. Devendra Amatya	Dr. K.K. Shrestha	Mr. Javed Hussain

Presentation

PHASE I: Planning

1. **Goal :** To develop a common framework for biodiversity assessment and monitoring.
2. **Objective :** Conservation of the Kanchanjunga Mountain Ecosystem.
3. **Activities**

   (A) Assessment

   1. Development of a Methodological Framework Output—Manual
      - (i) Standardisation of Nomenclature (Flora and Fauna)
        - (a) Scientific Names
        - (b) Common English Names
        - (c) Local Names
   2. Categorisation of Landuse Types
      - (a) Scale of Assessment and Monitoring (Flora and Fauna)
      - (b) Indicators (Flora and Fauna)
   3. Categorisation of Resource Use Types

   (B) Monitoring

   1. Monitoring Based on Assessment
   2. Mechanism: Technical Exchange Programme

(a) Identification of Resource Persons (from each country) and by required disciplines (June-Aug. 1997)

(b) Technical Workshop (Oct.-Nov. 1997) for development of a methodological framework, as mentioned above under Assessment

(c) Resource Person Exchange (Field Assessment)—(March-May 1998)

(d) Report Preparation—Distribution—Evaluation (July-Aug. 1998)

(e) Assessment and Planning Workshop (Sept.-Oct. 1998)

3. Budget:

1.	Development of a Methodology (Manual) Consultant and Consultancy	US $6,000/-
2.	Categorisation of Landuse Types / Categorisation of Resource Types Consultant and Consultancy	US $6,000/-
3.	Activities :	
	(a) Technical Workshop	US $15,000/-
	(b) Pilot Study	US $40,000/-
	(i) In-country	
	(ii) Trans-regional	
	(c) Workshop	US $20,000/-
	(i) Planning Phase Evaluation	
	(ii) Project Development and Planning	

Phase II

Implementation Phase

Operational and Management Cost	- US $13,000/-
TOTAL COST	- US $100,000/-

Outcome of Group II

The participants in Group II discussed "A Common Framework for Sharing Conservation Benefits with the Local People". The outcome of their discussions is provided in Table 1.

Participants

Ms. Fahmeeda Hanfee
Dr. Eklavya Sharma
Mr. D.D. Sharma
Mr. A.L. Joshi
Mr. Brian Penniston
Mr. Pradeep Mool
Mr. Ajay Rastogi

Goal: There should be a strong link between conservation and community benefits, so that local people would develop a purposeful stake of conservation and sustainable use of natural resources.

Approach: The perception of 'benefit' largely depends on the socioeconomic and cultural aspirations of the people. At the same time, all community benefit activities may not have positive linkages with conservation objectives and vice versa. Therefore, research needs to he undertaken to identify, plan, and implement the activities that fulfill the dual objective of conservation and development.

There are different types of development schemes currently being undertaken in the three countries. Sharing information on all aspects of these ongoing and future development programmes could prove very useful in initiating a comprehensive strategy that enables sustainable development of the Kanchanjunga Mountain System.

Preliminary discussions in the group led to the following Table (see Table) that outlines the activities and establishes community and conservation links.

APPENDIX A

Common Framework for Sharing Conservation Benefits with Local People in Watershads

Benefits	*Community*	*Conservation Link*
Tourism Domestic and Foreign	- lodge operators and campsite management • service—roads, guides, porters • indigenous products—consumables, handicrafts, etc. • transport-yak, dzo(s), jeeps, etc.	- sustainable use during tour souvenirs, etc. - minimum environmental impacts - alternative energy (kerosene, blankets, etc.)
Genetic Resources Animal and Plants Wild Domestic	Forests—fuel, fodder, construction, pasture/grass NTFP collection, med, plants, edibles - construction materials-bio prospecting-ornamental flowers, land races, plant stocks, domestic crops. • animal parts, genetic stocks, hunting pets, fishing • meat, daily, skinning + wool; transport, manure, ploughing construction materials	community managed forests - sustainable use (wild species) vegetable seed production - maintenance of genetic stock (plants and animals) - manure (soil conservation values)
High-value Natural Resources	extract mines, quarry-construction material minerals, precious stoness	- value-added (i.e., rings, etc.)
Water Power hazard control	- micro-hydel, employment subsidy and benefits, *ghatta*(s) (milling) - irrigation, drinking, soil fertility, land stabilisation, cash crops, flood control, lake management	- landscape/watershed protection/ downstream benefits.
Culture Antiquities (rugs) trade, etc.	religious centre + pride, tourism revenues, employment, festivals - ecological knowledge (avalanche safety, flood, etc.)	Cultural/religious prohibitions monasteries, etc. reverence for nature and cutlure historical importance, appropriate technologies (bamboo), medicinal practices. - (no hunting, fishing, etc.)
Aesthetic Fora and fauna	knowledge (guides, plants etc.) - services	direct payment for knowledge/services - fee sharing with community, e.g.,

Contd....

Benefits	*Community*	*Conservation Link*
viewing, photography Landscape—mountain climbing, viewing Lakes and waterfalls, Trekking	- attraction - fees (climbing, viewing, eating)	buffer zone
Global/Regional carbon sequestration global climate global genetic resources networking	- indirect, research/employment - info. sharing, networking, training indirect information exchange - indigenous knowledge system employment (*jhankri*, etc.)	- global links + long-term interests

Outcome of Group III

The members of Group III discussed 'A Collaborative Arrangement for Cooperative Management of the Kanchanjunga Mountain Ecosystem'. All the participants discussed the topic at length and came to the conclusions given below.

Participants	Mr. P.K. Basnet	Co-chairperson:	Dr. R.K. Rai India,
	Mrs. B. Zong	Dr B. Kattel	Mr. M. Sherpa
	Prof. P. Shengji	Mr. D. Rana	Mr. D. van Blitterswije

Presentation

Goal

All the three respective governments should endeavour to designate their respective areas of Kanchanjunga Mountain Ecosystem protected areas by the year 2000.

Regional Collaborative Arrangements

Approach

1. Conservation activities to be implemented by individual countries by maximising people's participation for poverty alleviation: micro-enterprises, income-generating activities, and a cooperative movement.
2. Establishment of a Coordination Forum for Conservation involving three regional countries with ICIMOD as the facilitator.
3. Tri-national expert coordination meetings should be held on a rotational basis to focus on the transboundary issues.
4. The responsible authorities should be mobilised through meetings to plan for local-level policies.

5. Collaboration can be carried out at three levels.
   (a) National Level
   - Pushing policy action collaboration
   - Making necessary regulations
   - Organising joint planning for management

   (b) Local Level
   - Establishment of management collaboration between offices and local bodies
   - Joint planning and implementation

   (c) Regional Level
   - Transboundary collaboration to establish a mechanism

Activities/Issues

(a) Protection of migratory species
(b) Illegal trade
(c) Fire control
(d) Exchange of information
(e) Capacity building and training

Assist in the implementation of international conventions.

Index

C

D

E

F

G

H

I

N

O

P

R